This book not only addresses how to succeed in college mathematics, but also how to succeed as a college student, regardless of one's field of study. Many non-mathematics issues are discussed that all college students face, and many suggestions given for studying mathematics can also be applied to studying other disciplines.

As a counselor and academic support person, many of my colleagues and I can attest to the fact that the foremost issue students come to us with, concerns problems they are having with their mathematics courses.

It is especially gratifying to see a discussion of vital psychological issues that affect so many of today's college students.

—**Monica Porter**, Ph.D.
Psychologist and Director of Women's Resource Center
University of Michigan, Dearborn

Students are finally made privy to what trained teachers receive; namely, what students need to do to learn mathematics and why they need to do it. Reason has to rule the day in the life of a college student, and it rules the content of this book.

The book is written for students and they will benefit greatly from it, but I would also encourage inexperienced and experienced mathematics instructors to use it as a resource to support their teaching and their students' learning.

Problem solving is a fundamental part of mathematics—many college students struggle with it and don't know what they can do to improve. I was a student in Professor Dahlke's problem-solving course and it has served me well. His two chapters on problem solving are comprehensive, lucid, and useful.

—**Daniel Buchanan**, M.S.
Mathematics Instructor
Henry Ford Community College, Dearborn

My job has just become easier with the availability of this book. Before, my mathematics colleagues and I had nothing to refer our students to that discussed the many ways they could improve their performance in mathematics.

It takes a lifetime of passion, caring, experience, and dedication to one's profession and students to write a book like this. Students need to do themselves a favor and read it, use its ideas, and realize its potential—in college and beyond. As a parent whose children are out of college, I wish this book had been available for them when they were ready to enter college.

—**Robert Fakler**, Ph.D.
Mathematics Professor
University of Michigan, Dearborn

How to Succeed in College Mathematics

A Guide for the College Mathematics Student

RICHARD M. DAHLKE, PH.D.

BergWay Publishing

How to Succeed in College Mathematics
A Guide for the College Mathematics Student

Copyright © 2008 by Richard M. Dahlke

ISBN 13: 978-0-615-16803-6

Library of Congress Control Number: 2007907747

BW
BergWay Publishing
P.O. Box 701785
Plymouth, MI 48170-0970

Ordering information address:
www.bergwaypublishing.com

Cover and interior design by Sans Serif Inc., Saline, MI
Printed in the United States of America, by McNaughton & Gunn,
an environmentally conscious book printing operation

A portion of the book sales goes to Haciendita Uno, a rural community in
El Salvador, and CRISPAZ, a faith-based organization dedicated to mutual
accompaniment with the church of the poor and marginalized communities in
El Salvador.

Dedication

To
Mary, on our 40th wedding anniversary

Andy, our son, and Laura, our daughter, and their families

The memory of my parents, Victor and Joan, and Mary's parents, Carl and Kay

Cousins—

> *Fr. Don, on his 60th year as a diocesan priest serving the people of central Wisconsin*
>
> *Sr. Eugenia, on her 61st anniversary as a Sister of Mercy of the Holy Cross, who served the sick as a nurse, among other ministerial duties*
>
> *Sr. Antona, on her 51st anniversary as a Franciscan Sister of Perpetual Adoration, and 29th year serving the people of El Salvador*

Sr. Peggy, on her 51st anniversary as a Sister of Charity, and 21st year serving the people of El Salvador

Fr. Peter, cofounder of Christians for Peace in El Salvador, on his 55th anniversary as a Carmelite priest, who has worked for decades in social justice ministry throughout Latin America and currently serves the people of Juarez, Mexico

Jim, a friend who continues to inspire me by a lifetime of service to those in need

All people who work for good

Contents

Acknowledgements

I am grateful to my mathematics colleagues who took time out of their busy schedules to critique specific chapters. These colleagues and their affiliated institutions follow:

University of Michigan, Dearborn—
   Robert Fakler
   Barbara J. Matthei
   Tim McKenna
   Helen Santiz
   Sandra Wray-McAfee

University of Michigan, Ann Arbor—
   Pat Shure

Henry Ford Community College—
   Daniel Buchanan
   Larry Smyrski

I appreciate the discussions I had with these staff members at the University of Michigan-Dearborn:
   Margaret Flannery, Counseling and Support Services
   William Keener, Director of Program for Academic Support
   Monica Porter, Director of Women's Resource Center (Dr. Porter also critiqued specific chapters and I thank her for that.)

All the above-mentioned professionals have been in academia for many years and their wealth of knowledge is vast. This is an opportune time and place to thank them publicly for their many years of dedicated service to college students.

I am indebted to Professor Ron Morash who collaborated with me in writing several subsections of this book. Ron and I have co-authored several academic papers, and years ago instituted a program for college mathematics students called the Keller Plan. This was a self-paced, mastery-oriented, and student-tutored program in Calculus I and II. The program proved to be highly successful and ran for many years. The knowledge I gained through teaching in this program is reflected in this book.

I am especially thankful to my wife, Mary, for her invaluable contributions to this book. She painstakingly critiqued all drafts of the manuscript. Her suggested changes made the book markedly more readable. I was fortunate that she has a teaching, writing, and mathematics background, which supported her efforts. I first met Mary in my first year of university teaching when she was my calculus student. Can anyone doubt that mathematics has its benefits?

I am grateful to the authors of academic articles and books that I have read over the years. Their writings helped form my educational philosophy, and I built upon them in this book. I stand on their shoulders.

My university students, unknown to them, were experimental subjects in my educational laboratory, namely, the mathematics classroom. Their influence played a major role in the selection of the book's content.

Mary Hashman made this a better book through her copyediting. Susan Kenyon and Barbara Gunia (Sans Serif, Inc.) were the book designers and a joy to work with.

It has been said that writing a book is easy for everyone except those who write one. How true that is. In talking with my colleague Professor Ron Morash at the outset of writing this book, I said, "Hey Ron, did you know that I am writing a book?" He replied, "Neither am I!"

Preface

Two key questions that you want answered about this book are:

1. Is this book written for me?
2. How creditable is the information in the book?

The remarks given here should help answer these questions.

1. Intended Audience

The primary audience for this book is college students who are or will be taking one or more college mathematics courses. It is of benefit to lower- and upper-division students in mathematics—those who take only one mathematics course in college to those concentrating in mathematics. It is also beneficial for high school students in college preparatory mathematics who are close to entering college. This book is not only meant for students having difficulty in mathematics, but also for those doing well in mathematics who want to learn more about what it means to understand mathematics, and how to better attain that understanding. Thus, there is valuable advice included for the strong student, the average student, and the struggling student. The student can select, from the many ideas and suggestions given, those that apply to him or her.

Even though mathematics is the discipline addressed in this book, much of the advice applies, directly or indirectly, to taking courses in other disciplines. This is especially true for courses in the hard sciences, such as physics, computer science, chemistry, biology, geology, and for those in the engineering and biological sciences.

Even though the book is written for college students, it is also a resource for beginning and experienced college mathematics instructors, not only because they are instructors, but also because they are advisors. Many college mathematics instructors have not been formally trained in educational pedagogy; therefore, what they learn about educational pedagogy in this book has the potential to positively affect how they teach mathematics courses. As advisors of students, mathematics instructors field questions on specific mathematics concerns, and on more general concerns facing college students. This book will help instructors become more effective advisors.

Inservice and preservice secondary teachers of mathematics can benefit from this book since many of the topics addressed apply to secondary mathematics students. Furthermore, these mathematics teachers need to inform college-bound students of what to expect in college, and specifically, in college mathematics.

This book can serve as a resource guide for college counselors, and for college academic support services and learning center personnel. Those I have spoken with mentioned that the typical student they see is one who is stressed out over the difficulties he or she is having in a mathematics course. Therefore, college support personnel can recommend this book to these students, as well as address specific sections of the book with them, either individually or in a workshop.

2. Support for the Suggestions

An educated person does not blindly accept suggestions, and is more apt to take advice if he or she understands the rationale behind them. Thus, *this book is replete with rationale for the suggestions that are given.* You need to understand this rationale to the extent that you can give it if asked. Sources used that lend credibility to the suggestions given include:

- Findings of educational psychology
- Reports and journals of professional mathematics or mathematics education organizations, including the American Mathematical Association and the National Council of Teachers of Mathematics
- Papers written by experts in the fields of mathematics, mathematics education, and education
- My formal academic training, research activities, and extensive and varied college teaching experiences in mathematics and mathematics education

  Throughout my teaching career, including the early years I was a high school mathematics teacher, I conducted classroom research on the problems associated with the learning and teaching of mathematics. At the college level I focused on teaching and research in the lower-division calculus-track mathematics program, and in mathematics courses for pre-service elementary teachers pursuing a minor or major in mathematics. I taught upper-division mathematics courses for mathematics concentrators including abstract algebra, linear algebra, and geometry from an advanced viewpoint. For many years, I also taught annual courses on problem solving in mathematics and on methods of teaching secondary school mathematics.

- Suggestions from successful students on what worked for them, and from unsuccessful students on what did not work for them (Their remarks are quoted throughout the book.)

- Comments by mathematics students that appeared on mathematics faculty evaluation forms in response to these questions: (1) "What did you like about the course and instructor?" and (2) "What did you dislike about the course and instructor?" (Their remarks are quoted throughout the book.)

- Suggestions from experienced college mathematics instructors on topics that should be included—some suggestions came from instructors who focused on teaching lower-division mathematics, and others from instructors who focused on teaching upper-division mathematics

- Astute quotations

    Quotations serve many purposes, including providing evidence for the relevance of ideas, giving credibility and authority to these ideas, and summarizing them. The wisdom that is gained throughout the years through experience is preserved by quotations. Look forward to reading them, learning from them, and being motivated and inspired by them.

Don't hesitate to discuss with others the suggestions and rationale appearing in this book, including your classmates and instructors. Hopefully, the give and take of these conversations will lend additional support, or lead to other ideas or suggestions. It's amazing what you can learn from conversation that includes good questioning, listening, and thinking.

Introduction

This chapter begins by identifying some mathematics and non-mathematics issues that affect performance in college mathematics, followed by suggestions on how to use this book, and a discussion on a key thing you must do to be successful. The chapter ends by addressing a compelling and empowering way for you to view college.

1. Two Fundamental Student Deficiencies That Affect Performance in Mathematics

You cannot be successful in mathematics without the ability to study mathematics, nor can you be successful in science without the ability to study science, and so on with other disciplines. Throughout all levels of education, the primary goal is to learn how to learn, and college affords you the opportunity to take this to a higher level. *Learning how to learn is the most valuable skill you can gain in college since learning is lifelong.* Many college students are deficient in effective study skills and many college instructors do not view it as their responsibility to teach them. A key objective of this book is to confront these obstacles.

It is one thing to be enrolled in a college-level mathematics course and another thing to complete it successfully. Receiving a respectable grade for the course does not necessarily mean that the content was learned, as it should have been. An A from one instructor may be a grade of B, C, or D from another instructor. *Two fundamental deficiencies of many students that negatively affect their performance in mathematics are: (1) not knowing what it means to understand mathematics, and (2) not knowing how to attain that understanding.* The percent of these students in a typical college mathematics course can be well over 50 percent.

It is unwise to equate intelligence with achievement. Many highly intelligent college students do not achieve well in mathematics, whereas many students of lesser intelligence do. This means that there are many other factors that affect achievement in mathematics, and foremost among them is whether a student has a viable method of learning mathematics.

Why do so many college mathematics students not know what it means to understand mathematics and how to attain that understanding? There is no definitive answer to this question, but there is little doubt that a major contributor is the lack of attention given to it by mathematics instructors. There are high school and college mathematics instructors who do not plan activities, to the degree required, that promote this understanding. Such activities include the nature of the content that comprises their presentations and class activities, the nature of their assignments, and the content for which they hold their students accountable. How many of your past high school and college mathematics instructors have consistently spent time giving sound and comprehensive advice on what you need to do to learn mathematics? And then followed up on that advice by having appropriate student expectations? Many instructors do, but some don't. If you have difficulty answering those two questions, you will be much better equipped to answer them after reading this book.

2. Non-Mathematics Issues That Affect Performance in Mathematics

Ask yourself this question: "What are the issues that affect my performance in a mathematics course?" You can quickly come up with some of these issues, but you would probably leave out many issues that affect your performance. Almost all that you are involved with as a college student is connected; that is, when your behavior changes in one area, one or more other areas will be affected. Your mathematics performance can be negatively affected by—

- Working too many hours on a paid job
- Reluctance to seek help from others

- Inadequate strategies for recalling content
- An instructor whose teaching style is incompatible with your learning style
- False beliefs on issues, such as your ability to learn, your responsibilities and those of your instructor, and your ability to make changes
- Lack of self-esteem
- Lack of motivation to succeed
- Poor management of time
- Taking a course in a shortened term
- Procrastination
- Anxiety
- Ineffective communication skills
- Transferring colleges without attaining vital information

I address issues that relate to you as *a college student*, and issues that relate to you as a *college mathematics student*. As a mathematics instructor and advisor for many years, I have been exposed to most issues that interfere with students' performance in mathematics. Many an office hour was spent discussing these issues with students. This book would be of limited value if it did not address the more important ones.

Once you better understand what mathematics is and ways to attain this understanding, you most likely will have to change certain study behaviors. Hence, the psychology of change is addressed, as well as other psychological issues that you may struggle with such as lack of confidence, lack of motivation, procrastination, and mathematics and test anxiety. Many "How to Study Books" avoid discussion of these difficult issues, namely, the psychological, attitudinal, and values issues. This is a deficiency of these books since these issues are *critical* to your success as a college student.

3. Using This Book

This is a *guide* or *resource* book. It contains a detailed Table of Contents and Index to help you locate information on a particular topic related to being a college or college mathematics student. The order of the chapters generally follows the order in which you have to make decisions, beginning with your plan to enroll in a college mathematics course and as you progress through the course. There is something to be gained by reading the book from beginning to end because of this logical progression, but this is not necessarily recommended. Most likely you will look at the Table of Contents or Index and start with a specific chapter, chapter section, or page, where you can find the information you need on a specific topic.

As you read on a topic, you will find that you are sometimes referred to other sections of the book for further information. Most topics associated with learning mathematics are linked together in a variety of ways forming an assortment of chains, and the disturbance of one link in a chain will affect other links in the chain. The book is organized so as to be sensitive to this linkage; thus, when you are reading about one link in a chain, you are often referred to other links of the chain. Hardly a section of the book stands alone, yet most sections can be read out of order. This is perhaps best illustrated by the primary organizational scheme of the book: A key topic is presented in a thorough way, and is often referenced again when other related topics are presented in thorough ways. Hence, this spiraling effect reinforces your knowledge of the topic.

Take brief notes as you read on suggestions that you want to make part of your study methods. Refer to these notes regularly, and each time you do ask yourself the question, "Have I been practicing these suggestions?" If your answer is, "No," then answer the question, "Why not?"

Choose What Is Best for You

You are unlike any other student; hence, what works for someone else may not work for you. You are the only one who can make the decision to adopt a specific suggestion. You are in the driver's seat and that should console you. There is a wealth of suggestions, ideas, and advice given here, to the point that you may feel overwhelmed by it. No one can come close to employing all of what is suggested, so don't even try. Some of the suggestions will resonate with you; others not. It is important to determine which ones are most beneficial for you. For these, the word "suggestion" takes on a more serious tone, similar to its use here: "When jumping out of an airplane, it is suggested that you wear a parachute." You have to read the suggestions, reflect on them, choose some to begin with that appear most fruitful, employ them, discuss them with others, and make wise decisions on whether to keep, revise or discard them.

Make Changes Cautiously

It is unwise to make wholesale changes quickly in your study methods; there is a chance you may become discouraged and completely abandon your efforts. Begin by working on one or two key changes, and relish any progress you experience along the way, as small as it may be. Making changes in your study methods is a process, not a single event, and you will have setbacks. You will eventually get where you want to be with slow and steady progress. Don't set unrealistic expectations on how fast you should progress. Some changes may happen within a day or two; others may take weeks, months, or even years. Well-entrenched study habits are slowly broken, and new study habits are slowly formed. Be patient and steadfast, work diligently, and anticipate the success that will eventually come.

4. Nothing Works Unless You Do

If you are expecting the advice given here to be a set of tricks and gimmicks that you can easily and quickly use to be successful in your mathematics courses, you will be disappointed. There is no royal road to learning mathematics, or to learning in any discipline. Getting educated is hard and often painful work. Understanding and applying a theorem may take you five hours, and be more difficult than digging a ditch three feet deep, two feet wide, and forty feet long. Learning mathematics is time consuming, often frustrating, most likely tiring, and can be downright depressing. But study becomes exhilarating when you know how to do it, know why you are doing it, and know that it will lead to success. Study also becomes easier since the more you know about something, the easier it is to learn more. But always keep this in mind as you pursue your goals: *nothing works unless you do*.

Working hard is not enough to be successful. Sitting for hours in a library or in your dorm room day after day working on mathematics does not necessarily equate to being successful in mathematics. Certainly success depends on the amount of time you expend, but more importantly on the quality of that time. Practicing quality study techniques in mathematics forms habits that lead to success in mathematics. Hence, to get the most out of your mathematics study, you have to work hard *and* smartly. You work smartly when you put to practice your understanding of what mathematics is and how to attain that understanding. You know what you need to do and then you do it. Getting schooled is serious business, and you either want to learn or you don't. *You cannot creditably say you want to learn if you don't do what needs to be done to learn.*

No doubt you have heard the rule of thumb that you should study two hours outside of class for every hour in class. That may make some sense if it is considered the average amount of study for all college students, regardless of the types of courses they are taking. But it may be totally inappropriate *for you,* and for being in a *mathematics* course. The time required depends on many things including the difficulty you have with mathematics, the complexity of the mathematics, and the demands of your instructor. You are well-served by throwing this rule of thumb out the window and basing the amount of time you need for studying mathematics on whether you are learning what needs to be learned. For many students, if not most, their mathematics study time needs to average more than two hours outside of class for every hour in class.

5. College as an End in Itself

Learn for learning's sake. Eliminate the mindset that you are in college only as a means to an end, but instead view it as *an end in itself.* Being a college student is your current profession so perform your work as a professional or responsible student would. Arise every morning with the attitude that you will be the best you can be. This attitude will serve you well in college and throughout your life. Enjoy all the benefits of being a college student, which includes your study.

College Mathematics Environment

When discussing the college mathematics environment, it is important to compare and contrast it with the high school mathematics environment since there can be profound differences. This chapter identifies many areas in which you may have to make changes, and this book will help you understand these changes and how to respond to them. My remarks apply to entering freshmen as well as to those with experience in college mathematics. I have known college sophomores who have taken two or three college mathematics courses and behave as though they are still in high school mathematics. *Your college mathematics courses will not adapt to you; any changes that need to be made have to be made by you.*

It is important to note that with some instructors, the differences between high school and college mathematics courses become blurred. That is, there are high school mathematics teachers who operate more like college mathematics instructors and vice versa. But it is not the norm for college instructors to behave like high school instructors, and for good reason. The college environment dictates a change in mathematics instruction from that of high school. This, in turn, dictates a behavioral change in college students to meet this changed instruction. The goal of this chapter is to identify the major behavioral changes you may need to make.

1. What You Can Expect

There are many differences between the college and high school mathematics environments. The sooner you know these differences, the sooner you can make the necessary adjustments in your college mathematics courses.

It is interesting and informative to compare the expectations of college students in mathematics courses with the reality of the college mathematics environment. You can make these comparisons since you will find in this section a presentation of key aspects of this environment, and in the second section the responses of college students to the question, "What did you dislike about your instructor?" Their responses to the question reveal their expectations of the college mathematics environment.

Descriptions of the differences between the college and high school mathematics environments follow.

More Work and Little Admonishment to Do It

Did you receive good grades in high school mathematics by doing a minimal amount of work? This may mean that you are an outstanding mathematics student. However, it may also indicate that your instructors had inappropriate or low standards, or your competition was weak. In the latter case, your instructors may have geared the course to the average student, causing your work to appear strong. Regardless of the reasons for your good grades, you could struggle in college mathematics because you lack the requisite knowledge needed for your introductory college mathematics course, or the study habits that would serve you well in college are not well-developed. As a college instructor, it was heartbreaking for me to see the reaction of students who had received A's and B's in the last two years of college preparatory mathematics but who did so badly on college mathematics placement tests that they were told to retake the content of this mathematics. At the very outset of their college years, they saw their desired career (e.g., engineer, chemist, doctor, computer scientist) in immediate jeopardy. If you developed poor study habits as a result of receiving good grades in high school with a minimum amount of work, it might take a while for you to realize this has to change when in college, and even longer to make the necessary changes.

In a college mathematics course you are likely to find that you are among high achieving students due to the selectivity of the college admissions process. The minimal amount of work that made you successful in high school will not suffice in college. What will separate you from your college classmates in achievement is your method of learning.

In high school, many of your teachers were almost like a father or mother to you. They may have continually pleaded with you to work harder, to get help if you needed it, to be more diligent about your assignments, and to study more for examinations. They probably answered all questions you had on the assignments and lectures, and reminded you frequently about upcoming examinations and when specific homework was due. Don't expect the same from your college mathematics instructors. It is not their job to be on your case if you are treating your mathematics course irresponsibly. Most likely your mathematics instructor will not take roll to see if you are in class. However, in small classes your instructor will notice if you are periodically missing, and may make a passing comment now and then to you or to the class about the importance of attending class. Also, don't expect to be admonished for a poor performance on a mathematics test. Having said all this, there are college instructors who want to have a close relationship with their students and will support them in becoming more responsible students. This is admirable. However, taking the necessary initiatives to be successful in college is your responsibility, not anyone else's.

More Freedom and More Decisions to Make

You will enjoy the extra freedom that you have as a college student. If you carry an average full-time credit load of 16 credit hours, you will normally be in class 16 hours per week. Assuming your classes meet during the day, that leaves you on average, for a *five-day* week, about five (5) hours each day, excluding your evenings, to work on your studies, be involved in other activities of your choice, and have lunch. You also have evenings and the whole weekend to study. (This is ignoring a paid job that you may have.) "Wow," you may say, "I love having all this free time. College is going to be a lot of fun!" Even as a full-time student you will often have several hours between classes, or even as much as a day and a half between classes. You will say, "This is so different from high school." Temptations will abound! Your classmates will want you to play cards, go to a movie or baseball game, or just hang out in various places on or off campus. You are encouraged to take advantage of some of these opportunities, for that is an essential part of being in college. But there are good times and there are bad times to do these things, and you have to restrict how many extra-curricular activities you engage in and the length of time you spend on them. For example, the one hour you may have between two classes is often better used as study time rather than free time. You have to discern what will be detrimental to your learning, mental health, and other responsibilities. Manage your time wisely.

Yes, college can be fun. But it also requires a lot of work, far more than you may realize. If you knew upfront how much work it requires you would think differently about how much free time you have. You don't want this realization to come as a shock to you several weeks into a semester. Focus on your studies from the outset of a term. Learning mathematics as it needs to be learned, is in itself a formidable task and if you don't know this now you will after reading this book.

College can also be daunting because of the numerous decisions you will have to make, academic and otherwise. There is far less hand-holding in college than in high school. You will have to decide many things and that may be terrifying for you since there are most likely no guidelines to read that specifically address what is best for you. For example, you will have to—

- Choose an area of concentration
- Decide on living accommodations and transportation
- Decide whether to have a job, what type of job, and the best hours for work
- Select courses and instructors, and schedule your courses
- Discern how to best use your time
- Map out when, where, and how you will study for each course

This book assists you in making many decisions.

I asked this question of an experienced director of introductory mathematics courses at a major university: "Why don't more freshmen students do better in introductory mathematics courses?" She said: "Because they are first semester freshmen. It takes them awhile to know this. They don't know what studying is. It is basically teaching yourself. Also, they don't seek out help."

If you are a freshman or are still acting as a typical high school student in your college mathematics courses, please read this chapter carefully.

Faster Paced Courses and More Content Details Left to You

The pace in a college mathematics course can be two to three times faster than that of a high school mathematics course, and in a shortened spring or summer term it can be four to six times faster. For example, in a typical college term of 14 weeks, you have about 56 class sessions for a four-credit precalculus course versus about 180 class sessions for a somewhat comparable high school mathematics analysis course (fourth year of high school mathematics).

The fast pace of college mathematics can be mind-boggling. It will not take very long into a course, perhaps only a few weeks, before any irresponsibility on your part such as missing classes, not seeing your professor for help when you need it, or procrastinating on your assignments may cause you to get too far behind. Mathematics departments and instructors have decided on the topics that comprise specific courses. Instructors of these courses will not slow down for you to catch up. Get too far behind and you will have no choice but to drop the course or receive less than an acceptable grade.

It is a challenge to adapt to the pace of college mathematics. Not only will your work habits have to change, but they will have to change quickly. You do not have the luxury of waiting a few weeks to get your act in order. In my mathematics department you have the first two weeks of a term to add a mathematics course. I have counseled many students against adding my courses if most of the two weeks had passed, since adding the course would have put them at a distinct disadvantage due to the amount of content already covered. I allowed some students to enroll only when their background suggested they had a reasonable chance to catch up.

You will be less "spoon-fed" in college mathematics courses. That is, your college instructor will not present in great detail all of the material you are expected to learn and will not discuss all of your homework concerns. He or she will refuse to be an oral textbook by repeating the content of the textbook before your eyes, dotting every "i" and crossing every "t." There is no time for this, and even if the time were available, it is unwise to do this. Your instructor views his or her role as a facilitator of your learning, and expects you to work diligently outside of class to fill in the gaps. You will not experience success in college mathematics if you refuse to adapt to this expectation of your instructor. Ways of filling in the content gaps left by your instructor comprise a substantial portion of this book's content.

Less Grading of Assignments and Less Feedback in Class

In many lower-division mathematics courses you will not be handing in homework to be graded other than perhaps a few assignments or specific exercises, now and then. Due to other commitments, most instructors do not have the time or the help to grade them. This should not be too disconcerting to you since the benefits from frequent grading of assignments are minimal. Having an assignment graded may motivate you to do it and to do it more carefully, but you need to learn to do this without this extrinsic motivation. Once again, you are an adult and should expect to be treated as one. Feedback on your assignments can be attained in ways other than having them graded by your instructor or his or her assistant. You can discuss assignments with your classmates and instructor, and you can obtain feedback in class when the previous day's assignment is discussed. However, not all assigned exercises will be discussed in class and answers or solutions will not necessarily be available for those not discussed. Some textbooks have solutions manuals that you can purchase, containing solutions to some of the exercises. Most importantly, you need to develop methods for determining whether you have correctly worked problems. This will serve you well in lifelong learning. The bottom line is you need feedback and it is up to you to get it.

Not only will assignments not be graded in many courses, but some instructors may not even make specific assignments. They may say something as casual as, "Work enough of the exercises from the section in the textbook that we are discussing so that you have a good grasp of the section." I disapprove of this since most students are not equipped to select the various types of exercises that they need to work. There are vast numbers of exercises in a section, and a number of cognitive skills that need to be employed to work them (e.g., computation, comprehension, application, and analysis). In general, only a small percent of the exercises need to be worked, but a more knowledgeable and experienced person must carefully select them, not students who are being exposed for the first time to the ideas addressed in them. This is the responsibility of your instructor, and he or she should give at least a minimal assignment. Beyond that, you are encouraged to select additional exercises, and by all means more challenging ones if you can handle them.

No In-Class Reviews for Examinations and
No In-Class Feedback on Graded Examinations

It may come as a shock to you to realize that, in the main, college instructors leave reviewing for an examination up to their students. This is not the case with many, if not most, high school mathematics instructors. High school instructors often set aside a class session to conduct an in-class review, prior to an examination. Why don't most college instructors conduct in-class reviews? For two reasons: There is not time in the course to do so, and even if there was, reviewing is not the responsibility of the instructor. Why should your instructor treat reviewing any differently from the assignments that you are given? They are both learning experiences for you; thus, they are your responsibility. There are better things an instructor can do in class such as beginning the next unit of study. That being said, your instructor has the responsibility to periodically summarize content throughout a unit. He or she may hand out a study guide for the unit test, and suggest some methods that you can use to conduct your review. But the bottom line is this: Reviewing is your responsibility, not your instructor's. You may not like this and believe your instructor is behaving irresponsibly, but try to understand the reasons for it.

Going over your examination once it is completed is your responsibility, not your instructor's. Many, if not most of your instructors, will hand out a solutions key to the examination. Regardless of what your instructor does, you have to take initiative to get feedback on your examinations by making use of what your instructor provided you, by talking with other students, or by attending your instructor's office hours. There are a variety of important things you can do with your returned examinations and they are addressed in this book.

More Lecturing, Less Discussion, and
Less Knowledge of You

College mathematics departments have instructors with varied academic backgrounds. Many have doctorates in pure or applied mathematics or in mathematics education and are in *tenure-track* positions (i.e., more permanent positions). Tenure-track mathematics instructors can be found teaching at all levels of mathematics—from the freshman level through the graduate level. *Adjunct* instructors are also employed, part or full time, with their contracts renewed each semester or after several semesters (i.e., they are in temporary positions). You are most likely to find adjuncts teaching only freshman or sophomore mathematics courses. Many of them do not have doctorates, but almost all have master's degrees. They may or may not have pre-college teaching experience. In large universities you will also have graduate students working as *teaching assistants* under the guidance of other faculty. These students are often pursuing a doctorate in mathematics. Usually these individuals teach freshman mathematics courses.

You can ascertain from the previous paragraph that college instructors of mathematics typically have more formal training in mathematics than high school instructors. This can lead to differences in content focus. High school teachers are apt to focus more on mathematical computation and factual information. College instructors will focus more on mathematics theory, which is not surprising considering their more extensive background in mathematics. This means that you will be expected to focus more on learning mathematics theory. You will be expected to demonstrate your understanding of theory in your assignments and on examinations. If this is something you have not done, then you have to learn and employ study methods that will help you do this. In learning mathematics theory you will learn more about what mathematics is and what it means to learn mathematics.

Most instructors of college mathematics have not been *formally* trained to be teachers. That is, they have not been in teacher training programs, as were your high school teachers. It is assumed that their extra mathematics knowledge compensates for their lack of courses in areas such as educational philosophy, educational psychology, and techniques of teaching mathematics. This is certainly debatable, but there is no doubt that their extensive and higher level of mathematical knowledge is of great benefit to you, especially as your mathematics courses become more advanced. Don't expect many of these instructors to be familiar with educational jargon and some of the more recent techniques of teaching (e.g., use of in-class student groups). College instructors who are beginning their careers may be better equipped to employ more current teaching techniques if they themselves experienced these techniques as students in pre-college or college mathematics. Recent techniques of learning mathematics are addressed in this book. You can employ them, regardless of what your instructor does.

In college there is far less opportunity for your mathematics instructor to get to know you for the following reasons:

- You are not in class each week as much as you were in high school and you may only have your college instructor for one semester. For example, in a one-semester college precalculus course, you may be with your instructor in class for about 56 meetings. Compare this to a yearlong high school precalculus course where you may have about 180 meetings.
- College instructors are engaged in a variety of activities outside of class, including research, and serving the department, college, or university, as well as their local and professional communities. They may not take the time to learn who you are and many will struggle learning your name right up to the end of the course.
- College mathematics instructors are notorious for walking into class, lecturing, taking a few questions, and then walking out. Engaging in "small talk" is not common for many of them; hence, less chance to get to know you. If you want more attention from your instructor, then you have to take the initiative to see him or her outside of class.
- Introductory college mathematics courses can have very large enrollments. In some colleges you can have 40, 80, or even 120 students in a mathematics course. Not only will the instructor not learn your name, but he or she may not recognize you throughout the term as being a student in his or her course if they happen to meet you outside of class.
- In large-section mathematics courses, your instructor may not grade your assignments and examinations, leaving this work to an assistant.

Are you beginning to feel alone? To be less isolated from your college mathematics instructor you can ask and respond to more questions in class, and make more use of his or her office hours.

Lecture is often emphasized more, and therefore discussion less, in college mathematics courses than in high school mathematics courses. The reasons for this are varied and include these:

- Less class time available means more lecturing since lecturing allows the instructor to move faster through the content (but not without you paying a price)
- Less knowledge by college instructors of the benefits of discussion and how to employ it
- More opportunity for students to have discussions with others outside of class

Discussion is critical for learning mathematics; hence, you have to take the initiative to see that it happens. Outside of class you can meet with your classmates and instructor, as well as carry on discussions though email and phone. Also, you will learn to rely more on your reading, which can be viewed as carrying on a silent discussion with the authors of your textbook.

More Complex Content

There are differences in focus between high school and college mathematics content. These differences make college mathematics courses more formidable. In college mathematics you will experience—

- More concepts
- More sophisticated concepts
- Greater emphasis on relating concepts
- Greater use of symbolic form
- More applications and greater diversity of applications
- Higher levels of cognitive thinking
- More emphasis on justifications
- More difficult computations

You are expected to have, at the outset of a course in college, the prerequisite mathematics for the course. For example, before enrolling in a first course in calculus you should know well the four years of high school college preparatory mathematics. It is not enough to say that you passed your pre-college courses with an acceptable grade. You have to know the mathematics and it is unfortunate that grades do not always reflect this. Your college textbook may have a review of some of the prerequisite content, and your instructor may do a quick review of some of it in class, either at the outset of the course or when needed in the course. But this review will be done quickly and you may have to spend considerably more time in personal review, especially on past topics that you learned less well. A "look-back" at a topic can only be called a review if the topic was learned before.

I cannot stress enough the importance of having the prerequisite knowledge for a course. If you don't have it, you will not be successful in the course. Learn well the content of each mathematics course, and approach your study knowing that you will be held accountable for it in subsequent courses. This mindset will help you learn it well and retain it. You will be successful in any mathematics course if you have the prerequisite knowledge and employ appropriate study methods. If one or the other is lacking, you will be in serious trouble.

There are *many avenues for receiving help* when in a mathematics course in college, and you have more time to avail yourself of this help. Besides the normal means of obtaining help such as your instructor, classmates, and a tutor you may hire, help is available from the mathematics department, from the college or university learning center or program of academic support, and from professional student organizations.

Greater Necessity to Read the Textbook and Take Notes

Saying that you must be able to read a college mathematics textbook with understanding is an understatement. You should not be able to survive in college mathematics without doing this. If you are able to survive without doing this, then you have to question the quality of the course you are taking. What your instructor does in class should be done with the assumption that you are reading your textbook. It is not uncommon for high school mathematics students to boast that they were able to get a respectable course grade without reading their textbook. If that is your experience then you probably had instructors who were "oral textbooks." That is, they may have over-emphasized lecture, basically writing the textbook in class before your very eyes.

This behavior breeds in you the attitude that you cannot learn mathematics until it is first presented to you by your instructor, and that all you need to know is the content presented and discussed in class. This is not preparing you for college mathematics to say nothing of lifelong learning. For you to make the necessary changes at the college level after 18 years of pre-college schooling requires an attitude change, lots of work, and the ability to learn from reading your mathematics textbook. There is significantly more writing in college mathematics textbooks, and for good reason.

You may have taken notes in high school mathematics but note taking was probably not emphasized to the extent that it is in college mathematics. I have had students in a variety of college mathematics courses who did not take notes, even though I frequently mentioned the importance of doing this. This would boggle my mind since I knew that they would struggle without notes. I do my thing, the textbook authors do theirs, and students need to have records of both. Did I approach specific individuals outside of class and try to convince them to take notes in class? Absolutely not! I treated my students as adults, giving reminders when appropriate, but stopping far short of holding their hands.

Fewer, Longer, and More Difficult Examinations

In some courses, there may be a few short quizzes in a unit to help you monitor your progress in the course. The first examination you see will probably be an hourly test, given about three to four weeks into the term, and covering a significant amount of content. You might say that three to four weeks of content is not that much. It may not be at the high school level, but it definitely is at the college level because of the pace at which the course moves. In a three-credit course you may only have three hourly exams over the course of the semester, and a comprehensive final examination. Hence, you can see the weight that each hourly exam carries. Now assume you did badly on the first unit test, and the second unit test is coming up in another three to four weeks. This content may be more difficult and most likely requires a good knowledge of the content of the first unit. You now have a very short time to learn the content of the first unit as well as that of the second unit that depends on it. This is a tall order indeed. It is not difficult to see that your situation can get out of control very quickly. The moral here is this: Work diligently to understand the content of your mathematics course from day one. Having to remedy significant deficiencies in a fast-moving course is problematic.

Since an examination covers a huge quantity of content it will be less clear as to what your instructor will ask you to do on the examination. The content that appears on the examination is only a *sampling* of the content you have to know, and know well. This means that you have to know a lot of content well to have a chance of performing well on the limited content appearing on the examination. It is not that un-common for instructors to make their examinations long, which means that you have to know the content thoroughly because you cannot take much time to respond to examination questions.

Finally, many of the examination questions are different from those in your assignments. They can be more difficult, or require you to be more creative in responding to them. The wording of an examination problem may be quite different from the wording of your textbook authors' problems because the purpose of the examination is to test your *understanding* of the content. Learning suffers if the type of instruction you receive is one of, "Watch what I do, practice what I do, and spit it back to me verbatim on the examination."

Less Opportunity to Improve Poor Test Scores, and Lower Course Grades

If your mathematics class does badly on an examination, the chances are exceedingly high that your instructor will not give another examination on the same unit. Similarly, your instructor will not give you some other project to do so you can make up for a low grade. Most likely you have to suffer the consequences of a poor examination grade; therefore, work as hard and as wisely as you can throughout the unit so you can do the best job possible on your examination. In all probability, you have one chance at it. You also need to know that your examination scores will most likely comprise most of your grade for the course.

College instructors are freer to give lower grades than high school teachers, *and they do*. It is not atypical in a college precalculus or calculus course of 30 students to see a grade distribution of 3 A's, 7 B's, 5 C's, 5 D's, 3 E's, and 7 W's (withdrawals from the course). Here we see that half of the students withdrew from the course or received a course grade of D or E. Some college instructors' grade distributions will contain no A's; others may have less than one third of the grades above C. Grades given by college mathematics instructors are generally lower than those given by instructors of other college disciplines. In my college, students concentrating in mathematics had the lowest *mathematics* grade point average compared with the grade point average of other concentrators *in their discipline's courses*. I have no doubt that this is typical for most colleges. These statistics are not meant to frighten you, but to prompt you to take your mathematics study seriously.

If you don't know what to expect in a college mathematics environment, you may be too slow to respond to the changes you need to make in order to succeed. It is better to be forewarned than to be surprised. Awareness empowers you to act.

Is it becoming clearer to you that much of the "free time" you thought you had as a college student is vanishing? It is time mainly needed for study so you can meet the demands of your mathematics course, as well as those of your other courses.

2. Students' Expectations of the College Mathematics Environment

The expectations of students of the college mathematics environment were determined through analyzing their responses to this question appearing on forms they used to evaluate their instructors: "What do you dislike about your mathematics instructor?" The students were enrolled in various calculus-tract courses. Many of their responses to this question indicate that the expectations they had of their instructor were not met. As you read them, decide whether a student's remark was unreasonable, based on what you read in the last section about the college mathematics environment. You might want to discuss your decisions with some of your classmates.

1. *"The instructor did not cover material in class that was on the test."* (Calculus I student)
2. *"Didn't always lecture on the material before assigning the homework."* (Calculus I student)
3. *"The material in the course is overwhelming. It seems that if you fall behind once, you never catch up."* (Calculus II student)
4. *"Some hard sections were covered too fast. He followed the syllabus when he should have followed the needs of the class."* (Calculus I student)
5. *"Lecturing was done on three sections from the next unit of study, before a test was given on current subject matter."* (Calculus II student)

6. *"I don't particularly like the way material was passed up after one day, even if we couldn't understand it."* (Calculus III student)

7. *"I found it difficult to keep up with the homework because I had a lot from other classes."* (Calculus I student)

8. *"No homework collected. This made it easy to fall behind."* (Calculus I student)

9. *"I think she should have reviewed before exams and tried to leave time for questions."* (Calculus I student)

10. *"My teacher was an evening teacher; hence no office hours were scheduled."* (Calculus I student)

11. *"He assumed you knew a lot."* (Differential Equations student)

12. *"We are not expected, in other math classes, to do a whole lot of reading on our own."* (Calculus I student)

13. *"He didn't tell us where in the book he was. It was hard to follow along."* (Calculus I student)

14. *"I sometimes was confused by the different formulas or methods the instructor used compared to those used in the book."* (Calculus II student)

15. *"He didn't do problems in the same manner as the book. Very confusing."* (Matrix Algebra student)

16. *"She uses techniques not shown in the book."* (Calculus III student)

17. *"His tests were unfair (sometimes) because he put problems on them that he had not explained in class. However, it did make me think and study harder."* (Calculus I student)

18. *"Not enough tests. Because there weren't enough tests, each test couldn't cover everything so some students studied for the wrong stuff."* (Calculus II student)

19. *"She went through the material very quickly and a lot of material was involved in each exam."* (Calculus I student)

20. *"On the test there wasn't sufficient time to check problems over."* (Calculus II student)

21. *"He put hard material on the exams (no easy stuff)."* (Calculus II student)

22. *"I wanted the instructor to grade on a curve."* (Calculus II student)

Outline of College Mathematics Courses and Programs

This chapter provides an overview of the variety of mathematics courses and programs that are available at most colleges, including discussion of distribution requirements, honors programs, mathematics concentrations, and how mathematics serves concentrations outside of the mathematics department. Descriptions are given of some courses, especially those most widely selected by students (e.g., those comprising the calculus track). These descriptions include their prerequisites, differences between being traditional or innovative, typical backgrounds of instructors, challenges, and pitfalls to avoid when enrolled in them. The information available in this chapter addresses many of the concerns that you may have about mathematics courses or programs. It is not a substitute for what you can get at the college you are attending or are considering attending. It is generic in nature and should help you place in context the mathematics offerings and programs of your college. The discussion begins with lower-division mathematics courses, progresses to upper-division mathematics courses, and ends with a focus on the mathematics concentration program.

1. Lower-Division Mathematics Courses

Mathematics courses in college mathematics departments are categorized as lower- or upper-division courses. Freshmen and sophomores typically take lower-division courses, whereas juniors and seniors typically take upper division courses, although exceptions abound. For instance, you may change your concentration in your junior year, and thus have a need to backtrack to get prerequisite courses for a different concentration. If you advanced placed out of one or more calculus courses in high school you may be qualified to take an upper-division course as a sophomore since you may have the prerequisites for this course. Your decision to take an upper-division mathematics course as a lower-division student should be based on whether you have the content prerequisites and mathematical maturity for the upper-division course.

It is not possible to describe all lower-division mathematics offerings across the country. Some courses offered at one college may not be offered at another for a variety of reasons: the types of students being served are different; not all colleges have the same programs that mathematics serves; and philosophies vary among colleges, schools, and departments within a university as to the mathematics that is considered most appropriate.

There is a common core of lower-division mathematics courses that you will find at almost all colleges. These courses can be categorized by their clientele. A mathematics course may appear in more than one of these categories. The categories into which lower-division mathematics courses fall include mathematics—

- For students with pre-college mathematics deficiencies
- To satisfy college distribution requirements
- For honor students
- For mathematics, chemistry, physics, and engineering concentrators
- For business, management, social science, and biology concentrators
- For preservice elementary teachers

The courses comprising each of these categories are described below, along with the typical prerequisites, methods of instruction, and description of instructors. Additionally, some cautionary comments are given.

We begin the discussion with mathematics courses for students with pre-college mathematics deficiencies. In most colleges these courses are considered *remedial* (i.e., not college level); hence, they do not carry regular college credit. Nonetheless, they are offered in most colleges, and are important to the students who need them.

Mathematics for Students with Pre-College Mathematics Deficiencies

Some students who are admitted to college do not have the mathematics prerequisites to enroll in a regular college mathematics course. To remedy these deficiencies, they enroll in one or two remedial mathematics courses offered at their college, most likely an introductory or intermediate algebra course. These courses are somewhat comparable to the first and second years of high school algebra. Some colleges may offer one course comprising the content of these two courses. Some colleges encourage students to enroll in a summer session to take remedial courses, prior to their first semester of college. Some very competitive colleges do not offer remedial mathematics courses because almost all of their students don't need them, and those that do can take them off campus (e.g., at a community college or in an adult community education program).

Mathematics entrance requirements

In the recent past the minimal high school mathematics needed for admittance to college was three years of college preparatory mathematics, including a year of algebra and a year of geometry. This has been strengthened since many states are now requiring four years of high school mathematics. More states are likely to follow this increased requirement. For many colleges, the preferred mathematics entrance requirement is four years of college preparatory mathematics. Some college engineering programs prefer their entering freshmen to have successfully completed a year of calculus in high school. There are variances among high schools in what constitutes their college preparatory mathematics, especially with the increasingly wider use of integrated mathematics programs.

Why are some college students not ready to take college-level mathematics? There are a variety of reasons, including the following:

Some students get admitted to college when they have not met, on paper, the minimal requirements for entrance, which includes the area of mathematics. They may not have the required courses, or may not have received acceptable grades for them. To be admitted, these students have to show potential for academic success and are admitted into a special academic support program. Many of these students have to take remedial mathematics.

Some students do not perform well enough on college mathe-

matics placement tests even though they have been admitted to the college, having satisfied, on paper, the mathematics entrance requirement. They either do not understand well enough the content of these high school courses, even though they received an acceptable grade for them, or too much time has passed since they had these courses and they need an extensive review of them.

Once these remedial courses have been successfully completed in college, these students then take further mathematics, either to satisfy the college mathematics distribution requirement, or to satisfy the mathematics requirements for a concentration they are pursuing. An intermediate algebra course is a prerequisite for college algebra, college trigonometry, and precalculus. Precalculus is a prerequisite for many concentrations. Elementary algebra is a prerequisite for introductory statistics, which is often required for some concentrations (e.g., sociology, psychology).

What if you don't know geometry?

High school students struggle more with geometry than they do with algebra, yet you have to look long and hard to find a remedial geometry course in a four-year college. This is mind boggling to me since the content of precalculus and calculus courses is filled with geometric concepts and relationships, and assumes that you are able to think geometrically in solving problems and providing justifications. There is some geometry in remedial algebra courses but it is minimal at best. If your geometry background is weak and you desire to pursue precalculus and calculus courses, it is in your best interest to enroll in a geometry course somewhere (e.g., in an adult education program or community college). As difficult as it may be to find a geometry course outside of high school, it is more difficult to find an appropriate one. *At the very least, get hold of a high school geometry textbook and use it as a resource.* You may need a personal tutor to help you, perhaps a high school mathematics instructor. There are several geometry textbooks written for college students. These may be more appropriate for you to study since they focus on what you need in geometry to succeed in college mathematics. If your goal is to enter a precalculus or calculus course, it is wise to reflect on this inscription that was written over the door of Plato's Academy: "Let no one ignorant of geometry enter here."

Instructors of remedial mathematics courses

Most likely the instructors of remedial mathematics courses are part-time or adjunct instructors. They could be full-time lecturers or instructors, not in tenure-track positions, or instructors who are currently secondary school mathematics teachers. Not every college instructor is prepared to teach remedial mathematics. College instructors of remedial mathematics need to be skilled in teaching remediation, and minimally have taught college algebra and trigonometry or a precalculus course. These are courses that most students in remedial mathematics will be taking after their remediation is completed, and their instructors need to know well the demands of these courses. Many tenure-track instructors with doctorates in mathematics are not well suited to teach remedial mathematics courses at the college level because they have never taught courses that are prerequisite for calculus.

How to approach remedial mathematics courses

You have to approach remedial mathematics courses with a commitment to work diligently and with a commitment to understand the content. *It is not enough to get a passing grade.* To have any chance to succeed in a college algebra, college trigonometry, or precalculus course, an understanding of introductory algebra and intermediate algebra is absolutely critical. You also need to have well-developed mathematics study skills. A college precalculus course is a huge step up from an intermediate algebra course. There is almost no comparison between these two courses in terms of expectations. In a precalculus course (normally a one-semester course), the number and complexity of concepts and problems, and the pace at which they must be learned, is taxing even for the well-prepared student.

There are many students in remedial mathematics courses who don't know why they are in college, much less why they are in these courses. Many of these students want to be successful even though they behave no differently, socially or academically, from when they were in these courses in high school. They believe that all they need to do is show up for class and the instructor will do all the work. This didn't work for them in high school, as evidenced by having to take remedial mathematics in college, and it certainly won't work in college.

Lecture versus self-paced instruction

The method of remedial mathematics instruction will vary from college to college, and even within a college. Some remedial mathematics courses will be primarily lecture courses, some will be lecture courses coupled with substantial small group work, and others may be self-paced (perhaps with some lectures now and then). In self-paced courses each student decides when to take a unit test (perhaps within certain time limits), and has the opportunity to take an equivalent test later if the test performance is unacceptable. There are pros and cons to this method of instruction. What is most important is how well the instructional method is carried out, and whether you are self-disciplined. Remedial courses cannot be strictly computational or mechanical. They must require students to be intimately involved with understanding concepts and relationships, and engaged in problem solving. Furthermore, sufficient one-on-one help must be available. It is unwise to enroll in a self-paced course if you are not unusually self-disciplined.

Where to take remedial mathematics courses

It is normally best to take remedial mathematics courses at the college where you will receive your degree, if they are available, because they are constructed to meet the prerequisites of the college-level mathematics courses offered at your college. For example, if you are in a four-year college, you may not be well served taking remedial courses at a two-year college, or through an adult education program, because there may be a diverse class of students enrolled, many of whom are not headed for the same mathematics courses or degree as you are (e.g., some may be headed for an associates degree, which may not require calculus-track courses, but you may be headed for a four-year degree that does). The content of the remedial course, and the standards of the course, may not be what you need, so talk to the instructor before enrolling in the course.

Comments by a university director of remedial mathematics

A university director of remedial mathematics said that he sees two groups of students in remedial mathematics: (1) those who think they know more than they do and get upset when things don't go well, and (2) those who say that they can't do well in mathematics and believe the placement test placed them too high. He continued, saying that "Most students in the former group get angry when things are marked incorrect, and are very adamant about how good they are." Both groups need an attitude change. Those in the second group need to be more positive about their abilities. When he was asked what advice he would give to students who need remediation in mathematics, he said he would tell them the following:

- *"Make sure you are in the right course."*
- *"Don't be angry over the course in which you were placed."*
- *"Realize that just because you had the content in the past does not mean that you understand it."*
- *"Do the work."*
- *"Get help when you need it."*

If you find that you have to take remedial mathematics in college, do what is needed to remedy your deficiencies, and then move on. It is important not to be angry with the instructor, the class, or yourself.

Mathematics to Satisfy College Distribution Requirements

Schools or colleges within a university can differ on how their students can satisfy the mathematics component of the campus distribution requirements. Colleges want to ensure that all their students have certain skills, competencies, and knowledge from various disciplines, including mathematics. It is expected that students will satisfy this mathematics component within their freshman and sophomore years.

Satisfying a mathematics distribution requirement occurs automatically for students enrolled in specific concentrations. For example, the mathematics that engineers have to take to be prepared for their engineering and math-related courses will most likely satisfy this mathematics requirement. The same can be said for business students, students pursuing an elementary or secondary teaching certificate, or students pursuing a concentration in mathematics, physics, or chemistry. For students who will not have mathematics prerequisites for courses they will be taking in their concentration, there may be a variety of courses identified by their college or school that they can select to satisfy the mathematics distribution requirement. For example, courses that are normally accepted include precalculus, calculus, statistics, and a course whose content is generally referred to as liberal arts mathematics. This liberal arts mathematics course covers a variety of mathematics subjects, and has minimal mathematics prerequisites.

It is important for you to know the courses that will satisfy the mathematics distribution requirement for your college, and their mathematics prerequisites. *A word of caution:* Some students, leery of their ability to succeed in almost any mathematics course, put off satisfying the mathematics distribution requirement of their college until they near graduation. *This is unwise.* I have known students who have had their graduation delayed for one or more semesters because they have not successfully satisfied the mathematics distribution requirement after several attempts to do so.

Mathematics for Honors Students

If you have outstanding ability and accomplishment, and are highly motivated, you may wish to be part of an honors program. You can apply to enroll in a college-wide honors program or in the honors program of a specific discipline. Honors programs are comprised of courses and seminars that are more stimulating and challenging then regular college courses.

College-wide honors programs

College-wide honors programs are comprised of a series of honors courses, tutorials, and seminars across an assortment of disciplines. Honors courses are frequently team taught, and interdisciplinary. They may or may not include mathematics. The mathematics course or courses in a college-wide honors program are frequently constructed and taught by mathematics instructors. Examples of topics appropriate for a mathematics honors course include the history of mathematics over a specific period of time, mathematics modeling, the nature of mathematics, mathematics and technology, unsolved problems in mathematics, great theorems of mathematics, and mathematics in science. Mathematics honors courses are typically not part of the regular mathematics curriculum.

Honors courses in the mathematics department

Apart from a college-wide honors program, many mathematics departments have honors courses in mathematics at the lower-division level, and perhaps also at the upper-division level. For example, there may be one or more honors calculus courses. In some mathematics departments, honors courses are offered beyond the standard mathematics curriculum in developing fields such as biomathematics or financial mathematics. To enroll in honors mathematics courses you have to satisfy the criteria set up by the mathematics department. You may be able to enroll for honors credit in specific upper-division mathematics courses that are part of the regular mathematics curriculum, but you will have obligations to satisfy that the non-honors students in the course do not have.

The benefits of being in an honors program or honors course are many, including the quality of the discussions that take place, the fostering of critical thinking and problem solving, the depth of understanding you can gain, and a greater opportunity to improve your oral and written communication skills. Some mathematics departments have an honors program for mathematics concentrators. Check out what is available at your campus.

In some mathematics departments there is opportunity for high achieving and motivated students to work on a research project with a mathematics professor, which is a unique and privileged experience.

Mathematics for Students with Concentrations in Mathematics, Chemistry, Physics, and Engineering

In general, lower-division mathematics for students with concentrations in mathematics, chemistry, physics, and engineering—known as the *hard sciences*—consist of the following:

- Three or four semesters of calculus
- Introductory differential equations
- Matrix algebra or introduction to linear algebra (somewhat similar courses)
- Bridge course to higher mathematics and proofs (many colleges do not have this course)

  A bridge course is a specially constructed mathematics course to better prepare students for specific upper-division mathematics courses. It covers basic mathematical concepts and techniques that are pervasive in many upper-level mathematics courses and includes concepts such as set theory, logic, and methods of mathematical proof. Rather than have each instructor of an upper-division course take the time to cover these techniques, it is deemed best to do it in one course.

This collection of lower-division mathematics courses is often referred to as *the mathematics prerequisites for the mathematics concentration*—a program comprised of upper-division mathematics courses and mathematics related courses referred to as *cognates*. The number of semester credits for these prerequisite courses varies from college to college, but is typically 16 to 18. The packaging of this content can differ significantly across colleges. For example, what is covered in three courses in calculus in one college may be covered in four courses in another. Matrix algebra may be packaged with the last course in calculus or with differential equations, or differential equations may be packaged with the last calculus course. In some colleges, the last calculus course may be different for engineers. For example, in one college the last lower-division calculus course was labeled Calculus III and was a three-credit course for engineering students (Math 205) and a four-credit course for other students (Math 215). The four-credit course covered more topics. (If you intend to enroll in upper-division mathematics courses, do not sell, or otherwise fail to keep, your lower-division mathematics textbooks once the course is over. They will serve as an invaluable resource. Also, keep your course notes and other course materials.)

You may receive advanced placement credit for one or two calculus courses if you received an acceptable score on the Calculus Advanced Placement Examination, or on other examinations acceptable to your college. At a minimum, students who are bound for concentrations in mathematics, chemistry, physics, and engineering should take four years of college preparatory mathematics with the goal of achieving sufficiently well so that upon entering college they qualify for the appropriate introductory calculus course. However, this is not a perfect world. You may be placed in a college course called precalculus, which carries college credit and is the prerequisite for introductory calculus. More details about a precalculus course are presented here because so many students take a precalculus course in college, and because solid knowledge of its content is critical for success in calculus.

College precalculus course
There is not enough space to present the possible scenarios for the types of non-remedial mathematics courses offered at various colleges that precede the first introductory calculus course required of concentrations. The listings of these courses boggle the mind in quantity and description. To make matters easier for you, when I refer to a precalculus course, I am talking about a course (or equivalent course or courses) whose content is *the* prerequisite for introductory calculus. That is, this is the course that immediately precedes introductory cal-

culus. The prerequisites for this precalculus course are normally introductory algebra, high school geometry, intermediate algebra, and an
introduction to trigonometry. Its content is typically included in the
fourth year of high school college preparatory mathematics.

Precalculus is a formidable course containing a significant amount
of content. This content includes some review of basic algebra and analytic geometry concepts and procedures, with emphasis placed on linear, quadratic, rational, logarithmetic, and trigonometric functions, as
well as on their inverses. The graphs of these functions, solutions to
equations involving them, and their applications receive detailed attention. It is typically a one-term course in college, about 14 weeks in duration. A relatively complete treatment of analytic trigonometry is
covered within the course in about five to seven weeks. Imagine the
pace of this development.

A major problem many calculus students have is a lack of precalculus knowledge. Make sure you know precalculus well before enrolling in calculus. There are no "ifs," "ands," or "buts," about this. If
you have any doubt as to whether you are prepared for calculus, see
section 3 of chapter 4 on "How to Determine Successful Completion
of a Prerequisite Course."

There are different types of precalculus instructors. They range
from part-time instructors to tenure-track professors, to those who
want or don't want to teach precalculus, and to those who are or are
not experienced in teaching precalculus. At most four-year colleges,
the chances of getting a tenure-track professor as your precalculus instructor are small. For this course, and perhaps for any mathematics
course, what is important is not tenure-track versus non-tenure track
instructor, but whether the instructor wants to teach the course, uses viable instructional techniques, and has appropriate expectations and
standards.

Calculus sequence

An introductory calculus course can be described as a student's first
"real" mathematics course. Moving from precalculus to calculus is a
big step upward. The power of calculus is undeniable since it is used to
solve important problems in natural science, computer science, engineering, business, economics, medicine, and many other areas. It
should be expected that a subject that is so useful has concepts that are
more sophisticated and more challenging to learn. There are three

major calculus concepts, each related to the other two, that have to be learned and applied; namely, limit, derivative, and definite integral. The power of calculus stems from these concepts. Other concepts and their relationships are outgrowths of these major concepts, and they come fast and furious in a calculus course. The pace of a calculus course, much like that of a one semester precalculus course, is rapid. Wear your track shoes!

Calculus is not easy. It is built on a powerful and elusive idea, namely, limit. You can manage calculus if you have the desire, the prerequisites (not just on paper, but in understanding), and the necessary study habits.

Are subsequent calculus courses more difficult than the first one?

If you had the prerequisites for the first calculus course, and understood it well, then there is no reason why you shouldn't do well in subsequent calculus courses. You may have to spend more time in subsequent courses, but that is completely in your control. Once again, it is about meeting the prerequisites of subsequent courses and then employing appropriate study techniques. For example, in Calculus III you need to understand well the content of the first two calculus courses since it makes extensive use of this content. It should be clear to you by now that just getting a grade of B or better in a course does not necessarily mean that you thoroughly understand the content of that course, or that you have good study techniques. Meaningful grades have a lot to do with the appropriateness of your instructor's expectations and the standards to which you are held accountable. If you have a choice, choose your mathematics instructors wisely. How to do this is addressed in chapter 13.

Instructors of calculus

The background and experiences of instructors teaching calculus courses vary considerably in many if not most colleges. They could be—

- Graduate students working on a masters or doctoral degree in mathematics (i.e., teaching assistants)
- Part- or full-time non-tenure track instructors with or without doctoral degrees
- Full-time tenure track instructors most likely having a doc-

torate in pure or applied mathematics, statistics, or mathematics education

The quality of instruction can vary significantly depending on the instructors' formal and experiential knowledge of teaching, and commitment to their work. In later lower-division mathematics courses, perhaps beginning with Calculus III, the instructors are typically tenure-track faculty with doctoral degrees, except perhaps in some community colleges. Instructors of upper-division mathematics courses are typically tenure track professors, perhaps teaching courses in their mathematics specialty.

Traditional versus innovative calculus courses

Some precalculus and calculus textbooks, written in recent years, are a big departure from traditional textbooks. Not only is there a different focus on the method of learning or teaching used in each approach, but also on the content that is included and how it is packaged. There is more emphasis on the use of technology and applications, and less emphasis on paper and pencil computations and formal proof. Colleges and universities using these textbooks have significantly minimized in-class lecturing, which is being replaced with in-class small group work. Also, significantly more emphasis is placed on out-of-class group projects. If you have the option of choosing an innovative calculus course, determine the advantages and disadvantages in doing so by talking to instructors of innovative and traditional calculus courses. In some colleges you have the option of choosing these innovative courses. Many, if not most colleges, do not have them, and in some colleges that is all you can elect for specific courses.

Problems can result if you take an innovative course followed by a more traditional mathematics course, and vice versa. You might have to switch types of courses because you are transferring to another college, or taking a course in a summer term in a college nearer to your summer home so you can progress faster through your college mathematics. If you are considering going from a traditional course to an innovative course that has the traditional course as a prerequisite, or vice versa, see if this is going to be a problem by talking to instructors of both courses. Many instructors, whether teaching traditional or innovative courses, see this as a significant problem.

Elect subsequent calculus courses in a timely manner

It is unwise to wait more than one regular term to take a subsequent calculus course. For example, if you take a first course in calculus in a winter term, then all efforts should be made to take the second course no later than the following fall term. The second course could be taken in the summer term following the winter term, but before making this decision read section 2 of chapter 6 on discerning whether to elect mathematics courses in a shortened term. Strong students may be able to wait a year before taking a second calculus course, but success becomes more problematic as time passes. There is a wealth of material in a calculus course and more of it is forgotten as time passes.

Mathematics for Students with Concentrations in Business, Management, Social Science, and Biology

In many college and universities, there are mathematics courses specifically constructed for business management, social science, and biology students. There may be only one course required consisting of a smattering of mathematics topics, including calculus. Or, there may be one or two calculus courses required. These calculus courses are typically less formidable than those for students with concentrations in mathematics, chemistry, physics, and engineering. That is, the prerequisites are more relaxed (e.g., trigonometry may not be required), there is less formal theory, and fewer complex computations. At some colleges, a student's performance in these courses may be used to determine admittance to specific concentrations (e.g., business management).

Perhaps the main difference in content of these calculus courses, compared with those for the hard sciences, is that the calculus content and the applications that are chosen are *geared to the specific concentrations for which the courses are required.* Hence, some schools or colleges recommend that students take the calculus courses that are specifically constructed for these concentrations. Many, if not most of them, will allow courses from the sequence of calculus for the hard sciences to satisfy the mathematics requirement for these concentrations. *A word of caution*: If you are pursuing a concentration in business management, the social sciences, or the biological sciences, and wish to take more mathematics at the undergraduate or graduate level, beginning with the third calculus course constructed for students in the hard sciences, realize that the two calculus courses specifically constructed for these concentrations do not adequately prepare you for this third calculus course. In that sense, they are terminal courses. The first two calculus courses for the hard sciences concentrations are the prerequisites for the third calculus course. Hence, you may want to take these two hard sciences calculus courses, even though they are not the ones recommended for your concentration (but most likely allowed as substitutes). Consult the mathematics department or advisors in your school or college.

The backgrounds of instructors for the calculus courses constructed for business management and the biological and social sciences are similar to those for the hard sciences calculus sequence, as discussed earlier.

Mathematics for Preservice Secondary School Teachers

In many colleges, preservice secondary school mathematics teachers are in the same mathematics concentration program as other students. Their mathematics electives are often chosen from courses that are more relevant to teaching mathematics such as number theory, probability and statistics, and history of mathematics.

I would be remiss if I didn't mention the importance to these preservice teachers of their professional sequence of courses from the school, college, or division of education. Most important among these are the methods course in teaching secondary school mathematics, and the student teaching experience. The knowledge gained from these is highly dependent on the qualities of the methods instructor and cooperating teacher(s) in the secondary school.

Mathematics for Preservice Elementary Teachers

College students pursuing a teaching certificate in elementary education have mathematics requirements to satisfy. At most colleges these will consist of two or more courses that have been specifically constructed for them. Common titles for these courses are Mathematics for Elementary Teachers I, II, and III, with the first course being a prerequisite for the second and the second course being a prerequisite for the third. Usually mathematics departments offer these courses but it is the school, college, or division of education that sets the requirements needed to meet the requirements of the State Department of Education.

The teaching of elementary school mathematics is critical to future success in mathematics of elementary school students. If you are pursuing an elementary education teaching certificate, realize the importance of your college mathematics courses to your success as a teacher of mathematics. You need to have quality courses, and you need to do well in them.

Purpose of the courses

The purpose of these mathematics courses is to provide elementary teachers with a perspective for understanding the mathematics they will teach in the elementary and middle schools. To achieve this, school topics are covered from a more sophisticated mathematical viewpoint. *Teachers need to know considerably more about the mathematics topics they teach than what they expect of their students.*

Topics addressed

Over the years, mathematics content that was first introduced in the secondary school, has filtered down into the elementary school. Hence, elementary school teachers are increasingly more responsible for expanding their mathematics background. A myriad of mathematics topics for preservice elementary teachers is addressed in college mathematics courses and most likely include these:

> Elementary set theory, numeration and estimation, the integer/rational/real number systems, elementary number theory, decimal numerals, ratios and proportions, two- and three-dimensional geometry of shape and measurement, geometric concepts of similarity and congruence, coordinate and transformational geometry, data analysis, and statistics and probability.

These mathematics courses need to be of high quality, and taken very seriously by those electing them.

Course emphasis

Emphasis in the courses needs to be on modeling real-world problems using an investigative approach involving problem solving, reasoning and proof, connections, and communication; and using calculator and computer technology to support the investigations.

Course prerequisites

The prerequisites for the first course vary from college to college, but professional mathematics and mathematics education societies recommend a minimum of three years of college preparatory mathematics, including a year of algebra and a year of geometry. If you don't have the prerequisites for the first course, talk to your instructor. Most likely you will have to satisfy them before enrolling in the course.

These courses must be suitably rigorous; hence, a good high school mathematics background by the participants is necessary. Also, they need to be more mathematics courses than methods or "how to teach" courses. This is not to say that methods of teaching specific mathematics, including the use of manipulatives, should not be part of the course—to the contrary. But they must not be the tail that wags the dog. That would be doing you and your future elementary students a disservice.

Significance of the courses

The importance of elementary teachers having a good mathematics background cannot be overstated. In a lifetime of teaching, the number of students who will depend on a specific elementary teacher to give good mathematics instruction is significant. If this is your chosen occupation, your actions will significantly influence your students' attitudes about mathematics, approach to learning mathematics, and knowledge of mathematics. What a weighty responsibility you have! You have within your power the ability to help or harm them in the area of mathematics. You must treat very seriously the mathematics courses for elementary teachers that you have to take. Do not view them as just requirements to complete, but as critically important to your success as an elementary school mathematics teacher. *You should not teach elementary school mathematics if you are not prepared to do so.*

Consider pursuing a major or minor in mathematics

You may want to consider pursuing a mathematics major or minor for preservice elementary teachers. If so, don't wait to begin taking the initial two or three core mathematics courses since they are most likely prerequisites for the other mathematics courses you will need to take. College years go by quickly and you need to fit in the remaining mathematics courses; hence, an early start is important.

Some schools or colleges of education believe so strongly in their preservice elementary teachers having a good background in mathematics, that they require all of them to have at least a mathematics minor. I urge you to consider pursuing at least a mathematics minor.

But you may say, "I am not strong enough in mathematics to pursue a minor." This is all the more reason to pursue a mathematics minor—to make you stronger.

Faculty for these mathematics courses

The backgrounds of faculty for mathematics courses for preservice elementary teachers vary greatly from college to college, and even within a college. These are not easy courses to teach—it takes years to feel comfortable doing so. Instructors need to bring considerable knowledge to the course, knowledge that is not found in the textbook. Instructors can range from part-time instructors with master's degrees, no experience teaching elementary school mathematics, and with little interest in teaching these courses—to instructors who have teaching experience in the elementary school, doctorates in mathematics education, knowledge of the importance of these courses to your chosen profession, and specialize in teaching them. Your instructor plays an important role in your success where success is defined by what you learn, not by the grade you receive. Choose instructors wisely if you have the opportunity.

2. Upper-Division Mathematics Courses

College juniors and seniors normally take upper-division or upper-level mathematics courses. Almost all mathematics departments offer similar basic upper-division mathematics courses. Beyond these basic courses are courses that are not found in all mathematics departments. For example, large colleges will have more upper-division mathematics offerings because of vast numbers of concentrations available at these universities that have specific mathematics requirements. Examples of upper-division courses include these, with perhaps some changes in titles at different colleges for similar courses:

> Elementary Number Theory, Modern Algebra, Linear Algebra, Stochastic Processes, Mathematical Statistics, College Geometry, Advanced Calculus, Functions of a Complex Variable, Numerical Analysis, Mathematical Modeling, Matrix Computation, History of Mathematics, Topology, Independent Studies in Mathematics, Fourier Analysis, Graph Theory, Integral Equations, Non-Linear Differential Equations, Linear

Partial Differential Equations, Operations Research, Founda-
tions of Mathematics, Combinatorics, Linear Programming
Methods, Differential Geometry, and specially constructed
courses for preservice elementary or secondary teachers who
are pursing a mathematics major or minor.

Prerequisites for Upper-Division Courses

Most upper-division mathematics courses, except those specially constructed for preservice elementary teachers, have as prerequisites most
of the lower-division mathematics courses, which include the calculus
sequence, differential equations, and matrix algebra. For mathematics
concentrators, these lower-division courses comprise what is called the
mathematics pre-concentration program. One criterion for success in
most upper-division mathematics is quality achievement in these
lower-division courses.

Transitioning from Lower- to Upper-Level Mathematics Courses

The most important transition to make from lower- to upper-level
mathematics courses is a basic change in mindset that all but about the
top 20% or so of students need to make. The attitude of the average
(and below average) student making the transition to upper-level mathematics is to think of all that has come before as a number of unrelated
topics, rather than looking for connections among topics. Perhaps this
is partly a fault of the teaching that takes place at the lower level.
These students learn the material well enough to make it to the examination, or at best to the final examination, and then forget what they
learned. I repeat a piece of advice to all students planning to take
upper-division mathematics: If you intend to enroll in upper-division
mathematics courses, do not sell, or otherwise fail to keep, your lower-
division mathematics textbooks once the course is over. They will
serve as an invaluable resource. Also, keep your course notes and other
course materials.

There is no end to the learning skills that will stand you in good
stead in upper-level mathematics courses. Besides the learning skills of
making connections and retaining what is learned, other learning skills
in mathematics include understanding, reasoning, communicating
(verbally and in writing), working diligently and in a timely manner,

persevering, reading, seeking help (from classmates, instructor, etc.), asking questions, having a positive attitude, and being facile in understanding and using symbolic form.

Unfortunately, there are students who manage to get very good grades in lower-level mathematics but struggle in upper-level mathematics, at least initially. In general, this should not happen, but at least two reasons could account for this, namely, a lack of many of the learning skills just presented, coupled with instructors in lower-division mathematics courses who have expectations that are too far below those of instructors of upper-level courses. Many upper-division mathematics courses are taught by instructors whose mathematics specialty is the content addressed by the course. Hence, you can expect more content expertise from these instructors, and that is a good thing. Upper-level courses carry fewer credit hours than precalculus and calculus courses, but don't be fooled into thinking that they are less work. On the contrary, it is highly likely that a three-credit upper-division course requires considerably more work than a four-credit lower-division course. But upper-level students are typically better equipped to handle the extra workload. Finally, in terms of differences, you can expect upper-level courses to have smaller numbers of students. This is definitely to your benefit.

Who Takes Upper-Division Mathematics Courses?

Students who take upper-division mathematics courses are those who desire more mathematics knowledge, or their fields of concentration require them to take these courses. In a typical upper-division mathematics course many of the students are mathematics concentrators and most of them are now taking two or more mathematics classes each term. Some of these wish to become teachers of secondary school mathematics, others may want to go on to graduate school in order to become college mathematics professors or to pursue an occupation in which knowledge of higher mathematics is required. Some want to primarily study mathematical statistics, or computer and computational mathematics. Others may wish to divide their program in some manner between mathematics and related fields such as physics, chemistry, or engineering. For example, at some colleges students majoring in specific engineering programs may pursue a concurrent bachelor of science degree in engineering mathematics.

Nature of the Content

Some upper-level mathematics courses are heavily laden with theory requiring students to have the ability to understand and construct proofs. Examples of these courses are: Modern and Linear Algebra, Advanced Calculus, Topology, Functions of a Complex Variable, Introduction to Numerical Analysis, and Survey of Geometry. Other courses are more oriented toward applications requiring students to have the ability to solve applied problems. The use of technology is customary in these courses. Examples of such courses are Applied Combinatorics, Dynamical Systems, Applied Statistics, and Matrix Computation.

Nature of the Competition

Students who considered themselves average in their freshman-sophomore college mathematics performance, but are attracted to mathematics, should not shy away from taking upper-level mathematics courses, or indeed from majoring or minoring in the subject, based on an inordinate fear of competition at that level. It is not just the mathematical geniuses of the freshman-sophomore level who elect to concentrate in mathematics or to take a number of upper-level mathematics classes. Many of the students you may remember as having been the strongest in your calculus classes will have gone on to major in a physical science, biology, engineering, or computer science. Granted, some of these will also decide to take some upper-level mathematics classes that appear interesting or useful to them so they will not disappear completely from upper-level mathematics courses. Invariably, however, many of the students who elect to go on in mathematics are from the ranks of those who had to work hard to earn their B grades in the calculus sequence and introductory linear algebra. They discover that mathematics is their primary interest or decide to pursue a career that requires a background in mathematics.

How any particular student compares with the competition will depend on that student's level of maturity, commitment, interest, and mathematical talent. A typical student should not be surprised if his or her relative standing, in terms of an informal percentile comparison to one's peers based on grades, did not change much in the junior and senior years from what it was in the first two years, even though the pool of competing students is now much different and, in particular, much smaller.

Working with Other Students

Because a higher percentage of the students in a given mathematics class at this level are mathematics concentrators, they have a serious and direct interest and motivation (even enthusiasm in some cases) for being there. It might be said that, for many of these students, mathematics classes are now an end in themselves and not a means to an end, as was often the case for the students' required freshman-sophomore courses. It also means that, as early as a student's first term at the upper level, he or she may notice that some of the same faces are starting to show up in more than one class. (Once this is noticed, it will likely continue through to graduation.) These factors point out that there is an advantage in, and an opportunity for, cultivating a compatible group of fellow students with whom to form an effective working group (which may turn out to be the basis for a social group as well).

Fellow students with whom you regularly discuss course content, work together on assigned problems, and compare notes on attitudes, opinions, knowledge, and goals related to the mathematics program or to careers based on mathematics, can be one of your most valuable assets. Consider joining whatever mathematics-oriented student groups or organizations your department supports (e.g., Pi Mu Epsilon chapter if eligible, or a local mathematics club), and consider participating in student competitions such as the renowned Putnam Exam. Try to attend whatever guest lectures, colloquia, or other special events your department sponsors.

Means of Evaluating Students

At the junior-senior level, evaluation is not likely to be based on in-class exams covering in-class content. Take-home problem sets or homework assignments are usually required and are graded, but in some courses, such as statistics or computational mathematics, there may be one or more large projects, often involving some computer-related work. In the latter instance, these projects may involve working in groups as well. An additional feature of these methods of evaluating is that they can be, and often are, used to extend the material that may be presented or discussed in class.

As a second point, no matter what means of evaluation is used, including in-class examinations, success in completing a problem or exercise is more likely to be based on the quality of the process of reaching the goal, rather than simply arriving somehow at the correct answer or desired conclusion. This should be true at the lower level as well, but may not be in many instances.

Majoring or Concentrating in Mathematics

If you like mathematics and do well in mathematics you might want to consider majoring or concentrating in mathematics. In most colleges, all mathematics majors begin their upper-division work by taking some required upper-division mathematics courses and then elect one of several tracks or options to pursue, which in turn partially dictates the remaining courses that are required. As examples, there may be an algebra or analysis track, or a pure or applied mathematics track. Some universities include a track for those students desiring to teach mathematics in the secondary school (call it an education track, if you will).

Required number of credit hours

The number of upper-division credit hours in mathematics required for the mathematics major varies across mathematics departments—the average is about 24 credits. Most likely several upper-division courses that use mathematics (e.g., engineering, physics, computer science), have to be taken, which are called *cognates*.

Mathematics prerequisites for the major

Generally, the prerequisites for the mathematics major consist of the calculus-track lower-division mathematics courses (through differential equations), and matrix or linear algebra. The average number of credit hours of these prerequisites ranges from 16 to 18.

Mathematics minors are also available

The number of mathematics credit hours comprising a mathematics minor varies from college to college. A typical mathematics minor consists of 12 credit hours of upper-division mathematics.

Secondary Teaching Major or Minor in Mathematics

At some colleges there are specially constructed upper-division mathematics courses for students pursuing a secondary *teaching* major or minor in mathematics. However, at many colleges, these students select from the same mathematics courses as do any mathematics major or minor. There are pros and cons for having special mathematics courses only populated by those pursuing a secondary teaching major or minor in mathematics. One advantage is that more effort is made to relate the content to the secondary mathematics curriculum. One major disadvantage is a "watering down" of the mathematics content to the extent that a weaker knowledge of mathematics is the result.

It is possible to get a secondary *teaching* major or minor in mathematics without getting a *regular* mathematics major or minor. The difference between a secondary mathematics teaching major or minor and a regular mathematics major or minor that allows you to teach mathematics, differs among various colleges. Suffice it to say that the requirements for the teaching major or minor have to be approved by the State Department of Education, and those for the regular mathematics major, *which allows one to teach*, have to be approved by the mathematics department (since it is their major) *and* State Department of Education. Students with bachelor's degrees in an area other than mathematics, many of whom are already in a professional career, may decide to become secondary mathematics teachers. These college graduates often choose to get a secondary *teaching* major or minor in mathematics since it requires fewer hours of upper-division mathematics than the regular mathematics major or minor. Make sure you understand the situation as it exists at your college.

Mathematics Prerequisites

If you are considering enrolling in a mathematics course, or are already in a mathematics course, an important question to ask yourself is, "Do I have what it takes in terms of prerequisite mathematics knowledge and study skills to be successful in this course?" If you do not have the necessary study skills then it is probable that you also don't have the required knowledge. Since this book's primary focus is on developing viable mathematics study skills, it will help you determine how you may be lacking in this area and what you need to do about it. It is possible to have good mathematics study skills, yet lack the mathematics content requirements for a course. This can be due to not taking the prerequisite courses, the prerequisite courses having inappropriate standards or content, or unsatisfactory achievement in the prerequisite courses (for whatever reasons). This chapter addresses the importance of having the prerequisite mathematics knowledge for a mathematics course, how you can determine if you have it, and if not, what to do about it.

The sections in this chapter are titled:

1. Importance of Having Course Prerequisites
2. Identifying Course Prerequisites
3. How to Determine Successful Completion of a Prerequisite Course
4. Prerequisite Mathematics Courses for Transfer Students

1. Importance of Having Course Prerequisites

Prerequisite knowledge for a course is knowledge that you are required to have so you are equipped to learn new knowledge in that course. It is probably safe to say that no one has perfect prerequisite knowledge for a course. Are there many calculus students who know all the content from the high school mathematics curriculum? Of course not, yet that knowledge is considered prerequisite for calculus. There are times when a small piece of knowledge that you may not understand is needed in a course. Most likely you can gain this knowledge quickly, so not having it upfront won't cause you any serious problems in the course. Identifying a prerequisite topic you don't understand, before you attend a class session that uses it, will give you time to learn it prior to class. This chapter addresses a lack of prerequisite knowledge so substantial that there will be little chance for success in the course. The pace of college mathematics is too fast to expect to learn this knowledge *and* the knowledge of the course that requires it. You would have to learn two courses while in one course, and in a manner that coordinates the learning of each topic of the prerequisite course with those in the subsequent course that uses it. This is virtually impossible to do.

Good instructors almost always conduct a brief review of important prerequisite knowledge that is needed for a particular topic. But it is important for you to understand the use of the word "review" in this context. *When a topic is reviewed it is done with the understanding that the topic was, at one time, understood by those receiving the review.* The review helps a student recall what he or she once knew. Reviews are conducted quickly; hence, what is done in reviewing a topic falls far short of what is done when the topic is taught for the first time. The latter is often done over a few days with appropriate assignments given, resulting feedback provided on them, all of which is then reviewed in preparation for a test on the topic. A *review* of a topic may be completed in a subsequent course in a few minutes. Here is an example from calculus: If at the end of a Calculus I course you are very confused on how to use the chain rule in finding derivatives of composite functions, you will struggle in Calculus II when you have to use it, not only on functions you have studied in Calculus I, but also on functions introduced for the first time in Calculus II. Rather than spending a few minutes reviewing the chain rule, you may have to spend hours over a span of days to learn it, perhaps with some tutorial help. If you didn't learn it when first exposed to it, why would you think it would take significantly less time to learn it in a subsequent course? In the meantime, the course you are enrolled in keeps "trucking along," and you are getting further behind.

If the content of a prerequisite course was understood at one time, then the question becomes, "Was it too long ago?" *If it was too long ago, a review might not be enough.* That is, you may have to relearn it by retaking the course. What constitutes "too long" varies from person to person and is a function of many variables, including the quality of the initial instruction and learning. A year may be too long for you, but it might not be for other students. Perhaps you can wait two years before taking the subsequent course, although that is highly unlikely. Adults returning to college after many years away, who need to take a precalculus or calculus course, may need to start by retaking elementary or intermediate algebra, regardless of how well they understood it when in high school. It is critical to begin in an appropriate level course. Some mathematics advisors believe that if it was at least two years since you had the prerequisite course, you should re-enroll in the course. There can be exceptions to this rule. If you had an excellent instructor for the prerequisite course, did very well, have strong mathematics ability and study skills, and are willing and have the time to review the prerequisite material as it comes up in the next course, than it may be a good decision to take the next course. Anything less than that requires serious thought before you decide to take the next course without repeating the prerequisite course.

In summary, it is a misuse of the word "review" to say that you need to review mathematics content that you never understood. *Don't enroll in a mathematics course unless you satisfy the mathematics prerequisites of the course.* The next section addresses ways to determine if you satisfy these prerequisites.

2. Identifying Course Prerequisites

You need to know whether you have the prerequisites for a mathematics course you want to elect, or for a mathematics course you are currently enrolled in that is causing you some difficulty. You can determine this by asking yourself two questions: (1) Have I *completed* the prerequisite course or its equivalent? (2) Have I *successfully* completed the prerequisite course or its equivalent?

Have You Completed the Prerequisite Course or Its Equivalent?

Who would think of taking a mathematics course if he or she had not completed the prerequisite course or its equivalent? More students than you would believe. Students who do so are either unaware of the prerequisite course or wrongly think they don't need it. They may also not understand, or don't want to understand, the interdependence of prerequisites and courses. For example, suppose a student wants to take an introductory calculus course, but first needs to take a precalculus course, which has an intermediate algebra prerequisite. After the student passes the intermediate algebra course with a good understanding of it, he or she enrolls in the precalculus course, but eventually drops it or receives a poor grade. The next semester the student enrolls in an introductory calculus course. Does this make sense to you? I hope you are saying, "No." Unfortunately, this is not just a hypothetical case—it happens, and far too often.

Here are typical prerequisites for some basic lower-division mathematics courses:

Course	Prerequisite Course
Algebra I (remedial)	Arithmetic (remedial)
Algebra II (remedial)	Algebra I (remedial)
Precalculus	Algebra II (remedial)
Calculus I	Precalculus
Calculus II	Calculus I
Calculus III	Calculus II
Differential Equations	Calculus III
Matrix or Linear Algebra	Calculus II

In basic upper-division mathematics courses, the prerequisites are normally one or more of Calculus II, Calculus III, Differential Equations, and lower-division (i.e., introductory level) Matrix or Linear Algebra.

Have You *Successfully* Completed the Prerequisite Course or Its Equivalent?

Suppose you received a passing grade for a course. Does this mean you are prepared for the next course that has this as a prerequisite? Perhaps! A grade of D is a passing grade, but that grade means you are not prepared for the next course. A grade of C should mean that you are, at best, *marginally* prepared; but often that is not what it means from the hands of an instructor who grades too high. Many instructors grade too high, which is known as grade inflation. A grade of A or B should mean that you are prepared, but don't bet your life on it. You may be happy to receive an A or B, but if it is an inflated grade, you will suffer the consequences in the next mathematics course, in a mathematics course further down the road, or in a mathematics related course.

I have seen many college freshmen with four years of college preparatory mathematics, who received all A's and B's in their last two years of high school mathematics, take a college mathematics placement test, and are told to repeat the equivalent of the last two years of high school mathematics.

It is not uncommon for instructors to have students, who received A's in their previous college course, struggle to attain B's or C's from them. Or, they had students with B's or C's in their previous college course, struggle to attain D's or E's in the course. This could be caused by a change in study habits, higher standards, or a more difficult subsequent course. However, it is often caused by previous instructors having weak or inappropriate content objectives or standards. Couple this with the students not knowing how to study more appropriate content objectives or rise to higher standards, and problems will arise.

I asked college students in calculus to tell me what advice they would give students entering college calculus. A Calculus I student said, "An "A" in Precalculus doesn't guarantee a good grade in Calculus I." Another Calculus I student said this:

> My biggest problem with Calculus I was with the trigonometry that was assumed to be known. As a freshman finishing my first term, I had precalculus in high school, which included a fair dose of trigonometry. I was aware that Calculus I required some knowledge of trigonometry, and judging by the placement test given at freshman orientation, I didn't think that I was that lacking in it. In calculus there were many identities that I never saw before, and we had to work with many trigonometric graphs. I wish my high school instructor had spent more time on these ideas for they are relevant and useful to calculus.

A Calculus II student said, "Knowledge of algebra and trigonometry is a must in order to do well in calculus."

In summary, *the grade you receive in a prerequisite course is not necessarily a reliable indicator of how successful you were in the course, or how successful you will be in a course that has it as a prerequisite.* Your question now should be: "If I receive a grade of A, B, or C in a mathematics course, how will I know if I am adequately prepared for the next mathematics course?"

3. How to Determine Successful Completion of a Prerequisite Course

Since respectable course grades may not be indicators that you have the prerequisite knowledge, you will need other means to determine this such as—

- A college placement examination in mathematics
- Your own assessment
- Consultation with an instructor to help you discern if you had a quality prerequisite course

College Placement Examination in Mathematics

If you are an entering college freshman, opportunity is typically provided for you to take a mathematics placement examination during college orientation, or perhaps even at another time if you missed it during orientation. If you are a transfer student, you can request to take this test. Typically, results of this test, along with other information about you, including your high school mathematics courses and your mathematics ACT or SAT score, determine your mathematics placement. If course grades could be taken at face value, there would be no need for such an examination. The fact that it exists should tell you something about the unreliability of grades.

How reliable is a college's placement examination in mathematics? It depends on whether the examination is well constructed. Does it test enough topics, and topics that are good predictors of success in specific courses into which you may be placed? Does it have subscores that address the content of particular courses, such as Algebra II, Precalculus, or Calculus I? The placement examinations at most colleges test high school mathematics content through precalculus. Your score, along with other information previously mentioned, is used to place you in one of several courses, perhaps Algebra II, Precalculus, or Calculus I. Typically, this examination is used to determine if you qualify for Calculus I, and if not, whether you then qualify for Precalculus. Some colleges give additional tests to determine your placement in a course lower than Precalculus.

Should you accept the results of your mathematics placement test? *You may be placed too high, too low, or just right.* That is your decision, although you may not have a choice if your college mandates your placement. If you have a choice, don't make your decision lightly. There is a tendency to ignore a placement if it is lower than what you expected, for that means you have to retake content, which is most likely an unattractive option for you. This is particularly difficult if you had calculus in high school and are being asked to take a precalculus or algebra course in college. Many do not accept this placement. This lack of acceptance can be detrimental to them. Some students are placed too low, and that can also be problematic.

Students who believe they are placed too high have to be sure that it is not unreasonable fear causing them to feel this way. For example, I knew a student who was at the top of her high school graduating class, received all A's in her four years of high school mathematics, and was placed into Calculus I based upon a college mathematics placement test. She had a difficult time accepting this placement because she wanted to start off her college career with good grades and was leery about getting a good grade in Calculus I. Unfortunately, she lacked confidence in her ability and background, and her less than stellar performance in the calculus course appeared to be a self-fulfilling prophecy.

How do mathematics instructors feel about students who don't accept their mathematics placements if they were placed lower than where they thought they should be? It upsets many of them for then they have to deal with students who are not prepared for their courses. It is not good for the instructor, the unprepared students, or their classmates. It seems to be happening more and more, causing more and more mathematics departments to make mathematics placement mandatory. How do I feel about mandatory placement? Again, it depends on the information that is used to make the placement. I am not in favor of mandatory placement if there is no opportunity to appeal the decision to a competent mathematics instructor, or to retake the examination. Unusual circumstances on the day the placement examination was given may have prevented the student from doing well, or it may have been an unreliable and invalid placement test. I have known students who were placed too low and it was disconcerting to me, and to them, to realize that they had to unnecessarily repeat a course.

There are additional ways to determine if you have the prerequisites for a course you wish to elect, or for a course in which you are currently enrolled and are experiencing unusual difficulty. For the latter situation, you may be in the course because the college suggested or required it for you; you may have declined their suggested placement; or you chose it without a placement recommendation. We now turn to another way to determine if you have the prerequisites for a mathematics course, and that is through your own assessment.

Your Own Assessment

You do not want to ignore how you feel about your placement, regardless of whether you feel you have been placed too low, too high, or just right. You can lend credence to your belief if it is based on the answers you give to these, and other questions that you devise *concerning the prerequisite course*:

- Did you have confidence in the standards of the instructor?
- Did you have confidence in the knowledge of the instructor?
- Did you know students who had other instructors who were teaching the course at a higher level?
- Did you get higher grades on your tests than you deserved?
- Were the tests surprisingly easy?
- Were you surprised at your course grade because it was higher than what you expected?
- Were you slacking off in the course, yet had little difficulty with the examinations?
- Were students missing a lot from class, yet received good grades on the examinations?
- Did almost nobody drop the course?
- Did the final course grades seem higher than they should have been?
- Were there frequent opportunities to do extra credit work to make up for poor performances on examinations?

- Were the examinations basically computational?
- Was it usual for class periods to end early?

Have you discovered that if you consistently think something isn't right, it usually isn't?

Consult with an Instructor to Help You Discern If You Had a Quality Prerequisite Course

If you intend to enroll in a mathematics course and question whether you have the necessary prerequisites, ask an instructor who has taught this course to look over your graded examinations from the prerequisite course to see if he or she believes you had a quality course and did well in it. It is not only your performance on the examinations that needs to be examined, but also the examination questions themselves. Were appropriate questions asked that would accurately characterize the content that should be in the prerequisite course? Realize that the quality of advice you receive, no matter the topic, is dependent on the expertise of the person giving it. It can help to get more than one opinion.

If you do not have your examinations from the prerequisite course, you can ask an instructor who has taught the course, to give you exercises to work in his or her office that will determine whether you understand the key prerequisites.

For example, if you want to enroll in a—

(a) Precalculus course, you can be asked questions from an Intermediate Algebra course that relate to linear and quadratic equations, including their graphs and algebraic manipulations.
(b) Calculus I course, you can be asked questions from a Precalculus course that relate to trigonometric functions, including their domains, ranges, graphs, and compositions.
(c) Calculus II course, you can be asked questions from a Calculus I course that relate to your understanding of the concepts of the derivative, definite integral, and the chain rule for finding derivatives of composite functions.

Have the instructor grade your responses with you present and discuss the results with you. As you are talking about the prerequisite course, the instructor can ask you additional questions about the course, including the way you approached your study. After all of this you should be equipped to make the right decision.

What If You Are Experiencing Difficulty in a Course and Question Whether You Have the Prerequisites?

If you are experiencing more than the typical difficulty that is part of any mathematics course, you may wonder if you have the prerequisite knowledge to continue on in the course. You question if it was a mistake to enroll in the course, regardless of whether you were placed there or you freely chose it. Ask your instructor to help you determine whether you are adequately prepared. The two of you can look at examinations that you have taken in the course to see if a lack of prerequisite knowledge is evident. Also, both of you can look at some of your homework to see if a lack of prerequisite knowledge appears there, and whether the homework was being done. Finally, just dialoguing in general with your instructor on the difficulty you are experiencing can be illuminating. Nothing beats a good conversation where both parties make comments and ask each other questions. Usually when I conversed with a student about his or her poor performance, I could diagnose the student's problem. In the main, it was due to one or more of the following: a lack of prerequisite knowledge, poor study habits (which can relate to a lack of prerequisite knowledge), or personal problems. Once your meeting takes place with your instructor, you will have to make a decision about whether to drop the course, and if so, to determine if you need to retake the prerequisite course. *Don't ignore the advice of your instructor*, but you are the one that has to make the decision. (See section 2 of chapter 25 on dropping a mathematics course.)

---

4. Prerequisite Mathematics Concerns for Transfer Students

If you are transferring to another college or university, you need to know how this change may affect the mathematics you will be taking at the transfer institution. The related questions you need answered include these:

1. How important is it to take all of my introductory or lower-level mathematics courses at one of the two institutions, rather than dividing them between these institutions?
2. If I divide the courses between the institutions, how can I determine what I need to know about the first mathematics course I will be taking at the transfer institution, and how can I better prepare myself for this course?

These questions, among others, are addressed in this section.

At Which Institution Should You Take Your Remedial or Lower-Division Mathematics Courses?

You may have little choice as to where you take your remedial or lower-division mathematics courses. For example, you may have made the decision to transfer to another institution after you had already completed some or all of your mathematics courses. But if you still have a choice, reasons are given here for—

1. Taking all of your remedial or lower-division mathematics courses at one institution
2. Dividing your remedial or lower-division mathematics courses between the two institutions

Reasons for taking all of your courses at one institution

The reasons for taking all of your courses at one institution include:

The sequencing of content may differ across institutions. For example, the content comprising the calculus sequence, the mathematics sequence for pre-service elementary teachers, and the remedial mathematics sequence, may be packaged differently across institutions. Continuity of content will be interrupted if not all are taken at the same institution.

The length of terms may be different at the institutions. Differences in lengths of terms can cause differences in the sequencing of content between the institutions (e.g., content covered in a first course at the transfer institution may not be covered in the first course at the other institution). Hence, a student taking a second course in the sequence at the transfer institution will not get this content.

Mathematics course goals and expectations may be different from one college to another. For example: the use of manipulatives, technology, or a focus on relating ideas in mathematics as opposed to a focus on computation, may play a major role in one college but not in another. Institutions vary in quality, resulting in different expectations from their instructors.

Textbooks used for sequences of courses may be different. Textbooks for a specific subject area may develop the content very differently. That may not be a problem since there are some safeguards put in place to help ensure that courses with supposedly comparable content are equivalent courses across institutions. Specific courses have to be accepted as equivalent to courses at your transfer institution before specific transfer credit is given for those courses. However, these decisions are often made by institutional staff outside of the mathematics department based on incomplete information gained from college bulletins. This can cause the acceptance of courses as equivalent to specific courses, when they are not. Being given transfer credit does not guarantee that you have an adequate background for a course at the institution you are transferring to that has this course as a prerequisite. Solutions for this are given later in the section.

You don't want to get behind in your education. Waiting to take mathematics courses until you are at the transfer institution may get you off the normal track of when these courses are typically taken. This could delay your enrollment in other courses that have these courses as prerequisites, which could delay your graduation.

Reasons for dividing your courses between institutions

The reasons to take some of your remedial and lower-division mathematics courses at your transfer institution include:

More appropriate development of the content may exist. Depending on the nature and quality of the institution and mathematics department, the transfer institution may provide a more meaningful and higher level development of the content, or a development that conforms better to the institution's concentration programs. The problem is being able to determine if the courses at the transfer college are more appropriate for you. There can be more variance in the quality of a course among instructors in the same college than there is in comparable courses across colleges.

Transferring to another college cannot be put off until all your remedial or lower-division mathematics courses are taken. For example, your family's move to another state may cause you to move with them and enroll in another college. Or, doing badly in a mathematics course may result in you retaking the course in a summer term at another college near where you live during the summer. Taking a summer course will get you back on your schedule for taking a subsequent mathematics course after you move back to the college in which you are enrolled.

Examples of Articulation Problems between Mathematics Courses at Different Institutions

For the *calculus sequence*, three examples of typical articulation problems follow:

Example 1. At College A, the last unit of the Calculus II course is on sequences and series. Its Calculus III course begins with a unit on vector calculus. At College B, the last unit of Calculus II is devoted to vector calculus and the first unit in Calculus III is on sequences and series. Hence, if you took Calculus II at College A and Calculus III at College B, you would repeat a unit on sequences and series and miss altogether a unit on vector calculus. Or, if you took Calculus II at College B and Calculus III at College A, you would repeat a unit on vector calculus and miss altogether a unit on sequences and series. The problem with each of these scenarios, and it is a major one, is not repeating a unit, but completing your calculus sequence missing a major unit of content. This is not a hypothetical example!

Example 2. At College A, Calculus I and II are taught in a traditional manner using a traditional textbook. In College B, Calculus I and II are taught in an innovative way using an innovative textbook. This means that there are significant differences in the way these courses are taught. The differences include computational emphasis, use of technology, role of instructor, use of class time, content inclusion, how the content is developed, and student projects. You can imagine the problems that result from taking Calculus I in one college and Calculus II in the other. Regardless of the college where the first course is taken, you will not be adequately prepared for the second course. You may be able to make this transition; but to be forewarned is to be forearmed.

Example 3. A community college's calculus sequence consists of three 5-credit courses. The college is on a 16-week semester calendar. A four-year college has a 14-week trimester calendar and its calculus sequence consists of three 4-credit courses. For coverage of basically the same content, the four-year college uses about 168 class hours and the community college uses about 240 class hours, which is 43 percent more class hours. This dictates that the pace of the course in the four-year college is significantly faster. One result is that its Calculus I course addresses the calculus of exponential and logarithmetic functions, whereas the community college addresses this in Calculus II. It is easy to see the problem that results from a community college student taking Calculus I at the community college and Calculus II at the four-year college. This student is at a distinct disadvantage in the Calculus II course, especially since the Calculus II course begins with content that uses what the student missed. It is very difficult to learn the missed content and at the same time be responsible for applying it to new situations. At a minimum, the student has to study the content missed *in advance* of entering Calculus II, which may or may not be sufficient preparation.

For *mathematics for pre-service elementary teachers*, an example of an articulation problem follows:

Example. College A, a community college, has a two-course sequence in mathematics for pre-service elementary teachers, totaling six credit hours. College B, a teacher training institution, has a three-course sequence for these pre-service teachers, totaling nine credit hours. Reviewing the course descriptions in both colleges, it appears that these sequences are comparable in content, but in reality they are far from it. Unlike topics covered at College A, the topics in the courses at College B are covered in much greater depth. They are also related in a more significant way to the mathematics of the elementary school and the means used to teach it. It is easy to see the difficulties that can arise if a student in College A has taken the first course there and will be taking the last two courses at College B. Similarly, problems will arise if a student has taken both courses at College A and the third course at College B. An articulation problem occurs with how the topics are handled.

For *remedial mathematics courses*, an example of an articulation problem follows:

Example. College A offers remedial courses, Beginning Algebra I and Intermediate Algebra, as does College B. The algebra courses in College A, a community college, focus on preparing students for an associate's degree since these students heavily populate the course. Preparing these students for the mathematics that they will need in their occupational programs is the foremost consideration. The primary focus of these courses in College B is to prepare students for a precalculus course that they will take prior to taking calculus. Suppose a student, bound for precalculus, takes one or both algebra courses at College A and then transfers to College B where he or she elects a precalculus course. The problem is obvious—the student most likely will struggle in the precalculus course because of insufficient depth in knowledge of algebra.

We now discuss how a transfer student can prepare to deal with potential mathematics problems that result from transferring colleges.

Solving Articulation Problems in Mathematics for Transfer Students

If you intend to transfer to another institution, or have permission to take a mathematics course at another institution, find out in advance what pitfalls await you. Before making decisions, know the various options that are available to you, and their pros and cons. This necessitates knowing how the mathematics courses you will be taking at your first college mesh with those you may want to take at your transfer college. To do this, you need course syllabi at both institutions to see how courses match in content and teaching methodologies (although the latter may be difficult to ascertain from a syllabus). You may need to speak with knowledgeable instructors at both institutions.

Who are the best college personnel to give advice? *In general, the individuals who can help you the most are in the mathematics department, not in the admissions office.* The most reliable sources are mathematics or mathematics education instructors, the mathematics department advisor, or mathematics department chairperson. Admissions counselors are good for some general questions (e.g., determining if a course is transferable and its equivalency at the transfer institution), but they are not that familiar with the intimate details of specific mathematics content and teaching methodologies. For this you need to talk to someone in the trenches. *Meeting the prerequisites of a course on paper and meeting them in reality are two very different things.*

You need to decide whether to take all of your remedial or lower-division mathematics courses at one institution. If you decide against this, then you need to determine what will help you be successful at the transfer institution if articulation in mathematics between the two institutions is problematic. Here are some suggestions:

Suppose you wish to take Calculus II at your transfer institution, but lack knowledge of a specific unit of content from Calculus I. Learning this content when enrolled in Calculus II can be difficult, especially if the mathematics studied at the outset of the course makes use of this content. The answer is quite simple: *do as much as you can to learn the content before the course begins.* You can learn this content if you are motivated, have the necessary time to learn it, and do what is necessary to learn it. For example, if you missed out on the calculus of exponential and logarithmetic functions, plan to spend about 15 hours learning this material prior to taking the course that uses it.

The question then becomes, "How can I learn this missed content?" The answer is that you need to do the things that you would normally do if you were enrolled in the course where it is taught. That is, read the sections in one or more textbooks, work a mixture of exercises, consult a solutions manual, and seek help from mathematics tutors or instructors. For this to work you must have the attitude that you can learn it, and the time to learn it.

Finally, what can you do to minimize problems at the transfer institution if the goals and objectives in the mathematics courses are different and perhaps at a higher level than those at the previous institution? The answer is not complicated: you have to work more diligently and make heavy use of the means available to support your work. It will be a challenge since the changes you need to make won't happen overnight. This extra effort of yours is not only directed to the content of the course you will be taking, but also to the content of the prerequisite course as presented at the transfer institution.

5

Examinations to Obtain Advanced Placement or College Credit in Mathematics

There are a variety of examinations you can take to show that you understand the content of specific college mathematics courses. Depending on the type of examination taken and the score received on it, you can place out of a course. That is, you will have satisfied the requirements of the course. You can also receive academic credit for the course, but this, and the number of credits you receive, is a decision of the college granting the credit. Different colleges will use different examinations or examination scores in making their decisions.

The means for placing out of a course or receiving academic credit for a course include these:

1. Advanced Placement Program (AP) of the College Board
2. College Level Examination Program (CLEP)
3. Examinations constructed by colleges
4. International Baccalaureate Program (IB)

These means are not all available at every college.

1. College Board Advanced Placement Program in Mathematics

The College Board Advanced Placement Program is a nationally recognized program of the College Entrance Examination Board (CEEB) that gives college credit, placement, or both to students while still in high school. In mathematics, the available subject areas are calculus and statistics. For calculus, there are two courses, Calculus AB and Calculus BC, and an examination for each. The credit or placement given is determined by each college based on its sequence of courses and the score received on an AP examination. Each AP calculus course is one year long. Calculus AB has some content on elementary functions, but Calculus BC does not, assuming this was learned well enough prior to the course. All the calculus topics of Calculus AB are included in Calculus BC, but the latter has additional topics. The Calculus BC examination has a Calculus AB sub-score. The depth of understanding required in both courses on common topics is comparable. These calculus courses are challenging, demanding, and taught like college-level courses. Students taking an AP calculus course should do so with the goal of placing out of a comparable college calculus course.

Prerequisites for AP Calculus

The prerequisites for AP Calculus are a demonstrated knowledge of the content of four years of college preparatory mathematics. Typically this prerequisite is accomplished by beginning with algebra in eighth grade and culminating with precalculus or mathematical analysis in the 11th grade. No matter how the courses are packaged and in what years they are elected, the content of these prerequisite courses has to be learned no later than the start of the 12th year. This raises a major concern: *high schools must be vigilant about offering AP calculus to only those students who have the prerequisites.* As a college instructor, I had students in an introductory calculus course who had AP calculus in high school in their senior year and did not do well enough to receive college credit or waive the introductory college calculus course. Unfortunately, they ignored their college placement results, which suggested they enroll in a precalculus course. They had precalculus problems in my calculus course to the extent that they had to drop it or received a poor grade in the course. They then had to take a precalculus course in their second college term.

The scenario just described is incredibly sad. These students had completed their four-year college preparatory mathematics program in the 11th grade. Their intent was to take calculus in their 12th year and receive college credit for it or at least place out of it in college. They knew if that didn't happen, they could retake this calculus content in college, pass it, and proceed to the next calculus course. This didn't happen. At the end of their freshman year in college they finally mastered precalculus, which they thought they had mastered two years ago. In addition, they had to delay taking non-mathematics courses that had calculus as a prerequisite (e.g., physics), to say nothing of the negative effect all this had on their psyches.

A scenario that is worse than the one just described would be one where these same students did not take precalculus in college in the term following their poor performance in college calculus, but re-enrolled in introductory calculus with the same results as before—either dropping it or receiving a poor grade. They then enrolled in precalculus after it finally hit home that they did not have the prerequisites for calculus. Finally, they successfully completed a precalculus course by the middle of their sophomore year in college. How demoralizing is that? And then we wonder why so many students drop out of college. When did this problem start? It started when *they enrolled in calculus their senior year in high school without having mastered the prerequisites.*

The coordinator of lower-division mathematics at a major university said this in response to my question about how well students follow placement advice at her university: "Advice is generally followed, but not likely by those who had calculus in high school and are being asked to take a precalculus course in college."

Scores Needed on AP Calculus Examination to Receive Advanced Placement or Credit

Colleges are interested in how well you achieved in Advanced Placement Calculus. The score you received on the Advanced Placement Calculus test best determines this achievement. A report of the Committee on the Undergraduate Program in Mathematics (CUPM) Panel on Calculus Articulation stated the following:

> Studies show that, overall, students earning a score of four or five on either the AB or BC Advanced Placement Calculus examination do as well or better in subsequent calculus courses than the students who have taken all their calculus in college.[1]

Consequently, if you received a score of four or five on the AB or BC examinations you should feel at ease if you are given one semester credit for these scores on the AB examination, and two semesters credit for these scores on the BC examination. You need to check the scores and the credit given for them at the college you are attending or want to attend. Colleges vary on these decisions. If there are circumstances concerning your performance in AP Calculus that you believe should keep you from accepting your placement, discuss them with a college mathematics instructor or advisor.

What if you received a score of three on either test and you were given college credit? What are your chances of competing favorably with those in Calculus II who took all their calculus in college? In this situation, the CUPM Panel on Calculus Articulation indicates that it is not clear how you will fare, overall, compared to those taking all their calculus courses in college. You have to decide whether to elect Calculus I or Calculus II. How to go about making this decision will be addressed shortly.

Since the Advanced Placement credit granted by colleges varies widely, no attempt is made to characterize it here. However, seven examples will illustrate this diversity:

1. Some colleges give no calculus credit for a score of three on either test.
2. Some colleges give four hours of credit for Calculus I for a score of three on either test.
3. Some colleges give four hours of credit for Calculus I for a score of three on either test and successful completion of Calculus II.
4. College W requires a four on the AB or BC test to receive four hours of credit, and requires a five on the BC test to receive eight hours of calculus credit.
5. College X has set up "phantom" courses Math 120 and Math 121, representing advanced credit for AB and BC examinations, respectively. It grants two hours of credit for Math 120 for scores of three, four, or five on the AB examination and another two hours of credit for Math 120 upon successful completion of Calculus II (but no credit is given if Precalculus or Calculus I is elected). This same college gives two hours of credit for Math 120 and two hours of credit for Math 121 for scores of three, four, or five on the BC examination (but no credit is granted if Precalculus or Calculus I is elected). No additional credit is given upon successful com-

pletion of Calculus II (assuming it is elected), but two additional hours of credit are given for Math 120 and two additional hours of credit are given for Math 121 upon successful completion of Calculus III. Whew!

6. College Y requires at least a four on the AB or BC examination to receive credit, but a score of three is subject to faculty review.

7. College Z considers giving placement and/or credit for scores from three to five on the Advanced Placement Examinations, but policies vary on the credit given depending on whether the student wants it applied to a major or to an elective area.

So what do you do if you receive a score of three on either test? Do you elect Calculus I (i.e., take the subject matter again), or do you elect Calculus II? This becomes a difficult question to answer even if you received Advanced Placement credit, unless you decide you don't need any more calculus in college beyond Calculus I. It is difficult to answer because it is quite likely that if you elect Calculus I you will resent taking it, and could develop a negative attitude that will affect your performance. If you elect Calculus II, you may not have satisfied the prerequisite of Calculus I, which indicates that you made the wrong choice. The CUPM Panel implied that you are between a rock and a hard place.

Retaking Calculus I in College with a Positive Attitude

It is not easy to retake a calculus course in college that you spent a year studying in high school. You have to combat resentment and harmful thoughts that can hinder your success in the course. You may have these thoughts:

1. "I know this material. I don't need to study as much, especially at the beginning of the course. It is boring and tedious to have to sit through it again in class, take notes, do the homework, and study for tests. I have better things to do with my time." (Perhaps an antidote to these thoughts is to heed the words of the well-known mathematician Paul Halmos as he describes an undergraduate experience: "I took a course in German the first semester, but I didn't like it. It was too easy. Therefore I didn't study, and therefore it was too hard."[2])

2. "I am overly anxious about how well I can do in this course. I will drop it as soon as I begin to have difficulty with it, and change my area of concentration to one that doesn't require mathematics."

3. "Some of my classmates know that I had calculus in high school. I will not discuss mathematics with them or volunteer to answer questions in class. I don't want to look ignorant."

Knowing that students who have to retake a course typically make comments like these can help you place them in the proper perspective. Reflect on the harm these thoughts cause you—what they will keep you from doing. If you have to retake a calculus course, your goal is to be positive about calculus, the course you are placed in, and your ability to succeed in this course. Do what needs to be done and do not entertain negative thoughts.

Pitfalls Associated with Taking a Non-Advanced Placement Calculus Course in High School

All high school calculus courses should be taught as college-level courses, as is Advanced Placement Calculus. If you were one of my college students, I would be concerned if you had taken a yearlong calculus course in high school that was not comparable to calculus as it is taught in college. What sense does it make for high schools to offer such a course? Not much. In such a course your chances of passing an Advanced Placement Calculus test or any college administered calculus test would be minimal, at best. If you are placed in Calculus I, you will probably experience the same problems as anyone else who has to retake a course, believing you had learned much of the content of the course the first time around. If by chance you happen to be placed in Calculus II, you will not be as prepared as most of the students enrolled, who were either placed there because they achieved well in Advanced Placement Calculus, or in Calculus I in college.

If your high school course was *one* semester of calculus, with the purpose of giving you a non-rigorous and non-detailed look at the main elements of introductory calculus (with the goal of preparing you for Calculus I in college), then the CUPM Panel and I believe that you were misguided in taking this course. (There are strong feelings among some college mathematics instructors that such courses should not exist at the high school level.) It will now be difficult for you in college to muster the motivation to work diligently and learn calculus as it should be learned. Professor Donald Sherbert on the CUPM Panel said, "It is like showing a ten-minute highlights film of a baseball game, including the final score, and then forcing the viewer to watch the entire game from the beginning—with a quiz after each inning."[3]

If you took a yearlong substantial high school calculus course that was not designated as Advanced Placement Calculus, then you can elect to take the Advanced Placement Calculus examination. It is the score you receive on the AB or BC test that colleges want to see, not whether the course was part of an Advanced Placement Program. However, the Advanced Placement Program was instituted to help ensure that quality courses are offered.

2. College Level Examination Program (CLEP)

Another vehicle for receiving advanced placement or credit in lower-division undergraduate mathematics courses at colleges is the College Level Examination Program (CLEP). It is likely that your college participates in CLEP, since most do. It is a good vehicle to use if you have attained specific mathematics knowledge, *regardless of how it was attained*. Some colleges may ask you to verify how you gained this knowledge. Perhaps you attained it through self-study, adult education, the military, high school, or a college mathematics course. You may not have received a high grade in a high school or college mathematics course, but rather than retaking the course—incurring tuition costs and considerable investment of time—you can pay a small fee to take a CLEP examination.

You will likely have to do additional self-study work before taking the examination. As resources for study, you can use textbooks, preferably several, that address the content of the test. CLEP does provide study guides for each subject, along with general information about the test and preparation advice. There are CLEP options available for military personnel at no financial cost. Go online to obtain more information.

In mathematics, the following CLEP examinations are available:

- College Mathematics
- College Algebra

- Trigonometry
- College Algebra-Trigonometry
- Calculus

A brief description of each examination follows.

CLEP College Mathematics Examination

This examination covers basic mathematics content that is often considered remedial mathematics at many colleges. Depending on the college, credit or exemption is given for a satisfactory score on the examination for a specific course or courses comprised of this content and offered at the college. At some colleges, an acceptable score may satisfy the college distribution requirement.

CLEP College Algebra Examination

This examination addresses content that is typically found in a college algebra course. Good sources to study, besides the CLEP study guide, are college algebra textbooks, preferably more than one.

CLEP Trigonometry Examination

This examination addresses content that is typically found in a college trigonometry course, with a focus on analytical trigonometry. Good sources to study, besides the CLEP study guide, are college trigonometry textbooks, preferably more than one.

CLEP Algebra-Trigonometry Examination

There are three ways to describe the content of this examination: (1) it addresses the same content as that addressed in the College Algebra and Trigonometry Examinations; (2) it addresses the content of a one-semester college algebra and trigonometry course; (3) it addresses the content of a precalculus course, which is prerequisite for an introductory college calculus course. Good sources to study, besides the CLEP study guide, are college algebra and college trigonometry textbooks, combined college algebra-trigonometry textbooks, or precalculus textbooks.

CLEP Calculus Examination

This examination addresses the content of a one-semester college introductory calculus course. Major topics addressed are limits, and differential and integral calculus. Good sources to study, besides the

CLEP study guide, are typical college introductory calculus textbooks for students with concentrations in mathematics, chemistry, physics, or engineering.

How Colleges Grant College Credit Through CLEP Examinations

It is usual for a college participating in CLEP to grant credit for acceptable scores on these four CLEP examinations: College Algebra, Trigonometry, College Algebra-Trigonometry, and Calculus. To find out how much CLEP examination credit you can receive for certain scores or what courses you are exempt from taking, check the CLEP policies at your college of interest. Here are examples of CLEP policies at two colleges:

1. College A offers three, three, three, and five credit hours for their courses in calculus, college algebra, college algebra-trigonometry, and trigonometry, respectively. This is the same amount of credit these courses carry at this college.
2. College B offers four, three, and four credit hours for their courses in calculus, college algebra, and college algebra-trigonometry, respectively. This is one credit hour less than the amount of credit these courses carry at this college.

For more information on CLEP go to the website of the College Entrance Examination Board.

3. Examinations Constructed by Colleges

Fortunately, many colleges, besides giving advanced placement or college credit in mathematics for one or more of the means already described, also give their *own* examinations to meet the needs of specific students. If for some reason you have not received advanced placement or credit through other means, ask if you can take an examination constructed by the college's mathematics department. If this is not done with regularity at your college, someone in authority in the mathematics department may grant your desire to take such an examination. A mathematics department might go out of its way to do this because it does not want a student to repeat a course or receive no credit for college level mathematics that the student knows.

But I would be remiss if I did not mention that the examination has to be comprehensive and well constructed. It must be constructed so that a passing score on the examination means that you know the content of the course that the test addresses, at least to the extent that it shows you have the major prerequisites for a course you want to take. An examination only samples the many content objectives of a course—it is not the course itself. You may be asked to tell how you learned the content being tested, before the mathematics department decides to give you the test.

4. International Baccalaureate Program

Another means of receiving advanced placement or credit in college mathematics is through the International Baccalaureate Program (IB), which covers the last two years of high school for students ages 16 to 19. This is a program of the International Baccalaureate Organization (IBO). This rigorous two-year program exists to accommodate geographic and cultural mobility by offering a curriculum that addresses the specificity of some national education systems (e.g., European) and the generality of others (e.g., American).

Subjects in this program, including mathematics, are offered at two levels, *standard* and *higher*. (In general, the mathematics courses for which you can receive advanced placement or credit are similar to those discussed in sections 1 and 2.) The higher level is comparable to the CEEB Advanced Placement Program. Students can earn, at many colleges and universities, advanced placement or college credit for certain scores on subject matter examinations acceptable to those colleges and universities. You have the option of enrolling in IB courses without taking the complete IB Diploma Program. If this option is chosen, you register for the subjects, at the standard or higher level, as Certificates. Graded IB exams receive a score ranging from one (very poor) to seven (excellent).

Placement or Credit Given by Colleges for IB Higher Level Examination Results

College policies for granting advanced placement and/or credit for IB examination results vary widely. Four examples follow:

1. College D grants four calculus credits for a score of five or six on the IB Calculus Examination, and eight calculus credits for a score of seven on this test.
2. College E gives placement credit for High-Level Pass only. Credit is not granted for the Subsidiary Pass or Diploma.
3. College F grants IB credit for Higher Level examination scores of four or better in all IB subjects.
4. College G grants General Education credit for courses in appropriate areas of study provided the student has scored five or higher on the Higher Level examination. Credit toward a major is recognized in some areas provided a score of five or higher was achieved on the Higher Level examination.

Contact your college to find out if it accepts IB examination results and, if so, what its policies are for granting placement or credit in mathematics. To obtain more information on the International Baccalaureate Program go to the website of the International Baccalaureate Organization (IBO).

6

Scheduling a Job, Courses, and Other Activities

This chapter focuses primarily on two issues and how they relate to each other. The issues are (1) scheduling a job (and other activities), and (2) scheduling your mathematics courses. If they are not properly attended to, they have the potential to raise havoc with your general performance in college and specific performance in mathematics. I have seen time and time again the academic problems that result from students' lack of forethought to these issues, especially how they coordinate with each other. The word "scheduling" here is used in an inclusive sense. It applies to the times during the day, the week, and the year that you choose to take specific mathematics courses or work a job, and to the number of courses and work hours chosen. There are potential pitfalls that should be avoided.

I cannot tell you how to schedule your mathematics courses, a job, and other activities. You know your specific situation and I don't. Perhaps you have overriding issues that dictate the schedule you need to follow. However, there are several issues you need to consider in making out your schedule that are discussed in this chapter. As you read through this chapter, ask yourself three questions: (1) Do I have scheduling options available to me? (2) If I have few or no scheduling options, how do I overcome the difficulties associated with the schedule I have? (3) Am I willing to make the necessary changes to minimize the negative impact that improper scheduling can have on my academic success?

Many choices you make as a college student affect other choices you have made or will make. Because of this, the decisions you make are often complex. This is especially true for scheduling issues. The following story illustrates a profound lack of forethought regarding scheduling: Mary was in my 9 A.M. mathematics class that met for one hour, three times a week. She was in her early 30s and was doing quite well in the course. As the semester progressed her grades began to slip. One morning before class, and about halfway through the semester, she came to my office and said that she was dropping the course. At that time she had a grade of B in the course. I asked her why she was dropping the course and she said, "Because I won't get an A in the course." I asked her why not. She said, "Because I don't have the time that I need to raise my grade to an A." (I won't get into my conversations with her on why she needed the A—that would be too depressing and not the reason I am bringing up this situation.) I asked her why she couldn't spend more time on the course. She said, "I work the midnight shift at an automobile company and come to your class directly from work." I asked her if she was a single mother, and she said, "No, my husband has a full-time job with the same employer." I asked her how many credit hours she was carrying and she said "fourteen." I asked if she had any children. She said, "I have three children less than six years of age."

Did this student have time for my course? Are her priorities in order? She had three full-time jobs! Perhaps if she had revealed more about her situation it would have been easier for me to understand, but as it stood, it was outlandish. There are countless ways to inappropriately schedule a job and courses that are nowhere as extreme as this example; however, they can still raise havoc with your academic success.

The sections in this chapter are titled:

1. Scheduling a Job
2. Scheduling Courses
3. Scheduling Other Activities

1. Scheduling a Job

Students enrolled in a non-commuter college are less apt to have a job or work fewer hours on a job than students enrolled in a commuter college. College students who commute may have less parental financial support, greater travel expenses, or are keeping a job they had in high school. It is common for college students to hold a job. You may have had a job during high school, and perhaps worked a considerable number of hours per week. Your academic work may have suffered because you worked too many hours or you scheduled your work hours during optimal study time. Receiving good high school grades under the job conditions you experienced may lead you to think that you could continue this schedule in college. This may or may not be the case, but be aware that college and high school are different institutions of learning with different demands placed on their students. Some of these differences are discussed in chapter 2.

If you plan to have a job while attending college, give thought to these related issues:

- Benefits of a job
- Knowing your priorities
- Feasibility of obtaining loans to reduce job hours
- Location of your job
- Suitable hours for your job

Benefits of a Job

There are good reasons to have a job. It can:

Provide money to meet necessary expenses. You have to meet your expenses; but if your job is interfering unduly with your academic work, then a loan (or a larger one) may be the best option (more is said on this later).

Provide health insurance. Most college jobs for college students are part-time and not professional level; hence, health insurance is not provided by their employers. But it may be imprudent to leave a job or reject a job offer if it provides health insurance, especially if you or your family need it and the job is your only option to obtain it.

Promote good mental health. A job can be a welcome break from academic work. There are times during the day or week when you may feel the need to get away from your studies. Perhaps an examination did not go well, or you need to be distracted from anything connected to your college life. A job may be a refreshing change or relief from intense and anxiety-provoking academic work. Regardless of the nature of your job, on certain days you might find yourself saying, "What a blessing it is to focus on this work, even if only for a few hours."

Provide a valuable learning experience. Any job is a learning experience. You learn (1) how to work with management and colleagues, (2) the cultural aspects of the workplace, (3) the importance of being punctual, and (4) specific job skills. Perhaps the most important thing you learn is that you wouldn't want to do that job forever, which can motivate you to finish your college studies.

Help organize your time. Another key benefit of a job is that it can help you complete your academic work in a timely manner. This may sound contradictory to you. How can spending time on a job help you get more accomplished in your studies? Won't there be fewer hours available for your academic work? Yes, but these hours will be better utilized. It is a truism that the more hours available to do a task, the more hours that will be wasted or used inefficiently. You often hear the maxim that "If you want something done, ask a busy person." By necessity, busy and successful people are organized. They take on a lot of work and are aware of the limited number of hours they have to get it done. They know that time cannot be wasted; hence, they work in a way that ensures the completion of each task in a timely manner. More is said on managing time in chapter 7.

Have a positive effect on your college experience. A job can increase your satisfaction with college. Richard Light reported on a study involving college students, in which it was found that "three-fourths of all working students say that working has a positive effect on their overall satisfaction with college."[1]

Please don't take this list of benefits as a strong endorsement for having a job in college. There are many factors to consider before you make your decision. A job can raise havoc with your academic success, but it doesn't have to.

Knowing Your Priorities

A frequent question that I ask struggling students who have a job that is interfering with their academic work is this: "What is your priority—your academic work or your job?" Few of them say that it is their job, yet their actions speak otherwise. If your priority is your academic work, then a job becomes secondary. It should not interfere with your academics to the extent that your career in college, to say nothing of your career after college, is jeopardized. Your job interferes with your college work if it does not leave adequate time for you to—

- Attend class regularly
- Meet with student study groups
- See your instructor for additional help

- Get your assignments completed in a timely fashion
- Study appropriately for examinations, and take them when scheduled
- Pursue recreational and other college activities
- Get adequate rest, exercise, and nutrition

Are you displaying some of these signs? If so, what do you intend to do about it?

In the first two-thirds of my teaching career, I had little difficulty scheduling office visits with my students. A student would ask, "When are you available to see me?" Within a day or two we were able to meet. This does not appear to be the situation today for many students. Perhaps this conversation that I had on a Monday morning with a student named John who had a job, represents the situation as it exists today with many students: "John, as you know, your performance on the test I gave back today was not good. I am very concerned about your welfare in this course. I would like you to come to my office so we can discuss the situation." As John looked at his calendar watch he said, "I might be able to fit you in Friday afternoon at 3:30."

Feasibility of Obtaining Loans to Reduce Job Hours

Attending college is a costly venture, with some colleges costing considerably more to attend than others. This causes many students to rely on student loans to pay their way. These loans add up to a significant amount over four years. Upon graduation, many students have college debts in the five- to six-figure range. Do you have to be concerned about college loan debt? Without doubt—as with most things in life, there needs to be a balance. You need to balance the number of hours you work at a job, the number of credit hours you take, and the loans you receive. You don't want a job to interfere unduly with your academics, yet you also don't want the size of your loan to be more than what you can reasonably pay off over a realistic period of time after graduation. Some students think it is almost sacrilegious to have any student loans. They work far too many hours on a job, which causes them to drop out of college or achieve minimally in their studies. There is good debt and bad debt. Good debt results from investing in your future, such as a home mortgage or student loan. Bad debt results when you spend money on things you don't need. Taking out student loans in a prudent fashion is one of the smarter things you can do.

Location of Your Job

The location of your job is also an important consideration. Travel takes time and money. You may have a job on or near campus, or at a considerable distance from campus. I have known full-time commuter students who lived about 20 miles from their work, and about the same distance from their work to college. Maybe it has to be that way, but perhaps not. I live in a metropolitan area 40 miles wide, and it seems that many people see medical doctors or dentists far from where they live. They could change doctors but they don't want to go through the hassle. If they only allowed themselves a few minutes of quality reflection, they would see that it is far less of a hassle to change. Time is precious, so why waste it traveling too far to a job if other arrangements can be made?

Suitable Hours for Your Job

If possible, choose hours on a job that have minimal negative impact on your studies. If that is not possible for your job, see if you can find another job. Perhaps you have some flexibility in the class times you choose for your courses. A good way to illustrate undesirable and desirable work hours is to relate two hypothetical work situations, one for Student A and one for Student B. Assume that each carries the same class schedule of 15 credit hours, and each works 20 hours per week. Their class times are labeled "x" and work times are labeled "A" (for Student A) and "B" (for Student B).

Time	M	Tu	W	Th	F	Sat
8–9 A.M.	x		x			
9–10	x	AB	x	A	x	B
10–11		AB		A		B
11–12	x	A	x	A	x	B
12–1 P.M.				A		B
1–2	x	x	x		x	B
2–3						B
3–4	x	x	B	x	AB	B
4–5			B		AB	B
5–6	A	A	B	A	AB	B
6–7	A	A		A	AB	
7–8	A	A			B	
8–9	A				B	

You can quickly see that Student B has the job schedule that better supports academic work. In terms of negative impact on course work, it is almost as though Student B has no job. The same cannot be said for Student A's job schedule. It has the potential to raise havoc with his or her academic achievement, and most likely will. Here is a look at each day for both students:

- Monday is a full day for Student A with nine hours of class time and work. Hardly any time is allowed for study, and no time for extended or group study. This is not true for Student B.
- Student A's schedule on Tuesday is also problematic since there is no time on Tuesday morning for studying for an examination (similar to the situation for Student A on Monday evening). Now two days have passed and Student A has had very little study time. This is not true for Student B.
- Wednesday is a great study day for Student A, and a good one for Student B. Student B is finished with his job early enough in the evening to find extended study time.
- Student A's schedule on Thursday is not that good, but perhaps manageable. There is opportunity for extended study time in the evening, but that might not begin until around 9 p.m. because of eating, travel time, etc. Fatigue may have set in by this time.
- Friday's schedule is fine for both of them since their Friday afternoon hours on a job do not interfere with any classes.
- Student B works his job most of Saturday, and Student A isn't on the job at all. Since both most likely can take one weekend day off from studying, I have chosen Sunday for this. Student B's job on Saturday is not interfering with study time, but may infringe on recreational pursuits.

To conclude this hypothetical situation relating the study times and hours working a job of Students A and B, *Student B has the best work schedule by a wide margin.* It allows Student B the opportunity to be significantly more successful in his or her academic work than Student A.

The main factors to consider in articulating class and job schedules include these:

- Free up times when you are most alert for extended study (two hours or more) every weekday (if possible). *It is unwise to put off studying for two or more consecutive days.*
- Reserve time shortly before classes to make final preparations for class.
- Reserve time shortly after class to review and adjust notes that were taken in class, while the content is still fresh in your mind.
- Reserve time during the day to seek extra help from your instructor, classmates, or tutors in a mathematics laboratory.

You can see the difficulty coordinating class and job schedules if more than 20 hours are spent on a job. Even 12 hours on a job can be problematic if they are not chosen carefully. Working a job from six to nine in the evening, Monday through Thursday, is fraught with difficulties. Contrast that with working four hours later in the day on Friday, and eight hours on Saturday. In a study I conducted, I asked calculus students to give advice to students on how to succeed in mathematics. Many of them mentioned how their job schedules impinged negatively on their success in mathematics. This can be a real problem, but early and careful planning can minimize the difficulty. If possible, find a job that better accommodates your academic schedule, or negotiate better hours with your employer. You may be in greater control of the situation than you think. Plan your job schedule carefully because it will have widespread effects.

If many of your hours on a job are fixed and scheduled at unchangeable times, and you know this before registering for classes, choose sections of your courses so their class times better articulate with your work hours. This may be difficult to do because of the limited number of sections of specific courses.

Making a good decision on the number of credit hours to elect in conjunction with the number of job hours to elect is crucial to your success. Abuses abound, but especially when making this decision for a shortened college term, such as a summer term. This is the first topic of discussion in the next section.

2. Scheduling Courses

These topics are covered in this section:

1. Discerning whether to elect mathematics courses in a shortened term
2. Minimizing the number of demanding courses in your course schedule
3. Spacing your classes and courses during the day, week, or year

Discerning Whether to Elect Mathematics Courses in a Shortened Term

During my years of teaching mathematics, there have been academic abuses by countless numbers of students who elect mathematics courses in a shortened term (e.g., an abbreviated summer term). These abuses are *costly* in more than one way. The advice in this section will provide a careful thought process on electing mathematics courses in a shortened term, and if you make the decision to do so, to not overextend yourself with other commitments.

A shortened term, often referred to as a summer term, is one whose length is shorter than a typical fall or winter term in a semester or trimester system. It is approximately six to seven weeks in duration. Many students who are enrolled in a mathematics course during a shortened term, and it can be a majority of them, perform poorly in mathematics because they mismanage their time. Dropping a mathematics course or receiving a poor grade costs dearly in terms of tuition, time, anxiety, motivation, confidence, and grade point average. Grades in other courses may also be affected by this mismanagement of time.

Reasons why you might want to take one or more mathematics courses in a shortened term include:

- To progress more rapidly through college
- To retake a mathematics course
- To attain the mathematics prerequisite for one or more upcoming courses
- To minimize loss of prerequisite knowledge recently attained by not waiting until a later term to take a course that has this course as a prerequisite

- To better attend to the demands of the course due to freeing yourself of other obligations

These are all honorable reasons to take a course in a shortened term.

Downsides of taking a course in a shortened term include these:

Class sessions are twice as long. A 50-minute mathematics class period can be wasted if confusion sets in at the outset of the class period and remains throughout the period. Now imagine what happens in a two-hour class period if you are confused at the outset and cannot get clarification at that time. There is potential to be confused and frustrated most of the two hours—the confusion will just compound. In a regular term, confusion arising during a 50-minute class can be cleared up before the next class session, which takes place on a subsequent day.

There are half as many hours available for completing assignments and preparing for examinations. In a shortened term there are about half as many weekdays and weekend days for the same amount of content that you would study in a regular term. In a shortened term, one-hour examinations come approximately every 10 days in a four-credit mathematics course (assuming four one-hour exams are given), to say nothing of having less time to study for the final examination. Also, the rule of thumb that says from two to three hours should be spent on assignments for each hour in class, means that in a seven-week shortened term, four to six hours should be spent outside of class for each two-hour class period. You cannot leave most of these study hours for the weekend; significant study time is required between classes. I always mention this to my summer term students and many of them are either shocked or don't believe me. They look at me as though they are saying, "Say this isn't true Professor Dahlke," or, "You are just trying to frighten us—we know better." Mathematics concepts and principles need time to ferment in your mind, and the time available for this to take place in a shortened term is about half as much as in a regular term. It is difficult and anxiety provoking to be behind in a mathematics course in a regular term. Imagine getting behind in a shortened term. Frightening, isn't it?

Consequences are dire if you miss class. Suppose you have to miss a class or two during a shortened term for acceptable reasons such as attending a wedding or funeral, illness, or a car breakdown. Missing a class period in a seven-week shortened term is equivalent to missing two consecutive class periods in a 14-week term. This also means that you may not be equipped to gain anywhere as much from the next two-hour class period, which is probably the next day. Hence, missing one class session in a shortened term could easily translate into missing the equivalent of one week's course content in a regular term. Imagine now missing two consecutive classes in a shortened term (which could easily happen since most class days for a four-credit course are consecutive). Alarming, isn't it?

A tendency to over-schedule yourself with a job and courses during the summer months. At many colleges, 12 credit hours in a regular 14-week term are viewed as a full-time load (although many students take several credits more than 12). Assuming this, six credit hours in a seven-week term is viewed as a full-time load, and a four-credit mathematics course is viewed as a two-thirds load. Since you may only be taking one course, you might decide to work more hours on a job. You might say, "After all, it is summer time, I am a part-time student, and this is my chance to make more money."

The following chart reveals, among other things, how much time on average you have to spend per week day involved in various activities in a seven-week term as compared to a 14-week term, assuming these conditions: (1) five weekdays are used; (2) you are enrolled in a four-credit mathematics course; (3) outside of class you study the recommended amount of two to three hours per each hour of class, and (4) you work 20 hours per week.

	7-week course	14-week course
A. Job hours per week	20	20
B. Number of class hours per week for four-credit course	8	4
C. Study hours per week for four-credit course	16–24	8–12
D. Study hours + class hours + job hours per week (A + B + C)	44–52	32–36
E. Average number of hours, per week day, devoted to study hours + class hours + job hours (D ÷ 5)	8.8–10.4 Equivalent to more than a a full-time job	6.4–7.2
Total number of study days in the term	49	98

The important row to reflect on in this chart is the second last one. If you restrict your job and study efforts to the five week days, you will devote from 8.8 to 10.4 hours per day to accomplish what is mentioned there (if you spend two to three hours in study outside of class per hour of class, respectively). This is a very demanding schedule, and does not include the non-sleeping time you may spend during these five days eating, driving to and from work and school, exercising, going to a movie now and then, watching television, spending some time with family and friends, attending events that are unique to summer, etc. This schedule becomes much more bearable during a 14-week term, but still very demanding. You have more than five days during the week to work your job and study; hence, using some of the weekend days helps, especially if most of your job hours are on weekends. Nonetheless, a significant amount of study has to be done during the week because mathematics study cannot be postponed until the week-end.

Imagine the scenario of enrolling in more than one course or working more than 20 hours per week. Would there be enough hours in the day for you to accomplish what needs to be done without significantly hurting yourself? Do the arithmetic, and the answer becomes obvious. There are other possibilities to consider involving more or fewer work hours or credit hours. Work out the last row of the chart (E) for the options you are considering. Think carefully about (1) whether you should enroll in a shortened term, (2) whether you should have a job, and (3) the number of credits or job hours to choose. Plan thoroughly, wisely, rationally, and realistically. Much is at stake.

Students who do not choose wisely suffer the consequences. Here is a typical financial result from an unwise decision: Mary needs to make money over the summer. She takes a summer job for 20 hours per week and makes $10 per hour. Over seven weeks she makes $1,400, ignoring any deductions from her paycheck. She also enrolls in a four-credit mathematics course with a tuition fee of $1,000. Her job interferes unduly with her course work to the point that she ends up dropping the course and losing her tuition, or she receives a D or E in the course. She has a net gain in monies of less than $400, receives little or no academic credit for the course, may have lowered her grade point average, and is demoralized. Wouldn't it have been wiser for Mary to take fewer credit hours, work fewer hours on her job, or work her job hours at times that coordinated better with her academic work?

In summary, my mathematics colleagues and I frequently say that taking mathematics in a shortened term is problematic; hence, we are hesitant to recommend it to almost anyone. We especially discourage students who are weak in mathematics (i.e., less than a solid B), due to the pace of the course. *However, we recognize that it may serve your needs to take a course in a shortened term, especially if you are a good mathematics student, have the prerequisites for the course, and have an appropriate amount of time to study.* Perhaps the fast pace of the course can be somewhat offset by being able to focus on one course if that is all you are taking. Remember, in a regular term a college mathematics course moves at a fast pace. In a shortened term it moves almost twice as fast. Imagine taking a very brisk walk for two miles. You will do some huffing and puffing, and wonder if you can keep up your pace as you near the two-mile mark. Now think about walking the two miles at a pace that is twice as fast. Are you up to it? Are you sure? Many students think they are, but find out later they were sadly mistaken.

Minimizing the Number of Demanding Courses in Your Course Schedule

Mathematics and science courses, especially science courses with a laboratory component, such as chemistry and physics, demand a lot of time. In any term it is wise to restrict the number of these courses you take with your mathematics course or courses. You may have little choice but to take several of these rather difficult and time-consuming courses in the same term. In that case, cut back on the number of credits, the number of courses, or the number of hours on a job. Even a strong student will have struggles if he or she elects, for example, three or more of these highly demanding four-credit courses in the same term: Calculus II, General Chemistry II, Introductory Physics II, and Introductory Economics.

Spacing Your Classes and Courses During the Day, Week, or Year

You may be tempted to schedule your courses close to each other, either during the day or over fewer days, so you can free up half- or full-days for your convenience (e.g., not having to be on campus on certain days) or for other activities, including studying or working a job. Your reasons for doing this may be understandable, but there are drawbacks.

For example, scheduling four courses over fewer days means that you will have more classes on some days, and less spacing between these classes. Four or five classes on a day can be problematic. If you are sick for a day you may have to miss all these classes. Or, imagine having three or four examinations on the same day. Suppose you schedule three of these courses to meet over three consecutive hours. There is no time to make last minute preparations for each course, to review or rewrite your notes after each class, or to see your instructor after class. Additionally, comprehending three different subjects in three consecutive hours is exhausting, and if class goes a few minutes longer than the stated time, you may be late for the next class. Now suppose that all of your courses are at most one hour apart. This leaves no time for any extended study between classes. These are serious and realistic issues to contemplate.

In a *shortened* term, some four-credit mathematics courses meet four consecutive days, Monday through Thursday. This gives a three-day weekend, for faculty and students, which is probably the reason for this scheduling. This is attractive, but a price is paid for no spacing between the meeting days. From an academic standpoint, scheduling the course on Monday, Tuesday, Thursday, and Friday might be better because you have more time on Wednesday to study before the next two class meetings on Thursday and Friday. Four hours of class will have passed after the first two class days, and extra time to learn this content will help you learn the content of the last two days. Contrast this with four consecutive days of class, where eight hours of class time has passed. You have the next three days to focus on those four days of classes, and that is good, but mathematics has to be studied *between* classes. Unless you are very disciplined, a three-day weekend will probably not give you any more study time than a two-day weekend since more time is wasted when more time is available.

On the other hand, too much spacing between mathematics classes can hinder learning. For example, two-credit courses are often spaced in a variety of ways. Here are some possibilities: Monday and Wednesday, Tuesday and Thursday, Monday and Thursday, Tuesday and Friday, or just on Wednesday (a two-hour class session). What is the best choice from a learning standpoint? I would eliminate the single class period on Wednesday. It is a two-hour time block and you know the problems associated with that. You have class and then it is a week before class again. The classes are spaced too far apart. Previous content will not be fresh in your mind for the next class period, and it will be difficult to bond with your instructor and classmates. Two of the scheduling possibilities (Monday and Wednesday, Tuesday and Thursday) give a one, and then four-day spacing between classes, and the other two scheduling possibilities (Monday and Thursday, Tuesday and Friday) give a two or three day spacing between classes. The spacing of two or three days between classes is a better choice because it minimizes the longer space.

Late afternoon or evening classes are often problematic. During those times, a four-credit mathematics course is typically offered twice a week; on Monday and Wednesday, or Tuesday and Thursday, and each class session is two hours. They are offered this way to minimize the days spent on campus to accommodate people who have a full-time day job. A problematic situation arises when a three-credit mathematics course meets once a week in the evening for three hours. There appears to be a trend of college mathematics courses meeting less often, which means the class sessions have to be longer. I understand the convenience of this, but learning suffers! Don't forget this when you choose your course schedule.

After completing a course, how long is it reasonable to wait to take a subsequent course that has this course as a prerequisite? *Don't wait too long.* For a complex course with detailed content that makes use of considerable content from a prerequisite course, take it as soon as you can after completing the prerequisite course. This is especially important if you are a student who struggles with mathematics. Consider this example from the calculus track: If Calculus I is completed in the fall term, it is best to take Calculus II in the winter term. If this is not possible, is it better to take it in a shortened term (about four months later) or wait until the fall term, which is about a year after completing the prerequisite course? If you wait too long (and a year may be too long for you), you will have to retake Calculus I. And if you do not adequately handle the pitfalls of taking a course in a shortened term, that can cause even more harm. Reflect on the pros and cons of both options, perhaps seeking counsel, and make a decision that is best for you.

3. **Scheduling Other Activities**

Some of the benefits of having a job, such as promoting good mental health, providing a valuable learning experience, helping you organize your time, and making your college experience more satisfying, *are also benefits of engaging in activities other than a job.* These other activities include volunteer work, playing intramural sports, and participating in college organizations. It is important to feel that you are part of a university community. Engaging in study groups and college extra-curricular activities can ameliorate a feeling of isolation. Richard Light reported on a study, which found that a student feeling isolated from the college community is indicative of problems to come in the student's college experience.[2]

The advice given in this chapter for coordinating your job schedule with your class schedule also applies to coordinating your schedule of other activities with your course schedule. Whatever you engage in—a job, your course work, volunteer work, a college sponsored activity or organization—it is important to *coordinate all this with your class schedule* to give you the best results.

Managing Time

I am hard pressed to come up with something that is more important to you as a college student than how you manage your time. No matter what other study skills you have, consistently misusing your time lessens their value. This is supported by a study reported by Richard Light, which involved two groups of sophomores who were interviewed thoroughly. One group was comprised of students who did well socially and academically as freshmen, and the other group were freshmen who did not. The interviewers found that the key factor accounting for this difference was how the students spent their time. Light reported:

> Sophomores who had a great first year typically talked about realizing when they got to college, they had to think about how to spend their time. They mentioned time management, time allocation, and time as a scarce resource. In contrast, sophomores who struggled during their first year rarely referred to time in any way.[1]

Light also stated that the importance of managing time was supported by comments made by seniors who were asked to give advice to freshmen. It was not atypical for them to mention time management. You may struggle with managing your time at the outset of your college career, but you can't wait very long to get a handle on it.

There are many issues related to managing time. In this chapter the focus will be on managing your time so that (1) you have an adequate amount of time for study, (2) your study is done at suitable times, and (3) your study is appropriately structured within any study session. Other issues discussed include motivating, focusing, rewarding, and evaluating your study time, as well as ideas for saving time.

The sections in this chapter are titled:

1. Scheduling Daily and Weekly Activities
2. Choosing an Environment for Individual Study
3. Judicious Use of Study Time
4. Evaluate Your Study Time
5. Commitment to Study

1. Scheduling Daily and Weekly Activities

It is wise to construct a chart that shows how you intend to spend your time each day of the week. Some of your activities need to be completed daily and will be at fixed times (such as your class times); for others you will have decisions to make as to when they are scheduled and for what duration. One major benefit of charting this information is that you don't need to spend time each day reconstructing your calendar for the next day. You can quickly look at your chart to see what you have scheduled for the next day, and quickly make changes to accommodate modifications. Carry your chart with you so you can look at it throughout the day. Looking at your schedule of daily activities places pressure on you to perform them at their designated time.

When creating your weekly schedule, list the days of the week at the top of the chart, and list the hours of your day, perhaps in half-hours or hour segments, along the left side of the chart. You may want to do this for six days of the week and start your day at 7 a.m. and end it at 10 p.m. Then fill in the time slots with your activities for each day. Include some or all of these: class, work, individual and group study, weekly review, travel, meals, and recreation. For class and study time slots, fill in the names of the course or courses to be studied. Use ink to fill in the fixed activities for the semester, such as class times, and use pencil for those that may change. It is prudent to also have a *long-range* time schedule that displays dates of examinations and due dates for papers. Not planning for these can raise havoc with your weekly schedule.

Building recreation time into your schedule is crucial for your mental health. It can be a reward for being responsible in your activities. Scheduling recreation time makes it as important as other scheduled activities (and when you are recreating, you won't have the guilt feelings that you should be studying). In a sense, by scheduling these fun times you are building procrastination from your studies into your schedule, which may be just what the doctor ordered. It seems reasonable to disengage from your studies for one day of the seven-day week. A six-day workweek for a student is sensible.

Example of Daily Schedules

At the end of your day is a good time to revise your schedule, as needed, for the next day. Examples of daily schedules for two consecutive days are given. These are the types of daily schedules that I would have constructed as an undergraduate student. It might not come close to what works best for you. The important thing is to have a schedule and revise it when you find it not working for you. You will discover that you need less study time for some courses and more for others, as well as the best times to study.

Monday's Schedule

6:15–6:45	Rise, shower, and dress
6:45–7:30	Read the section in Math 115 textbook on content that will be developed in today's class
7:30–8:00	(1) Review questions on yesterday's Math 115 reading and problem assignment that I will ask in today's class (2) Review yesterday's Math 115 class notes
8:00–8:30	(1) Eat breakfast (2) Email my Chem 115 professor requesting an appointment
8:30–10:00	Work Chem 115 assignment
10:00–11:00	Math 115 class
11:00–11:30	Review and rewrite today's Math 115 class notes
11:30–12:30	Lunch with Andy and Laura
12:30–1:30	Work Chem 115 assignment
1:30–2:00	Review Chem 115 notes from last class
2:00–3:00	Chem 115 class
3:00–4:00	Work Lit 101 assignment
4:00–5:00	Lit 101 class
5:00–7:00	Shop for groceries, eat dinner, and free time
7:00–8:30	Work Math 115 assignment
8:30–10:00	Work Hist 203 assignment
10:00–11:00	Free time
Sleep	

Tuesday's Schedule

6:15–6:45	Rise, shower, and dress
6:45–8:00	Work Chem 115 assignment
8:00–8:30	Eat breakfast
8:30–10:00	Weekly meeting with Math 115 study group
10:00–11:00	Math 115 class
11:00–12:00	Hist 203 class
12:00–1:00	Eat lunch and discuss project with Judy
1:00–3:00	Chem 115 lab
3:00–5:30	Work job
5:30–7:00	Eat dinner and free time
7:00–8:30	Meeting with campus newspaper staff
8:30–10:00	Work Math 115 assignment
9:30–11:00	Do laundry and free time
Sleep	

There are many issues to consider when you are scheduling weekly activities that are not fixed for the semester. As you progress through this book you will learn educationally sound activities to practice, and the best times for doing so.

Follow the Schedule

Any weekly schedule of study sessions is subject to change because of unforeseen events like the sudden announcement of an examination. *A word of caution:* Be careful what you call an unforeseen event. These examples are not unforeseen events: "I just don't feel like studying at this time," or, "I think I'll go see a movie this afternoon." There will be times when you follow these urges and that might be good for your mental health. But these exceptions should be minimal; otherwise you are defeating the purpose of a schedule. In the same vein, if you have scheduled two hours for studying mathematics on a specific day, you might be tempted to quit half-way through if things are not going well, or, you may be progressing so well toward your study goals for that time period that you reward yourself by quitting early. This is generally not a good thing to do. You might want to leave a challenging mathematics problem for the time being—that can be wise—but it most likely is not wise to leave your study of mathematics.

Being disciplined in your study implies that your study plans have become *habit*. That is, carrying out your study plan is second nature to you. Habit keeps you faithful to a plan. *Learning can take place whether or not you feel up to it.* Work on following your study schedule so it becomes habit. Janet Erskine Stuart says it well: "The great thing, and the hard thing, is to stick to things when you have outlived the first interest, and not yet got the second, which comes with a sort of mastery." If your study plan is not working, revise it or consult a college counselor for assistance.

Short Periods of Study

The disciplined student makes use of short periods of time for study purposes, perhaps sessions of an hour or less. Opportunities during the day for short periods of study include studying between classes, while riding a bus or train to and from college or a job, while on break from your job, as you eat, waiting to see a doctor or dentist, or immediately upon rising in the morning. Even five to fifteen minutes of review can pay enormous dividends, especially if it is near a class session. You can participate more completely and successfully in a class session if you prepare for it shortly before it begins, reviewing content that you knew yesterday, but may have forgotten today. Reviewing class content and rewriting class notes shortly after a class session reinforces it and helps fill in what you missed. Here is a list of work that can be completed in a short study period:

- Prepare for a class session by reviewing previous class notes.
- Get your homework questions in order that you will ask in class.
- Read an overview of the textbook section that will be discussed in class.
- Read your notes from the textbook covering the content of yesterday's class.
- Continue to work on a problem, concept, or principle that is difficult.
- Seek help from a classmate on any aspect of the course.
- Review specific material for an examination.
- Conduct a weekly review of a course.
- Reflect on, or revise, your calendar of activities for the day or for tomorrow.

- Work a small part of an extensive project (e.g., write a summary of a laboratory assignment).

If you need to get in five to six hours of study each day, small study intervals can add up to a significant percentage of this time. When study time is broken up frequently with other responsibilities, which often happens in college, your sense of using time wisely has to come more from your use of small intervals of time (five minutes is not too small), than from big ones.

Sustained Periods of Study

As valuable as short periods of study are, longer or sustained periods of study are needed to accomplish certain learning goals. Learning new or complex material requires *sustained* periods of study. In Richard Light's report on the behavior of college students, he reveals a finding that may surprise you. He said:

> The single biggest trouble with time use for nearly all students who struggle is their pattern of studying in a series of short bursts. Instead of spending sustained periods of time engaging with their coursework, they squeeze in twenty-five minutes between two classes. They stop by the library for thirty minutes on the way to dinner. . . . They come home from a full evening of being with a drama group, or sports practice, etc. They begin to work on the next day's classes and are tired before they start.[2]

Light is not saying that there should be no short study periods. He is saying that *there needs to be an adequate number of longer periods of study.* For in-depth learning to take place, you need to spend sustained time with a few short breaks added. For example, through reading, it may take a few hours to understand the development of the mathematics in a section of the textbook. Perhaps at a later time you may have to spend another few hours applying this knowledge to your problem assignment. Then, with the knowledge you gained through working the problems, you may have to reread portions of the section to gain a better understanding of the section. Understanding the section helps you work the problems, and working the problems helps you better understand the section. These study sessions cannot be fruitful if they are a series of short study spurts. You will be leaving things unlearned that are needed to learn other things. Bertrand Russell said, "To be able to concentrate for a considerable time is essential to difficult achievement."

How many long study periods in a week do you need for a particular mathematics course? I recommend from four to five study periods per week, with 1.5 to 2.5 hours per period, assuming shorter periods of study are also used throughout the week for the course. *It should be a rare occurrence not to have at least one sustained study period between two consecutive class sessions of a mathematics course.*

Don't Mix Academic Work and Recreational Pursuits

Perhaps some of the best guidance I can give you is this: *Don't mix academic work and recreational pursuits.* That is, when you are engaged in academic work don't be thinking about recreational pursuits, and vice versa. If you are engaged in one area and a thought arises from the other area, do not allow yourself to entertain it—dismiss it quickly. Recreational pursuits, whatever form they take, should be times to *recreate* yourself, and that will not happen to the extent it should if you are simultaneously anxious about your pending academic work. When recreating, you need to feel that you are away from your academic work, and when doing academic work you need to leave recreation by the wayside so you can focus on your study. When you have studied as you should have, reward yourself by engaging in recreational pursuits. Don't mix work and play, for you will shortchange both.

2. Choosing an Environment for Individual Study

The environment you choose for individual study has a bearing on how successful you will be in your study. The degree of importance of this to your success depends on the subject matter. Content that is interesting and not too difficult can be learned quite well under less than ideal learning conditions, although most likely not in the most efficient manner. Learning complex or uninteresting content, or any content for that matter, is best mastered in a first-rate study environment.

Choose a Place with Minimal Distractions

There are general guidelines for choosing places for individual study. At times you cannot be too choosy, especially if the study sessions are short and coupled with another activity. For example, you have limited choices if you ride a train or bus to and from college and want to study during your ride. Or, when looking over your notes while you are eating at a college delicatessen, or when you take a 10-minute break from your job. You most likely have more control over the places where you do most of your sustained study, and it is wise to choose them carefully. Usually you will have a place on campus to study, as well as where you live. It may take you a short while to adapt to a more appropriate study environment since you will miss the distractions that are part of your current study environment. Here are suggestions to help you determine a quality study environment:

Avoid individual study in a room where you frequently hear people engaged in conversation. These conversations may take place in person or they may be on a television or radio. You may say, "That is when I study the best." Are you sure, or are you saying this because you have not tried, for a reasonable period of time, other study environments? Learning new and complex knowledge requires full concentration, and your attention to this should not be competing with your attention, consciously or not, to the conversations of others.

College and public libraries are typically good places to study. Not all areas within a library are suitable for study. Walk around the library to find a good study location. I always looked for an isolated carrel among the stacks of books. This is a sterile environment, but it is what I needed—no refrigerator—no television—no people—nothing except an isolated student or two browsing the stacks of books. In the carrel there was nothing to distract me. One of my rewards for a good study session was leaving that sterile environment to rejoin the glitzy world.

Look for an empty classroom. Many times during the day, especially in the early morning and late afternoon, classrooms are vacant and can be used for study purposes. They may be better than the campus library and closer to your next class. If needed, you can find out what times specific rooms are open by contacting the office on your campus that deals with room assignments.

Find a study area that has good natural light during the day and good artificial light in the evening. It should be well ventilated—not too hot and not too cold. It should have a suitably sized desk and chair conducive to work.

Here are two responses I received on a questionnaire asking students to give me advice on what I should say to students who read this book: (1) "My study skills were much improved when I changed from doing my calculus homework in bed to working on the fourth floor of the library." (2) "Don't relax. Sitting in a chair is better than a recliner. When you study you need a minimum amount of tension to concentrate. Don't begin studying immediately after a heavy meal as the body tends to be relaxed."

Enhance the Study Area

You can do things to enhance your major study area. Have the necessary study materials at your fingertips (computer, calculator, pencils, pens, paper, dictionary, etc.) Make it attractive by eliminating clutter and other items that can distract you from the task at hand. Newspapers, magazines, novels, and a television are items of distraction. It is too tempting to turn to them when you want to avoid studying. The desire to watch television is especially great. You may say to yourself, "I'll just spend a few minutes flipping through the channels to see what is on and then turn it off." Even if you turn it off within a few minutes, what you saw will linger with you for a while and interfere with your study. You are likely to watch a program far longer than you intended. It is an entirely different matter if you have planned on rewarding a good study session by watching some television. If visitors distract you, place a sign on your door indicating that you do not want to be disturbed, perhaps displaying the times you are not available. It works in hotels and motels. To mask noise, play background music.

When you see or visualize your study space, it is important to think, "This is where I study." You will not find the perfect study place, so don't expect to. As important as a study space can be, it is of little value to you if you don't use it. Take the study time you need to be successful.

3. Judicious Use of Study Time

A necessary condition for having a good individual study session is to know how to select and structure the activities you engage in during the session. Many college students do not do this very well. Ideas for selecting and structuring study activities comprise this section.

Set Goals for Each Study Session

Suppose you have scheduled a mathematics study session of two hours. Before the session begins, it is beneficial to set realistic goals for the session by writing them down. Having goals organizes the session, guides your focus, and gives you a sense of accomplishment as you achieve them. Here is an example of goals and associated time limits for a two-hour session:

1. Review class notes (10 minutes).
2. Do a quick reading of the problems assigned, and determine the number of pages in the textbook section that are to be read (5 minutes).
3. Gain an overview of the textbook section (15 minutes).
4. Carefully read the textbook section (40 minutes).
5. Work the problem assignment (50 minutes).

For a difficult task, break it into reachable subgoals. A good time to take a short break in the session is when you move from one goal to another, assuming a reasonable amount of time has passed. For example, you might want to take a break before starting goal 4, and one before starting goal 5. This breaks your study session into three parts, somewhat equal in length. These are natural breaks to take for two reasons: (1) You are most motivated to study at the outset (the first 30 minutes), less motivated the next 40 minutes (as you begin to tire), and then will regain your motivation after a respite, with the end in sight (the last 50 minutes). (2) The activities are different in each of the three parts, although related. For this reason, you are not disrupting your work until you have completed a specific task, and you will be motivated after your break since the next task is different.

Choose the Courses to Study in Any Given Study Session

If you intend to study more than one course in a given study session, there are some general principles that will help you plan your study.

Begin with your most challenging subject, and then proceed to the next most challenging subject. You have more energy and a greater focus at the outset of your study session to apply to difficult subject matter. Most likely what keeps you from studying your most difficult subject first is the thought that you will not be successful. We

procrastinate when we expect to fail or experience considerable difficulty. I often do that, whether it's repairing a faulty furnace, replacing
a broken window, or figuring out how to apply a computer application.
But almost always, I find out that when engaged in the task it is not as
bad as I thought and chastise myself for delaying. Furthermore, my
success or partial success in doing a task motivates me to spend more
time on the task, either then or at a later date. Success with challenging
tasks gives you confidence to tackle other tasks.

**Do not place the subject of least interest to you at the end of
your study session.** If this subject is placed last, chances are you will
short-change it because you didn't get to it (perhaps because you
didn't want to get to it), or you are tired and study it halfheartedly. You
have to take special steps to succeed in a subject that is non-motivating
to you. In your professional work you will have many activities to perform that you won't like. Gain confidence that you can do them well
by doing well in your least-liked college subjects.

Do not study like subjects consecutively. Changing the nature or
difficulty of the content studied is motivating, less tiring, and minimizes interference. Studying mathematics followed by physics or even
chemistry is not recommended, and neither is studying French followed by German. If you must study similar subjects consecutively,
take a short break before going to the second subject.

Stay Focused in Your Study

It is difficult to stay focused in your study session for long periods of
time. Only you know how long you can stay focused. Is it 45 minutes,
60 minutes, or 90 minutes? There are many variables that determine
how long during a specific study session you can stay focused. These
include the nature of the subject, your interest in it, your energy level,
and whatever else might be on your mind at the time. Here are some
guidelines to help you focus during your study session:

Schedule short breaks. Knowing that you will take a break after
45 or 60 minutes into your study will motivate you to stay focused for
this length of time. The break is your *reward* for staying focused. You
will be refreshed by the break, and motivated to return to your study.
Plan to do something during your five to ten minute break that re-
freshes you. It may be no more than walking around the library, treating yourself to a drink or snack, browsing through a magazine, or
talking to a close friend. If you study as long as 90 minutes without

taking a break, then you might want to take a longer break lasting around 15 minutes.

Initially, don't schedule study sessions that are too long for you. Is two hours, with short breaks, too long for you? It may be better for you to schedule an hour that you manage well than it is to schedule two hours and waste most of the last hour because you are tired. You can gradually increase the length of your study sessions.

Other Issues That Interfere with Your Study Focus

Allowing your mind to wander as you study is inefficient learning. Working on two things at one time guarantees that neither one will be done well. If you can't stop daydreaming while you work, pause and *seriously entertain your dreams.* Once you have taken care of them, start your work again. Keep doing this. Eventually you will tire of this and get down to serious work. Similarly, if you have a personal problem that is anxiety-provoking and interfering with your study, put this problem aside for a half hour or so, and then devote your full attention to it. Worry, worry, and worry over this problem if you have to—get it out—then devise a plan to confront it. Now go back to your study, knowing that you will carry out your plan later. We often don't like it when others interrupt us, so why tolerate the times you interrupt yourself, which are significantly more numerous?

Study is also inefficient if you are sleepy or hungry. Get at least eight hours of sleep per night and catnap when you find yourself sleepy as you study. Why struggle for an hour trying to stay alert as you study. Perhaps you will accomplish 20 minutes of learning through your misery. Nap for 15 minutes and then accomplish 45 minutes of more enjoyable learning. You know what to do if you are too hungry to study efficiently. You also need to exercise regularly—it makes you more alert in your study and it helps relieve stress.

Use Study Time Productively

We all waste time when we work, even though we may not think so. Consider these suggestions:

Don't dash from subject to subject as you study. Do you do a little work on one subject, then a little work on another, and a little work on a third, before you go back to the first subject and start the process over again? This is an inefficient technique, to say nothing of the interference caused by mixing different subjects. Bite the bullet and stay with a subject.

Early morning hours are excellent for studying after a good night's sleep. Upon arising you are renewed, refreshed, and have quiet time—all ingredients for productive study. It is a great way to start a busy day since you are buoyed by what you have already accomplished. Find an hour or two before 8 A.M. You will be amazed at what you can accomplish. You may reject this advice, viewing yourself as a night owl. Perhaps for years your study time has been working late into the evening and extending into early morning hours. It doesn't have to be that way—it may just be habit. Making the necessary change may take time. It may not be a good idea to study an hour or two before 8 A.M. if you worked late the previous evening, say as late as 11 P.M. or midnight. Mid- to late-evening is a great time to relax for the day and get to bed so you are equipped to face the next day. Studying after 9 P.M. is not that productive after a full and tiring day. I know there are many night owls who will not agree with this last comment.

4. Evaluate Your Study Time

Occasionally keep a time log for a week on how you spend your time. For each day of the week, name each activity, the time you began it, and how long you stayed with it. Your daily log hours should total 24. Analyze your log in a variety of ways, but answer these three questions: (1) When and how have I wasted time? (2) What am I going to do about the time I wasted? (3) What other valuable information did I learn?

5. Commitment to Study

The key ingredient to making your study successful is your commitment to study because "Nothing works unless you do." I was conscientious in my mathematics teaching. I always put it at the top of my professional responsibilities. I worked diligently on attending to the many duties associated with teaching. These duties included careful class preparation using innovative approaches, giving meaningful assignments, writing quality examinations, and valuing the time I spent with students in my office. Teaching well is hard work and I was committed to it. The same can be said for many of my colleagues. Unfortunately, some of my students lacked this work ethic. There was a mismatch between my work ethic and that of these students. Success in mathematics is highly dependent on a strong commitment to study. I wanted my students to be successful so I worked hard at doing my part, but I was only part of the equation. For my un-committed students I might as well have been a telephone pole in class, for my commitment and expertise was of little value to them. It was wasted on them. Your learning is enhanced when there is a partnership among you, your instructor, and your classmates. You have an important role in this partnership, which includes commitment to your study.

Answer these questions:

1. Am I making the best use of my instructor's commitment and expertise?
2. Does it make sense to enter a mathematics course without having the will to succeed?
3. What do you think is the main reason there are many unsuccessful talented people?
4. Am I committed to my mathematics study? What is my evidence?
5. How can I be committed to my mathematics study when the course is a requirement and I am not interested in the content?

This chapter concludes with a comment by Tyron Edwards, which characterizes the importance of time management: "Have a time and place for everything, and do everything in its time and place, and you will not only accomplish more but have far more leisure than those who are always hurrying."

Nature and Evolution of Mathematics

You cannot adequately learn mathematics unless you have a good sense of *what mathematics is*. Having this sense helps you know *what* to study, and you need to know what to study before you can learn it. By a good sense of mathematics I am not referring to knowing what distinguishes geometry, algebra, calculus, etc. from each other; instead, I am referring to what makes all of these specific subjects "mathematics." Once you have this sense, you will know more about the aspects of mathematics to study so that you are learning mathematics in a more comprehensive and useful way.

No doubt you have an understanding of what mathematics is, but if you are like many students, it is most likely more limited than it should be. I say this because of what I have noticed that students study in mathematics. I say to myself, "They don't have an adequate understanding of what mathematics is, because if they did they would address more in their study than the few aspects of mathematics that they do." What you need to learn in mathematics, and hence study in mathematics, must reflect in an appropriate way what mathematics is. Otherwise you are limiting your ability to use your mathematics in a meaningful and empowering way.

Your mathematics instructor should help you determine *what mathematics is* through his or her presentations, class discussions, assignments, and examinations. Some instructors fall far short of doing this by overly stressing one aspect of it (e.g., computation), at the expense of other vital aspects. Your next mathematics course can be a struggle for you if you have an instructor who expects you to work diligently to attain a more comprehensive understanding of what math-

ematics is. You will struggle initially in gaining this understanding because you have not learned, heretofore, what you need to do and how
to do it. Sadly, the course you are in keeps moving along.

What should you do if you have an instructor who does not expect
you to learn mathematics as it needs to be learned? If you determine
this early on in the course, perhaps you can drop the course and choose
another instructor. (A better option would have been to know more
about your instructor before selecting him or her. See section 4 of
chapter 13 on choosing your mathematics instructor.) If you decide to
stay with your instructor, then *you* need to do what needs to done to
learn what should be learned. This latter option will require some extra
motivation on your part since your efforts will not be reinforced by
your instructor's expectations and actions. You might get a good grade
in the course without doing anything extra; however, your motivation
for the course has to be more than receiving an acceptable grade. This
book will help you understand what mathematics is and what you need
to do to learn it.

1. Describing Mathematics

It is not possible to define a subject as complex as mathematics in a
few sentences or paragraphs. However, a good sense of mathematics
can be attained by looking at it from many perspectives *and* doing
some of the work that mathematicians do. Each of these points of view
of mathematics describes mathematics, yet fall far short of *defining*
mathematics.

Shortcomings of an
Abbreviated Definition of Mathematics

Suppose you were asked to give a one-sentence definition of mathematics. What would you say? Most likely you would say what many
students say: "Mathematics is computation." And why wouldn't you
say this since computing answers is what you have been doing, almost
exclusively, since your mathematics learning began. It is what is
taught, practiced, and "spit back" on examinations. Consequently,
since this may be your view of mathematics, it is what you primarily
studied in your quest to learn mathematics. My experiences as a mathematics instructor led me to conclude that the major problem of most

diligent, but *unsuccessful* students in mathematics, was their narrow view of what mathematics is, which dictated what and how they studied. Robert Lewis, a college mathematics instructor, addressed these issues:

> The great misconception about mathematics—and it stifles and thwarts more students than any other single thing—is the notion that mathematics is about formulas and cranking out computations. It is the unconsciously held delusion that mathematics is a set of rules and formulas that have been worked out by God knows who, for God knows why, and the student's duty is to memorize all this stuff. Such students seem to feel that sometime in the future their boss will walk into the office and demand "Quick, what's the quadratic formula?" Or, "Hurry, I need to know the derivative of $3x^2 - 6x + 1$." There are no such employers.[1]

Any attempt to give an abbreviated definition of mathematics does not serve you or mathematics well. This discipline, and many others, is too complex to be captured by a few brief statements. There have been attempts to give such definitions and some of them are stated in a general fashion so as to include much of what mathematics is. But they are so general that they are of no help in directing your mathematics study. A better way to get at the nature of mathematics is to give a series of short statements, each of which describes an aspect of the nature or essence of mathematics, but individually do not characterize it. The more descriptions given, the greater understanding you will have about the nature of this discipline. When these descriptive statements are viewed as a whole and incorporated into your mathematics study, the result is an adequate understanding of the essence of mathematics.

Suppose I want you to know who my friend Bill is; that is, I want to "define" Bill for you. The best way to do this is to describe aspects of Bill. I might say that Bill is a college graduate, is married, is 40 years old, has cancer, is employed in a service industry, was passed over for promotion at his job, likes boating, and is very active in his church. The statements about Bill that *relate* to other statements about him help define him further. For example, suppose you know that Bill was motivated to become an active volunteer for an organization that provides housing and support for developmentally disabled adults because his sister Mary is developmentally disabled. Virtually endless relational statements could be made about Bill. The more aspects of Bill that you know, and the more you relate them, the more you get to know who he is. All of this is enhanced if you can also spend time with him. Analogously, the more aspects or descriptions of mathematics that you know, relate, and work with, the more you will know what mathematics is. It is as simple as that.

If you incorporate into your mathematics study many of the aspects of mathematics that help characterize it, you will then be approaching your mathematics study somewhat how a professional mathematician approaches his or her study. You might say, "Surely I should not be expected to operate in my mathematics courses similar to that of professional mathematicians." To this I respond, "If you want to learn mathematics as it should be learned, you do not have the option of drifting too far away from how they approach their study of mathematics." The bonuses you receive include seeing mathematics as a way of thinking, as evolving historically (driven by applications and by intellectual curiosity), and as having a much greater scope than what you have been led to believe. Furthermore, you are more equipped to see the usefulness of mathematics and the ability to apply it in other academic fields and in your career.

Aspects of Mathematics That Help Define Its Nature or Essence

There are many aspects to mathematics, and the more you work with them as you study a specific subject area of mathematics, the greater will be your learning. They include the following:

- *Mathematics is experimentation.* (Collect and analyze data.)
- *Mathematics is discovery.* (Use experimentation, tables, pictures, and logic to make discoveries.)
- *Mathematics is concepts.* (Know and apply concepts.)
- *Mathematics is generalizations.* (Know and discover general statements that can be proven true.)
- *Mathematics is numbers.* (Know and work with the different categories of numbers.)
- *Mathematics is algorithms.* (Know, develop, and apply algorithms.)
- *Mathematics is formulas.* (Know important mathematics formulas and derive formulas.)

- *Mathematics is shape.* (Know and work with shapes of various dimensions.)
- *Mathematics is pictorial.* (Use geometric figures, graphs, or diagrams to understand and relate concepts.)
- *Mathematics is pattern.* (Know, discover, and apply patterns.)
- *Mathematics is order.* (Know what and how it orders.)
- *Mathematics is computation.* (Perform simple and complex computations—mentally, with paper and pencil, or with judicious use of technology.)
- *Mathematics is abstract.* (Know and work with various levels of abstraction.)
- *Mathematics is logic.* (Know and use logic to reason and justify.)
- *Mathematics is deduction.* (Reason from the general to the specific.)
- *Mathematics is induction.* (Reason from the specific to the general.)
- *Mathematics is a logical structure of connected concepts.* (Understand, appreciate, and apply the logic that is used.)
- *Mathematics is intuition and insight.* (Experience instantaneous insight and understanding without conscious use of reasoning.)
- *Mathematics is justifying generalizations.* (Prove generalizations using formal and informal reasoning.)
- *Mathematics is a game played by following specific rules.* (Know the rules and play the game.)
- *Mathematics is change.* (Know how changes in quantities affect changes in related quantities.)
- *Mathematics is solving problems.* (Solve a variety of problems.)
- *Mathematics is applications.* (Apply mathematics in an ever-increasing number of disciplines.)
- *Mathematics is simulation or modeling.* (Simulate real-life situations by constructing mathematical models of them.)
- *Mathematics is symbolic language.* (Understand and use symbols of mathematics.)
- *Mathematics is spoken language.* (Speak and discuss mathematics.)

- *Mathematics is written language.* (Write mathematics using symbols and words.)
- *Mathematics is abstract.* (Know, appreciate, and work with various levels of abstraction.)
- *Mathematics is precise.* (Know the exactness of what is being said, and be precise in what you are saying.)
- *Mathematics is concise.* (Use an economy of words without overly compromising clarity.)
- *Mathematics is compact.* (Understand and appreciate compactness, and express ideas in a relatively small space.)

This list of aspects of mathematics should broaden your view of what mathematics is, and what you need to address in learning it. You can see that mathematics is significantly more than computation. The materials you use in your courses, primarily your textbooks, will address most of these aspects of mathematics. You don't have to go looking elsewhere for exposure to them. That being said, it is important that you address them. You may have to go to other sources to address some of them in greater depth. You will focus on some aspects of mathematics more than others, but if key aspects are grossly unbalanced in the attention given to them, by you or your instructor, you are not learning mathematics, as it needs to be learned. Many of these aspects of mathematics, as they relate to your study of mathematics, are addressed in subsequent chapters of this book. It is there that I focus in detail on those for which you may need more guidance and background knowledge (e.g., speaking mathematics, writing mathematics, working with symbolic form, proof, and problem solving).

The following aspects of mathematics are less tangible than those in the previous list; however, awareness of them can be enlightening, inspiring, and empowering:

- Mathematics is beautiful.
- Mathematics is history.
- Mathematics is a product of human thought.
- Mathematics is dynamic (i.e., its history is continually being made).
- Mathematics is inspired by nature.
- Mathematics is the language of the universe.
- Mathematics is the foundation of all other sciences.

- Mathematics is the language in which scientific models are written.
- Mathematics is power.

2. Evolution of Mathematics

Do you believe that mathematics is a static discipline? That all the mathematics in existence today is from the past? That it was discovered many years ago with no more to come? Do you also believe that all *uses or applications* of mathematics were discovered many years ago? None of this could be further from the truth, and as a mathematics student and user of mathematics, it is important for you to know this. Mathematics is a dynamic discipline—discoveries are continually being made, as are uses for them. We now turn to how mathematics discoveries come about; that is, how mathematics evolves.

People develop mathematics. It was, is, and will continue to be invented or constructed by humans. I know most students understand this, to some extent, but students' methods of studying mathematics do not always support this. Their methods portray an unrealistic view of how mathematics comes about. How mathematics develops is related to how it should be learned—it doesn't develop miraculously and *neither is it learned miraculously*. Hard work is involved for the developer and the learner, and the practices of one are similar to the practices of the other. They both have to experiment—to try this—to try that—to perhaps start over—to perhaps learn more about a different area of mathematics that can then be applied to the current situation—etc. The history of mathematics development reveals a lot about how mathematics has developed, and consequently how it has to be learned. This development highlights the interplay between mathematics discoveries and the need to solve real world problems. Mathematical theories or discoveries evolve from the need to solve practical problems, or from existing theories or discoveries.

Mathematics Theories and Discoveries Can Evolve from the Need to Solve Practical Problems

The development of mathematics in its early years was almost exclusively motivated by practical problems that needed to be solved. Solving the same type of practical problems led to abstracting a method used to solve them. For example, based on a specific practical need, a method was discovered for finding the third side of *specific* right triangles knowing the other two sides. The method was then generalized, resulting in an abstract formulation that applied to any right triangle. You know this as the Pythagorean Theorem [$a^2 + b^2 = c^2$, where c is the length of the hypotenuse and a and b are the lengths of the legs]. In summary, more than a few thousand years ago, work on solving practical problems of a certain type gave rise to discoveries that can be used to solve similar practical problems, now and in the future.

These days, if you are to learn the Pythagorean Theorem and some of its applications, it is not enough for an instructor to say, "Here it is, now use it on these problems." That is like the original discoverers finding the theorem under a rock, which did not happen. It has to be developed with understanding, and this development should (somewhat) mirror the experimentation—the trial and error—that the original developers went through. As a student, you have to be treated as a discoverer—not for everything, but for some things. Fortunately, instructors can help speed up the process of learning by using carefully designed learning or discovery activities. If you are struggling in your college mathematics course with specific mathematics content, doing more discovery or laboratory activities that relate to it can most likely help you. Possible sources for these activities are your instructor (ask for them during his or her office hours), fellow students, or specific presentations or exercises appearing in your textbook, or in other textbooks. Your instructor can construct some of these activities on the spot—just indicate your confusion to him or her, and that you need more laboratory or concrete work to aid your understanding.

Here is an example of a current day practical problem, which scientists or engineers may want to solve: Design a windshield wiper apparatus that will best clean a specific type or shape of windshield. These individuals may, perhaps with the help of mathematicians, find existing mathematics that will help solve the problem. Or, they may need to develop new mathematics that can do the job, again enlisting the help of others, including mathematicians. The need to solve real world problems assists in the discovery of *some*, but not all, new mathematics.

This statement, by the well-known mathematician Halmos, attests to the fact that specific examples and problems, whether applied or mathematical, lead to general mathematical theories and discoveries:

The heart of mathematics consists of concrete examples and concrete problems. Big general theories are usually afterthoughts based on small but profound insights; the insights themselves come from concrete special cases.[2]

New mathematics is being discovered daily, and *another means of mathematical discovery*, namely, using existing theories or discoveries, is now discussed.

Mathematical Theories and Discoveries Can Evolve from Existing Theories or Discoveries

Using laws of logic, and existing theories or discoveries, other mathematics theories or discoveries are developed. Many current and past mathematics theories or discoveries came about this way, without any idea of how they could be used to solve a real world problem. They are developed from existing mathematics abstractions. They may not have an apparent application to the real world at the time of their development. However, it is quite possible that an existing mathematics discovery can be used to solve a real world problem that is first posed many years after this mathematical discovery. New applications of specific mathematics discoveries have been found hundreds of years after their discoveries. Nikolai Lobachesky, a Russian mathematician (1792–1856), saw this early on. He said, "There is no branch of mathematics, however abstract, which may not some day be applied to phenomena of the real world." Tobias Danzig, a mathematician, used an interesting analogy to make a comparable statement:

> The mathematician may be compared to a designer of garments who is utterly oblivious of the creatures whom his garments fit. To be sure, his art originated in the necessity for clothing such creatures, but this was long ago; to this day a shape will occasionally appear which will fit into the garment as if the garment had been made for it. Then there is no end of surprise and delight.[3]

Mathematics discoveries that come about by staying within the abstract field of mathematics do not necessarily come about through a nice, neat, and formal deductive argument. There is considerable trial and error involved as the mathematician works in a disorderly way with the other abstractions at his or her disposal. Halmos expresses how this works:

> Mathematics—this may surprise some—is never deductive in its creation. The mathematician at work makes vague guesses, visualizes broad generalizations, and jumps to unwarranted conclusions. He arranges and rearranges his ideas, and he becomes convinced of their truth long before he can write down a logical proof . . . The deductive stage, writing the results down, and writing its rigorous proof are trivial once the real insight arrives; it is more the draftsman's work not the architect's.[4]

To repeat, mathematics theories or discoveries arise out of the need to solve practical problems, or from applying laws of logic to existing mathematics theories or discoveries.

A Myriad of Mathematics Areas Has Evolved Over the Years

A few thousand years BC, the best mathematics minds would use pages of work to solve a simple arithmetic problem involving fractions. It was difficult work because the subject of fractions was not well developed—certain fraction concepts and principles that would have made their work easier were not yet discovered. Today, a typical fifth grader, having the benefit of good mathematics instruction, would solve the problem almost instantly—one that taxed great minds thousands of years earlier. The beginnings of elementary algebra, geometry, and trigonometry developed from a few thousand years BC to about 500 BC, through the work of the Babylonians, Egyptians, and Greeks. Additions were made to these subjects as time progressed, extending into the AD years. Beginning ideas of calculus came about in the BC years, but significant advancements in calculus first came in the 16th and 17th centuries. Mathematics subject areas continue to be invented, links between them continue to be made, and new theories or discoveries within a subject area are continually being constructed.

Perusing an inventory of mathematics subject areas shows how mathematics has grown over the years, to say nothing of reviewing a specific subject area to see the advancements made there. The growth in mathematics subject areas is illustrated by this partial list of mathematics subject areas, taken from the Mathematics Subject Classification scheme developed by the American Mathematical Society Zentral-blatt fur Mathematik:

> combinatorics and graph theory; order, lattice, ordered alge-braic systems; general algebraic systems; number theory; field theory and polynomials; commutative rings and alge-bras; algebraic geometry; linear and multilinear algebra; ma-trix theory; associative rings and algebras; nonassociative rings and algebras; category theory; homological algebra; k-theory; group theory and generalizations; topological groups; Lie groups; real functions and elementary calculus; measure and integration; functions of a complex variable; potential theory; several complex variables and analytic spaces; special functions including trigonometric functions; ordinary differ-ential equations; partial differential equations; dynamical sys-tems and ergodic theory; difference and functional equations; sequence, series, sum ability; approximations and expansions; Fourier analysis; abstract harmonic analysis; integral trans-forms, operational calculus; integral equations; functional analysis; operator theory; calculus of variations and optimal control; optimization; geometry, including classic Euclidean geometry; convex and discrete geometry; differential geome-try; general topology; algebraic topology; manifolds and all complexes; global analysis, analysis on manifolds; probabil-ity theory and stochastic procedures; statistics; numerical analysis.

Whew! Are you surprised, enlightened, confused, or frightened by the extensiveness and diversity of the list just presented? Perhaps you are thinking, "What is the nature of these subject areas?" I couldn't ad-equately answer this question for many of these subject areas, and even if I could, you would not understand much of what I said unless you had a background in higher-level mathematics. Two persons with a doctorate in mathematics, who focused their studies in different areas of mathematics, will most likely have difficulty understanding the most advanced knowledge in each other's focus area. You can spend many years, perhaps even a whole career, trying to gain an expert un-derstanding of one of the subject areas that are listed. I listed these mathematics subject areas to show you that mathematics is much more extensive than you may think, and will only become more so. This information will be of greater benefit to you if you have an interest in pursuing advanced mathematics, perhaps culminating in receiving a bachelor's, master's or doctorate degree in mathematics. You are probably thinking, "Just help get me through my current mathematics course."

How Mathematics and Technology Enhance Each Other

Mathematics and technology like each other very much. One field profoundly influences the other and that will only get stronger. Computers have a great influence on mathematics with perhaps the most significant change being the introduction of new problems and methods of work for mathematicians to grapple with. With the advent of computers and their increased power and speed, mathematics is more focused on finding fast algorithms rather than elegant formulae. With the continual increase in computing power, more mathematical results are used. Also, the language of mathematics and the way it uses logic to make its arguments is of great use in technology. For those of you who think that the need for mathematics has lessened due to the pervasiveness of technology, think again—it is just the opposite.

Understanding Mathematics

Mathematics teachers tell their students that they are to learn mathematics with *understanding*, or to study mathematics for understanding. The operative word here is "understanding." What does it mean to understand mathematics? This fundamental question, together with the topics of "thinking in mathematics," and "benefits of learning mathematics," is addressed in the next chapter.

Benefits of Learning Mathematics Through Emphasis on Understanding and Thinking in Mathematics

It is important for you to know the benefits of learning mathematics. Your motivation to study mathematics correlates highly with your knowledge of what mathematics can do for you in the short and long term. The general and specific skills attained from learning mathematics do not come automatically because you are enrolled in a mathematics course. They require knowledge of what needs to be learned, how to learn it, and considerable effort to learn it. Mathematics concentrators have an advantage over non-mathematics concentrators learning these skills because of the quantity and breadth of lower- and upper-division mathematics courses they take. Nonetheless, benefits of learning mathematics can accrue to a worthwhile degree from taking as little as one mathematics course.

Learning *specific* mathematics content is important since much of this content is prerequisite for subsequent mathematics and mathematics-related courses. But more important are the general learning skills that come from studying mathematics. Not only will they help you learn mathematics content, but they will also be extremely useful throughout your career, regardless of the direction it takes. It may surprise you to read what Jack Meiland, professor emeritus of the University of Michigan, says about general learning skills (he is not restricting his remarks to mathematics courses):

> professors are not overly concerned when students say that
> they forgot some of the content of the course soon after the

115

final examination; professors believe their courses to be worthwhile anyway because in courses students learn ways of thinking that stay with them even if they forget particular content. In order to learn how to think, you must think about some particular content. But the content is not the main point. Much of the content that you are taught in college will be outmoded or discarded anyway in ten or twenty years. Learning intellectual skills and attitudes is far more important.[1]

Jerome Bruner, a former Harvard professor, is not downplaying the importance of general learning skills when he says, "The first object of any act of learning, over and beyond the pleasure that it may give, is that it should serve us in the future. Learning should not only take us someplace; it should allow us to go further more easily."[2]

It was always my goal as a mathematics instructor to have my students learn how a mathematician thinks and works. To accomplish this, their work had to bear a reasonable semblance to the work of a mathematician—for then they would have acquired many general learning skills. College graduates, who have been in the work world awhile, regardless of their career path, value their general learning skills highly. They value their ability to think and learn more than any specific content knowledge they learned while in school. My experience talking with employers indicates that they also value more highly these general learning skills in their employees.

The sections comprising this chapter are titled:

1. Benefits of Learning Mathematics
2. What It Means to Understand Mathematics
3. Thinking in Mathematics

1. Benefits of Learning Mathematics

Your purpose in studying mathematics has to go well beyond just passing the course. You need to study knowing that what you learn in the course will be used in subsequent mathematics or mathematics-related courses, and in your after-college life. You learn the content of the course by applying general learning skills, and your general learning skills will improve because they are repeatedly used to learn content. However, the *focus* of your mathematics study has to be the use and improvement of your general learning skills. Mathematics is not the only discipline in which you can improve general learning skills. However, mathematics is viewed by many as the best discipline for improving certain ones. For example, it is second-to-none in improving analytical thinking skills. The same can be said about the value of mathematics in promoting the ability to work with many ideas at the same time. Many employers love to hire mathematics concentrators, especially if they come with knowledge in a related discipline (e.g., engineering, computer science, physics). They see attributes in these concentrators that they do not see in other graduates.

How to study in your mathematics courses in order to realize the benefits of your study will be obvious to you as you progress through this book. You already know that you need to include in your study many aspects of the nature or essence of mathematics, as addressed in the last chapter. Included among the benefits of learning mathematics appropriately are these:

Increased ability to be a self-sufficient learner. A primary goal should be to release yourself from dependence on your instructor (or anyone else) for your knowledge. Your goal should be to "learn how to learn." Achieving this goal prepares you for lifelong learning. Once you graduate you will not have the luxury to pull a teacher out of your back pocket to access his or her answers or guidance. You nurture self-sufficient learning by weaning yourself away from your college instructors.

Increased ability to work cooperatively and collaboratively with others, and understand the value of doing so. Working with others helps you learn specific mathematics content, as well as prepares you to use this skill in the professional world, where it is highly valued. Working with others is addressed in chapter 23.

Increased ability to understand the nature of a technical subject, how it evolves, and how it is learned. Attaining a historical and cultural perspective of mathematics was the focus of chapter 8.

Increased ability to understand what is to be learned. *Understanding* mathematics is the primary goal of mathematics learning—so much else depends on it. What it means to understand is the subject of the next section.

Increased ability to know, value, and employ problem-solving skills. There are specific problem-solving skills that are important to learn and apply. They are useful inside and outside the discipline of mathematics. Employers love problem solvers, and mathematics is an excellent discipline for developing problem-solving skills. Problem-solving skills are addressed in chapter 22.

Increased ability to know and employ the traits of a successful problem solver. This refers to traits like persistence, coming back to a problem at a later time, and having the right attitudes. Problem solving traits are addressed in chapter 21.

Increased ability to work with many ideas at one time. A wealth of ideas can be considered and related in mathematics because of the many ideas that are related, and because of the compactness of mathematics.

Increased ability to think critically and logically. Mathematics is an ideal discipline for developing analytical thinking skills. Think like a mathematician and you will think analytically.

Increased ability to reason deductively. Reasoning from the general to the specific is a step-by-step reasoning process in which logical arguments are made to justify or prove that something is true.

Increased ability to reason inductively. Reasoning from the specific to the general is discovering a general pattern by observing many examples.

Increased ability to make connections within a technical subject, across disciplines, and to the physical world. The ability to interpret the physical world in technical terminology is accomplished through constructing mathematics models of the physical world.

Increased ability to apply a technical subject. Apply mathematics content to solve routine and complex problems in an ever-increasing number of disciplines.

Increased ability to use appropriate technology to learn and apply a technical subject. Available suitable technology to use in your mathematics courses includes graphics and symbolic calculators, and desktop, laptop, notebook, and cell phone computers. This technology makes use of word-processing software, dynamic or motion software, symbolic manipulation software, spreadsheet software, and databases. Mathematics is also learned through the use of compact disks, tutorial websites, and email discussions. Using technology in mathematics is discussed in section 3 of chapter 18.

Increased ability to speak the language of a technical subject. This benefit is accomplished through discussion, and is the focus of section 4 of chapter 14.

Increased ability to read complex technical literature. Reading mathematics is discussed in chapters 15 and 16.

Increased ability to read, interpret, and manipulate symbolic form. Learning symbolic form is the focus of chapter 19.

Increased ability to write in a technical subject. Use the symbols and vocabulary of mathematics to write complex arguments and solutions to problems. Writing mathematics is the focus of chapter 20.

Increased ability to write *about* a technical subject. This is different from writing mathematics, although writing mathematics is sometimes used in writing about mathematics. It is *expository* writing that is used to demonstrate knowledge of concepts, principles, and their connections. Examples of this are writing a laboratory report or responding to an essay-type question appearing on an assignment or examination. Writing about mathematics is the focus of section 2 of chapter 14.

Increased ability to make calculations and appropriate decisions about the type of calculation to make. Perform simple and complex computations, using paper and pencil and electronic technology, and make decisions on the degree of accuracy needed.

Increased ability to predict and generalize through experimentation. Collect and analyze data to make discoveries.

Increased ability to visualize a technical subject. Represent concepts and relationships by using diagrams, graphs, and tables or charts. This topic for mathematics is the focus of section 5 of chapter 16.

Increased ability to understand logical structure, appreciate its beauty, and work within it. Mathematics is a logical and beautiful game played by following prescribed rules. This benefit is the focus of section 4 of chapter 16.

Increased ability to understand our information-laden world. The different modes of thought used in mathematics, such as modeling and logical analysis, give you the mind facility and versatility to be a critical observer and participant in a world that is becoming increasingly more technological and complex.

This list of abilities clearly shows that there is much more to learn in a mathematics course than learning specific mathematics content. It is important to pass the course to satisfy a requirement, and to *learn the content* so you can succeed in a subsequent mathematics or mathematics-related course. Let there be no doubt about that. That being said, it bears repeating that the general abilities or skills you gain are more important. With these general abilities or skills you have a much greater chance to succeed in subsequent college work and in your career.

2. What It Means to Understand Mathematics

If I asked you if you want to understand the mathematics content of your course, your answer would most likely be, "Yes." If I asked your mathematics instructor if he or she teaches for understanding, I would expect to get the same answer. Nobody likes to be confused and teachers don't want you to be confused. Understanding things allows you to understand more things, and not understanding things can be depressing, demoralizing, and cause you to disengage from whatever is causing the confusion. However, if I asked students and instructors what understanding mathematics *means* to them, the responses would vary considerably from person to person, and many would be deficient. The notable mathematician Paul Halmos, reflecting on his freshman year in college mathematics, said, "I didn't understand what it meant to understand something, and what one should do to get there."[3] No doubt part of his problem can be attributed to the dearth of explanatory writing in the textbooks of his era.

What it means to understand something is a complex issue since understanding is not an all or nothing matter. We continually evolve in our understanding, and many things that we do influence it. Prominent educational psychologists have grappled with this issue for many years. I would be presumptuous if I told you that I know *exactly* what it means to understand something. However, a description of understanding is formulated in the next section that will serve you well in your study of mathematics. This description is followed by applications of it—in and out of mathematics. The section concludes with a discussion on acquiring the mindset of looking for connections to better understand mathematics topics.

A Description of Understanding

A description of understanding that some educational psychologists, mathematicians, and I subscribe to is this: *Understanding something comes from connecting or relating that thing to other things you know.* You have probably heard a mathematics instructor discuss the importance of learning the *structure* of the mathematics you are studying.

What does that mean? It means learning how things are related or connected in mathematics. An analogy would be this: If you want to understand how to build a house, you need to learn about the structure of a house. You need to know the pieces of the house and how they relate to each other. Similarly, understanding the mathematics you are studying means learning the structure of the mathematics. You want to know the basic elements of the content *and* how they are connected. Many of the learning skills listed in the last section help facilitate this understanding. They help you connect or relate what you already understand to what you will come to understand.

Many things affect your understanding of something. You can always understand something better. *Understanding does not take place in a vacuum.* The more you can relate what you understand to what you are trying to understand, the better your understanding will be. To understand something new, you have to use what you already understand.

You know how difficult it can be to understand a new mathematics concept. When first exposed to it, you can be very confused and think that you will never understand it. But you find out that the more that is said about it, the more things that you can relate to it, and the more you work with it, the less confusing it becomes. Eventually, you understand it and wonder why you ever had trouble in the first place. You may forget your understanding of it, and that is why it is important to realize that *understanding needs maintenance.* Reviewing mathematics is performing maintenance; but it also enhances understanding, since concepts understood earlier are re-related to knowledge that has been better understood as time passes.

Mathematics students I have had over the years who struggled in my classes, and mentioned to me that mathematics was always difficult for them, appeared to have one thing in common: *They did not study mathematics for understanding.* They learned content rotely and viewed most aspects of mathematics as isolated from one another. Their main goal in their mathematics study was to memorize pieces of information without looking for connections among the pieces. They also manipulated symbols mindlessly. In a mathematics course, where the instructor stressed understanding, these students had no chance of receiving a good course grade unless they *radically changed their approach to studying.* That doesn't happen overnight. To succeed, they needed to make three changes: (1) have a strong desire to understand, (2) know what needs to be changed to understand, and (3) persevere in their efforts to make these changes. The revelatory look on the faces of these students when I told them how they needed to study mathematics was telling. More than a few of them said to me, "I didn't know that is how I need to study. No one told me this." For some of them who took my advice, the change in attitudes about themselves and mathematics was remarkable. They improved their performance, and would have improved significantly more in their courses if more time were available. College mathematics courses move very quickly—the pace doesn't slow for someone who needs to make changes.

Applying the Description of What It Means to Understand or Learn the Structure of Mathematics

To make the description of understanding that is used here more meaningful, three examples are given. The first two examples are non-mathematical.

Example 1. The parents of a 16-year-old teen named Robert want to determine if it is reasonable to allow him to drive his girl friend and another couple to the high school prom. Robert received his driver's license that week. To better understand the decision they need to make, they reflected on these events that bear on their decision:

- Robert's driver education teacher said that he was one of the best student drivers she ever had.
- Robert's mother rode with him a lot when he was driving with a permit and observed that he eventually drove like an experienced driver.
- Robert's father saw him driving very fast one day with a carload of people.
- The prom location is 35 miles from Robert's home.
- The young couple riding with Robert was arrested last year for having open alcohol containers in a car.
- After the prom, Robert intends to drive to a friend's home where a private after-prom party will be held for eight couples, and the only people in the home will be these couples.

Each of these statements will be used to help Robert's parents make the decision. No doubt many more events can be identified and connections made, more questions can be asked, and specific conditions can be demanded, but Roberts's parents are most likely well equipped with what is here to make an informed decision. They can relate or connect the various statements in their efforts to determine the decision they will make.

Example 2. You have a classmate John and were recently introduced to two of his friends, Agnes and Fred. You would like to know more about the story of this trio so you can relate better to them. You find out that Agnes and Fred are engaged, that Agnes has her eye on John, and John and Fred are cousins. This could go on and on. (This is the stuff romance novels are made of!) The point being made here is that when you know more about the structure of the relationship among these three people, the more you will be able to better function within this relationship. The same can be said about knowing the structure of mathematics—it too has its stories. Knowing the structure of mathematics makes mathematics inherently interesting, and the connections you make will enhance your understanding and ability to apply mathematics.

The third example is a mathematical one and comes from precalculus. It won't make much sense to you if you do not have the prerequisite mathematics. Nonetheless, you most likely will be able to understand the point that is being made.

Example 3. Determine if $y = f(x) = 2x - 1$ has an inverse function; if so, find it, and then verify that it is the inverse function.

To work this example, and *understand* why you did what you did, you have to understand these prerequisite concepts, principles, and techniques that are connected to the problem:

1. Definition of a function
2. Definition of a dependent variable of a function
3. Definition of an independent variable of a function
4. Definition of a one-to-one function
5. Principle: A function has an inverse function if it is a one-to-one function
6. Definition of the inverse of a one-to-one function
7. Technique for finding the inverse of a one-to-one function
8. Definition of a composite function
9. Principle: For a function f and its inverse f^{-1}, $f \circ f^{-1}(x) = x$

The story that solves the example follows:

The function, $y = f(x) = 2x - 1$, is one-to-one since, for different real numbers x_1 and x_2, $2x_1 - 1$ and $2x_2 - 1$ are different. To find the inverse of f, we apply the technique of statement 7, replacing y by x and x by y, yielding $x = 2y - 1$, and then solving for y in terms of x, getting $y = \dfrac{x+1}{2} = f^{-1}(x)$. We see, for the ordered pair $(2,3)$, that this inverse function does what it is supposed to do, since $f(2) = 2(2) - 1 = 4 - 1 = 3$, and $f^{-1}(3) = \dfrac{3+1}{2} = \dfrac{4}{2} = 2$. That is, f takes 2 to 3 and f^{-1} takes 3 back to 2. To show that this works for any x, we employ the principle of statement 9 to f and f^{-1}:

$$f \circ f^{-1}(x) = f\left(f^{-1}(x)\right) = 2(\frac{x+1}{2}) - 1 = x + 1 - 1 = x.$$

Don't be alarmed if you did not follow the solution to example 3, even if you were exposed to these ideas in the past. To understand the solution just given, each of the statements 1–9, and how they are related, has to be understood. If you were expected to understand this example in a course, you would have already learned statements 1–9. It is a simple matter to find the inverse function of $y = 2x - 1$. However, the point of the example is to show you what you need to know and relate so you can *understand* if it has an inverse function, how to find it if it does, and to justify that the technique used actually gives you this inverse. This is *understanding the solution* to the problem, versus just arriving at a solution. By reading this example you were privy to a mathematics story. It is important for you to realize that when you are doing mathematics you are telling stories, and that you need to bring in statements and relate them in your stories so your stories are understood by anyone who hears or reads them, including yourself.

Benefits of Understanding or Learning the *Structure* of Mathematics

I place high emphasis on learning the structure of mathematics because of what it can do for you. The short- and long-term benefits are enormous.

The benefits of learning structure include these:

Increased ability to reconstruct content (thus, aiding the recall of content). Learning the structure of mathematics increases your ability to reconstruct content. Memory loss does not mean complete loss. The little we may remember can be used to reconstruct what we lost. To the unsuspecting person it may seem like someone has recalled something from pure memory, but in reality that person may have quickly reconstructed some or all of the knowledge. This was possible because the person was able to relate or fit the knowledge that he or she *wanted* to recall to other knowledge that he or she *could* recall. Here is an example: Mary was asked if the sec x is defined at $\frac{\pi}{2}$. Mary did not know the answer, but knew that she could quickly give the answer if she knew the graph of sec x. She could not recall this, but knew that $\sec\frac{\pi}{2} = \dfrac{1}{\cos\frac{\pi}{2}}$. She couldn't recall $\cos\frac{\pi}{2}$ but knew she could get it from the graph of cos x, which she knew. From the graph (which is in her mind's eye) she saw that $\cos\frac{\pi}{2}$ is 0. Then knowing that $\frac{1}{0}$ is undefined, she knows that $\sec\frac{\pi}{2}$ is undefined.

Mary reconstructed this knowledge mentally, and virtually instantaneously. If you had not been privy to Mary's thought process, would you know if Mary gave you the answer to the question through memory, or through her ability to obtain the answer by reconstructing it? It would be difficult to answer this question. However, you were privy to her process of answering the question. Mary was able to answer the question because she is attentive to learning the structure of mathematics. Mathematics is a story to her, and if she forgets a part of the story, she can fill it in because she knows other parts of the story that relate to what she needs to know.

Knowing the structure of mathematics helps you better determine the meaning of an examination question. For example, students often complain that they didn't understand a particular examination question, criticizing their instructor for wording it badly. Frequently, the reason they could not understand the question was that they did not learn the structure of mathematics, as they should have. They did not learn enough connections to various items in the question to be able to place the question into a structure that makes it understandable. Understanding a question is highly dependent on knowing where the question fits in the content of the unit of study in which it appears. This has everything to do with studying to understand, which means studying to relate ideas or make connections.

Increased ability to transfer skills to highly similar content.
An understanding of mathematics skills allows you to more readily
apply these skills to learn highly similar skills. Because you have been
relating ideas, your skills are greatly enhanced and you are more experienced in relating ideas; thus, the foundation of understanding you
have built makes it easier to learn highly similar skills that are related
to what you have already learned. For example, understanding how to
graph the sine function by plotting points helps you understand how
to graph the cosine function by plotting points, or find the derivative
of tan by using the Chain Rule helps you find the derivative of $\sec\sqrt{\dfrac{x}{3}}$.
This learning serves the future.

**Increased ability to transfer learning to special cases of a gen-
eral principle or concept.** An understanding of a general principle allows you to recognize subsequent problems as special cases of this
principle or concept. Once again, this learning serves the future. The
most important content to learn in a mathematics course are the fundamental or basic ideas and principles that give structure to the course.
For example, in an arithmetic course they include these principles: as-
sociative and commutative properties of addition and multiplication,
distributive property of multiplication over addition, additive and mul-
tiplicative identity properties, etc. In calculus they include principles
that give meaning to the derivative and definite integral, and important
theorems such as the Fundamental Theorem of Calculus, The Mean
Value Theorem, etc. It is the most fundamental ideas that have the
greatest applicability to other ideas and problems.

Two examples are given of how understanding a general principle
allows for recognition of a subsequent problem as a special case of this
principle:

1. Knowing the Pythagorean Theorem will help you recognize
   a special application of it such as finding the third side of a
   right triangle when you know the other two sides.
2. Knowing that the first derivative of a function $y = f(x)$, with
   respect to x, is the instantaneous rate of change of f, with
   respect to x, will help you recognize that the velocity of a
   particle, whose movement as a function of time is given by
   $y = s(t)$, is $s'(t)$. (If you have not taken introductory calculus, you probably will not understand this example.)

Increased ability to transfer attitudes toward learning. To learn with understanding, you have to develop appropriate attitudes toward learning. These attitudes then become part of you and you will transfer them to your future learning. Included among the attitudes to have is realizing the importance to your learning of relating, questioning, guessing, persevering, and believing in your ability to learn.

Increased ability to think analytically. How could it be otherwise, for analytical thinking is relational thinking—making connections.

Increased ability to be more flexible in your thinking. The greater your understanding, the more connections you will make; hence, the greater your ability to pick and choose from what you understand to understand other things. You are not locked in to one approach but can choose from a variety of approaches.

Increased ability to solve problems. Problem solving can be viewed as finding linkages to parts of the problem that will result in a solution to the problem. This often involves analytical thinking and flexibility in your thinking. In your efforts to solve a problem you are relating aspects of the problem to what you know, and then selectively choosing from what you know to solve the problem. Hence, a necessary condition for being a good problem solver is to learn the structure of mathematics—how things are related.

If you are not focusing on understanding in your study of mathematics, the list of benefits just presented should be motivation to change your approach to learning mathematics. Just imagine not attaining these benefits because you have a rote or mechanical approach to learning mathematics. In this changing and unpredictable world you will need these benefits and they come with understanding mathematics.

You can readily see why the renowned psychologist Ausubel said, "The most important single factor influencing learning is what the learner already knows."[4] When any student comes to me for help on specific content, the first thing I do is determine what the student knows that is prerequisite to understanding this content. I ask a series of diagnostic questions. What I look for in the responses is the student's understanding of these prerequisites. Until the prerequisites are understood, I cannot help the student with the issue that brought him or her to my office. My actions support what Ausubel said about learning, for he ends his statement with the word "knows," and by knows he means *understands*.

There is no doubt that learning the structure of mathematics makes the course more difficult. When you are learning structure, you are working at a higher cognitive level. It can be frustrating work, since understanding does not come immediately. Perseverance, a most admirable and critical trait in a student, has to be developed. Stay with your work for the rewards are many.

Studying with the Mindset of Understanding or Making Connections

If you do not approach your study with a strong desire to understand, then that lack of desire is the first thing you need to change, and that has to be a will of the mind until the satisfaction and joy that comes with understanding takes over. It will come with time. Desire, perseverance, and an inquisitive mindset are needed to understand. Approaching your learning from a questioning standpoint can take you a long way down the road to understanding. *Ask yourself and others these types of questions* as you read, attend class, and work with others:

What is this used for?
Where did this come from?
How does this follow from . . . ?
Where is this going?
What justifies this result?
Is there an easier way to do this?
Is this idea related to . . . ?
How does this fit with what I already know?
I already know how to solve this, so why are we learning a new method of solving it?
Is there a graphical representation of this that makes it plausible?
I wonder what I can do with this result?
Where can I find a formal proof of this result?
Can you suggest another book that gives a more thorough justification of this?
What are the textbook authors saying at the top of page 124?
What justifies the second last step of your argument?

You have to be continually inquisitive when you are in class, reading the textbook, doing the assignments, working with your classmates, getting help from your instructor, and studying for an examination. You have to want to know where ideas come from, why we have them, how they relate to what we already have, and where we are going with them. *To place content into a structure you have to be inquisitive, continually asking many questions.*

Writing *about* Mathematics to Gain Understanding

The most important thing I did as a mathematics instructor to improve my students' understanding was to have them write *about* mathematics. This came through exercises that I placed in assignments, study guides, and examinations. In most exercises of this type, you are working with the structure of mathematics; that is, you have to write about how things are related or connected. To respond to these exercises in a meaningful way you have to know how things are related. This forces you to focus on relating knowledge in your study. Textbooks need more of these exercises, and instructors need to supplement what their textbooks lack with their own writing-about-mathematics exercises. They have to test their students on this type of exercise. It has to be an instructor expectation that students accomplish these exercises. Short of this, students have to construct their own writing-about-math exercises, for themselves and perhaps for their study partners. If it is not an expectation of your instructor that you respond to this type of exercise, then you have to be motivated to do it yourself. There are times you have to go beyond your instructor's expectations.

Writing *summaries* of content developed in your textbook or in class is good practice in making connections. If you can do this in your words and still capture the essence of the development, then you have understood the development.

You should be able to defend what you believe about specific mathematics topics, and writing is an excellent vehicle for making your defense. Meiland equates an understanding of a belief to being able to defend that belief: "If one does not know how to defend a belief, if one does not know what counts as good reasons for a belief, then to that extent one does not understand the belief." [5]

The topic of writing about mathematics is the focus of section 2 of chapter 14. Examples of writing exercises are given here, but many more appear in that section. You may first realize that this work is no easy task as you attempt to respond to this type of exercise. If the basic content of any of these exercises from precalculus and calculus is familiar to you, then write your responses to some of them.

1. Explain why a function that is not one-to-one, will not have an inverse that is a function.

2. In a half page or less, summarize what you know about the inverse of a function.

3. For the functions $\dfrac{f}{g} + h$ and $(f \circ g) \circ h$, discuss how you would find each range value associated with the domain value b. In your discussion, you can use specific functions for f, g, and h.

4. Explain why the period of the tangent function is π and not 2π.

5. In simplifying $\sqrt{\dfrac{4}{9}}$, Tom wrote $\pm\dfrac{2}{3}$, and Karen wrote $\dfrac{2}{3}$. Who is correct and why?

   In solving $y^2 = \dfrac{4}{9}$, Tom wrote $\dfrac{2}{3}$ and Karen wrote $\pm\dfrac{2}{3}$. Who is correct and why?

6. Discuss why $\sqrt{x^2}$ is not necessarily equal to x.

7. In graphing the equation $y = \dfrac{4x^2 + 7x - 2}{x + 2}$, Tom graphed

   $y = 4x - 1$ (obtained by reducing the right side of the first equation). Was Tom's graph the graph of the first equation? Explain.

8. Argue that if the graph of a function is symmetric to at least two of the x-axis, y-axis, and the origin, then it is symmetric to the third.

9. Given function $y = g(x)$, how would you explain to a classmate how to find g (-5) and g ($x^2 - 1$)? (You are not to use a specific function for g in your response.)

10. How do the graphs of $g(x) = (x + 2)^3 - 1$ and $f(x) = x^3$ compare? (Answer the question without constructing their graphs.)

11. Compare and contrast the Trapezoidal Rule and Simpson's Rule for approximating the value of a definite integral.

12. In a paragraph or less, describe the basic steps needed to solve an applied max-min problem.

13. Suppose the graphs of a function and its first and second derivatives are given, but are not identified. Explain how you can determine which graph goes with each function.

14. In words that accompany an illustration, give the geometric interpretation of the partial derivative of $f(x, y)$ with respect to x at (a, b).

15. Knowing that a function f is increasing over an interval (a, b) if its first derivative is greater than 0 over (a, b), explain why f'' is greater than 0 over (a, b) implies that f' is an increasing function over (a, b).

Have you been expected to respond to exercises like these in your mathematics courses? If not, do you intend to look for or create exercises similar to these to work? You might want to encourage your instructor to construct these types of exercises for you and your classmates.

3. Thinking in Mathematics

You cannot succeed in mathematics without the ability to think. Fortunately, thinking can be nurtured. This section begins by describing two modes of thinking, and finishes with a discussion on how you can nurture thinking in mathematics.

A Few Descriptions of Modes of Thinking

Analytic or logical thinking. This is thinking that advances step-by-step, where each of the steps is clear in your mind and you are aware of going from one step to the next. If you are thinking analytically you can tell someone exactly what you did, step-by-step. It is often a deductive process and logical in its progression. This type of thinking is widely used in mathematics. *Reasoning* is thinking logically or analytically.

Intuitive thinking. This thinking is very different from analytic or logical thinking. It is thinking that reaches a result or conclusion with little or no awareness of using steps along the way. Somehow, the intuitive thinker has a flash of insight of the result or of an idea that may prove fruitful in eventually obtaining the result. Ask an intuitive thinker how he or she obtained the result and the answer you most likely will receive is, "I don't know." Sometimes, by carefully asking an intuitive thinker a series of questions, you can help him or her to see what led to the result. These questions lead the thinker through an analytic thinking process that justifies the result attained intuitively.

Intuitive thinking is used in mathematics by students, but most likely not to the extent that analytic thinking is. The reason for this is that it is not as well developed in most students as analytic thinking. How to nurture intuitive thinking is not well known. It seems that a broad knowledge of mathematics content related to the problem under discussion supports intuitive thinking. Also, a good knowledge of problem-solving skills used to attack a problem supports intuitive and analytic thinking. Problem solving is important in its own right, but it has the added benefit of nurturing analytic and intuitive thinking. Chapters 21 and 22 are devoted to helping you improve your problem-solving skills.

Nurturing Thinking in Mathematics

From now on, whenever I use the word "thinking," it will mean thinking logically or analytically, or reasoning. The best way to improve thinking, as I have described it, is to practice it. Unfortunately, *thinking is silent.* No one can hear you think. When thoughts float silently and quickly through your mind it can be difficult to know what they are or how logical or illogical they may be. However, if you *vocalize* them they are no longer silent to you or to anyone nearby, which allows for better evaluation of them.

Practice in thinking can be nurtured in a variety of ways, many of which appear explicitly or implicitly throughout this book. But here I want to stress the importance of verbalizing (vocalizing) the methods and strategies you use in your thinking. Doing so enables you to become more aware of what you are thinking and what influences your thinking. This verbalization can be with yourself, classmates, and instructor. There is the typical verbalization that takes place when you discuss mathematics with others, and this is elaborated in section 4 of chapter 14. The specific verbalization technique suggested in the following subsection is something that is typically not done; yet it can be a very effective way to improve your thinking.

Verbalizing Your Thoughts to a Partner

To help you become more aware of what you are thinking in mathematics and what influences it, it is helpful to have someone listen and react to your thoughts as you vocalize them. Whimbey and Lockhead have written an excellent paper on how to use a partner to help you improve your thinking or problem-solving skills.[6] My remarks reflect their ideas.

Choosing your partner. Ask someone, most likely a willing classmate, to listen to your thoughts as you think aloud. This person may already be a study partner. The content of your thoughts may be a problem that you are trying to solve, part of a development in your textbook, or something that your instructor said in class. It can be anything that involves thinking. Your goal is to improve your thinking skills, and that will take some time; hence, you have to do this activity over a span of time. In this thinking activity, we'll call one person the "Listener/Supporter," and the other the "Thinker/Problem Solver."

Changing partners. It is a good idea to change partners regularly, since no two people think or problem solve the same way. Some partners will benefit you greatly; others may be of some assistance; and still others of little assistance. Reflect on whether the person you choose is working out for you, but give it some time.

Alternate roles. At times you will be the Listener/Supporter and other times you will be the Thinker/Problem Solver. Your thinking skills can improve by playing either role.

Advice to the Thinker/Problem Solver

The role of the Thinker/Problem Solver is simple. This person has to vocalize all thoughts that come to mind while trying to understand the content being studied. Here are samples of thoughts that the Thinker/Problem Solver will be saying aloud:

> *"I am not sure what this word means."*
> *"I had better see what the textbook authors say about this."*
> *"The statement of this problem makes no sense to me."*
> *"I think I had better draw a picture of the situation."*
> *"The answer I just obtained doesn't make any sense to me."*
> *"I have no clue as to how to justify this step."*
> *"This problem seems similar to one I worked before."*
> *"I will use the problem-solving method of working an easier problem."*
> *"This appears to be a related rates problem."*
> *"This step is justified by Theorem 4."*
> *"I have no idea where to start."*

Advice to the Listener/Supporter

There are several key things the Listener/Supporter needs to do:
Ensure that the Thinker/Problem Solver is vocalizing all his or

her thoughts. If a somewhat steady stream of comments or questions is not coming forth, you can be sure that many of the thoughts that arise are kept silent. A simple prompt like, "Are you saying aloud all of your thoughts," should suffice, at least for the moment. Even simple steps may need to be vocalized; however, very routine and minor written work does not have to be vocalized if you believe the Thinker/Problem Solver understands it.

Work on the problem situation at the same time as the Thinker/Problem Solver. You can't be of much help if you are not familiar with the statement of the problem and a sequence of steps for solving it. Work on the problem situation at the same time as the Thinker/Problem Solver. You should not do it separately, and neither of you should let the other get ahead. Don't hesitate to tell the Thinker/Problem Solver to stop progressing on the problem situation if he or she is getting ahead of you. It is clear thinking and accuracy that are the goals, not speed. The Thinker/Problem Solver should not take the next step until you check out the current step for accuracy and clarity of thought.

Point out errors in the Thinker/Problem Solver's thinking as they arise. This may be an error in making a drawing, deriving a computation, misusing a theorem, supplying a reason for a step, etc. The correction of any error needs to be made by the Thinker/Problem Solver, not by the Listener/Supporter.

Ensure that the Thinker/Problem Solver justifies every step. You can support the Thinker/Problem Solver in making a justification by carefully asking leading questions if need be, but you are not to make the justification.

Be kind in your remarks. The Thinker/Problem Solver needs to understand that your interest is in helping with his or her thinking skills. You goal is not to criticize or in any way put the person down. Therefore, be aware of the tone and charity of your remarks.

Don't solve any part of the problem for the Thinker/Problem Solver. If the Thinker/Problem Solver gets stymied at some point, suggest a next step, but don't perform the step. You are there to guide and prompt, not to "do for."

Suggested Remarks and Questions That Help the Thinker/Problem Solver Progress in His or Her Thinking

> *"What problem-solving skills do you know that might apply here?"*
>
> *"Are you sure of that thought?"*
>
> *"Are you sure of that step?"*
>
> *"Are you sure of that computation?"*
>
> *"Might there be another way to approach the problem?"*
>
> *"Does the answer make sense to you? Elaborate."*
>
> *"You misused that theorem. Do you know why it doesn't apply?"*
>
> *"What prompted you to start with that idea?"*
>
> *"Did the textbook authors work a similar problem?"*
>
> *"Do you know what the textbook authors cautioned about that situation?"*
>
> *"I am not sure you understand what you just did. Can you explain it again, giving more clarity to your justification?"*
>
> *"Are you sure you read the problem correctly?"*
>
> *"Have you read all the information in the problem?"*
>
> *"Have you used all the information in the problem?"*
>
> *"Is there any extraneous information in the problem?"*

The use of the thinking technique of this section to improve thinking can work regardless of where you and your partner are, mathematics-wise. Just performing this process with someone else is helpful. That being said, there may be other partners who can help you more, but that is dependent on many factors, including personality, willingness to help, listening skills, tutorial ability, and knowledge of mathematics.

When Can I Stop Verbalizing or Vocalizing My Thoughts?

Speaking your thoughts out loud to yourself, not necessarily to a partner, causes you to hear them. That alone will help you become more aware of what you are thinking and what influences your thinking. It will slow down your thinking, allowing you time to reflect on it. It is a good thing to vocalize, even without a partner. Your eventual goal is not to rely on a partner; however, a partner is useful as a helper in the early stages of your deliberate attempts to improve your thinking and problem-solving skills. As time progresses you can minimize or even eliminate a partner for stretches of time. You may periodically need to use a partner, especially when you are faced with thinking situations that are more complex or require new thinking skills. You may never want to completely stop vocalizing your thoughts to yourself.

10

Changing Beliefs, Attitudes, and Study Methods

An objective throughout this book is to identify beliefs that spawn attitudes and study methods that interfere with learning mathematics. There is no doubt that all students carry some of these beliefs. You need to know what they are so you can become a better mathematics student.

Changing well-entrenched beliefs that negatively impact attitudes and study methods won't happen easily or quickly. You will probably need support to make the necessary changes. One form of support is knowledge of the *dynamics of change,* which is discussed later in the chapter. Change takes place in stages—there is a beginning, middle, and end stage. You will be helped in progressing through these stages if you know for each stage (1) what to expect, (2) what to feel, and (3) what you need to do.

The sections comprising this chapter are titled:

1. Importance of a Positive Attitude for Success
2. Beliefs and Habits That Hinder Learning
3. Basics of Change
4. The Process of Change

1. Importance of a Positive Attitude for Success

We bring many attitudes into mathematics courses that have been shaped by our beliefs. Unfortunately, false beliefs can form negative attitudes, which lead to unproductive practices that are a source of exasperation. In turn, unproductive practices tend to strengthen negative attitudes. It is a never-ending battle until deliberate intervention takes place.

A positive attitude makes the trials and tribulations of life, which can be a mathematics course, easier to handle. It is a joy for me to work with a student who has a good attitude, for then I know the student has a chance to succeed. This student is not interested in pursuing "the blame game," but wants to improve and looks honestly at the things he or she needs to change.

My attitude has become increasingly more positive as the years have passed, which has increased the quality of my life. If there is only one quotation that I want you to take from this book, it is this well-known one by Charles Swindoll:

> The longer I live, the more I realize the impact of attitude on life. Attitude to me is more important than the past, than education, than money, than circumstances, than failures, than successes, than what other people think or say or do. It is more important than appearance, giftedness, or skill. It will make or break a company . . . a church . . . a home. The remarkable thing is we have a choice every day regarding the attitude we will embrace for that day. We cannot change our past . . . we cannot change the fact that people will act in a certain way. We cannot change the inevitable. The only thing we can do is play on the one string we have, and that is our attitude . . . I am convinced that life is 10% what happens to me and 90% how I react to it.

I urge you to copy Swindoll's quotation and read it often. If you want your life to be better you may need to make an attitude adjustment. Perhaps you need to be more realistic in your expectations of yourself and of others, look more on the positive side of things, and take constructive steps to make situations better for you and for others.

2. Beliefs and Habits That Hinder Learning

There are many false beliefs related to mathematics or mathematics study. There is hardly an area of a mathematics course free of false beliefs about it, including the content, instructors, classmates, textbooks, obtaining help, assignments, examinations, and grading. If you have false beliefs about some of the above areas, your performance in the course can be seriously compromised. The following subsections give examples of false beliefs about mathematics and the ways they are learned, many of which relate to the areas just mentioned. By no means is this an exhaustive list of false beliefs; however, it contains some of the more typical ones held by students.

False Beliefs about Mathematics Aptitude

Do you believe that you were behind the door when your parents passed out their math genes; hence, you don't have the mathematics aptitude to be successful in mathematics?

This belief can cause you to ignore most of the advice in this book. Since you "just don't have it" in mathematics, why even bother when the going gets rough. You think that it is not possible to understand a derivation, work a difficult problem, or understand an examination question that is not routine. It will be easy for you to give up because you know from past experience that any further effort is fruitless. Your motivation to succeed in mathematics, if there is any at all, will wane quickly. Furthermore, you won't feel guilty about giving up since all of this is out of your control. Do you agree that the behavior that can result from this false belief is destructive? What could be more likely the truth is that your natural aptitude in mathematics is less important for success in mathematics than your motivation to succeed, methods of study, belief that it is worth studying, and work ethic. Until you utilize good methods of studying mathematics, you will not know how successful you can be. Don't entertain any thoughts you may have on your perceived lack of mathematics aptitude and how it will hamper your achievement in mathematics. You most likely have more mathematics aptitude than you realize. Know what you have to do in your study to be successful in mathematics, and then do it.

False Beliefs about Poor Memory

Do you believe that you were not blessed with good memory? That you either remember something or you don't? That there are so many things to remember for a mathematics examination that you cannot keep it all straight? That many of your friends are blessed with good memories? That you study as hard they do but they can remember and you can't? That you forget mathematics almost as soon as you learn it?

A belief that your poor memory is innate leads to difficulties. You will believe that there is little or nothing you can do about it, except to work harder to memorize content without understanding its meaning or significance. You make more flash cards and underline more in your textbook or notes. You become discouraged in your study and fear taking examinations. You will continue to be discouraged until you realize that likely reasons for your memory problem are that you are not studying for understanding, or that you have unsound maintenance and review methods for what you do understand. For help in remembering mathematics, read section 1 of chapter 24.

False Beliefs about Proof or Formal Reasoning

Do you believe that it is unnecessary in mathematics courses to learn and construct formal proofs, or reason formally? Do you also believe that the only purpose this serves is to confuse, and that it provides little help solving problems?

These false beliefs can cause you to dispense with proof or formal reasoning in your study. Learning and constructing formal proofs is *overdone* if it does not allow sufficient time for other necessary mathematics activities to take place. However, knowing why certain principles are true is not wasted time. It is through proof or formal reasoning that connections are made among definitions and previously proven or accepted generalizations that justify a principle. Dispense with this too much and you are dispensing with substantial understanding. You need to understand the justification of certain principles, using proof or formal reasoning, and the justification of other principles using *informal* reasoning. Informal reasoning uses a variety of elements for its justifications including diagrams or pictures and results obtained through laboratory activities. It is less precise than formal reasoning, but a greater understanding can often result from its use. It is used more than formal reasoning in less advanced mathematics courses. At times it is important to know *precisely* what you are talking about and that comes through formal reasoning. Good understanding comes from *using a mix* of formal and informal reasoning.

If you have the belief that any formal reasoning is unnecessary, a belief that is most likely not shared by your instructor, you will probably not work diligently to understand many of the proofs appearing in your textbook (or provided by your instructor in class), or to work diligently on exercises that are asking you to construct proofs. Analogously, a similar lack of effort can exist if you feel informal reasoning is unnecessary. In either case, your understanding of mathematics will suffer and this includes your ability to apply what you learn to solving problems. You may wrongly think that all problems only require you to apply mechanical procedures to solve them, and if you can't quickly determine these procedures, not much can be done about it. Eventually you may end up thinking that these problems are not that important to solve. They could very well be problems for which you have to use reasoning, formally or informally, to reach a solution, and not working on understanding proofs can hinder your ability to reason when attempting to solve them. These problems are often your most interesting ones, and offer you an opportunity to learn more. They will require extra time and thought, but the exhilaration and benefits that come from solving them are worth the effort.

False Beliefs about Obtaining Help from Your Instructor

Do your believe that obtaining help from your instructor is a sign of weakness and should be avoided if possible? That you should not have to get extra help? That you will suffer consequences if you see your instructor for help, since he or she will probe more deeply and discover that there is a lot more that you don't understand?

Obtaining help from an instructor is a sign of strength and quality college instructors believe this. Yes, instructors will probe more deeply, but be thankful for that. Wouldn't you want a medical doctor to do the same? The vast majority of instructors will grade you on your performance on assignments and examinations, not on visits to their offices. If your grade is changed because of your office visits, it will most likely be a higher grade because instructors admire students who show they want to learn. Consulting with your mathematics instructor is important to do. To discover the benefits of office visits and how to prepare for such visits, read section 1 of chapter 23.

False Beliefs about the Role of Your Instructor

Do you believe that a major role of your instructor is to make learning as easy as possible? That he or she is to remove every bump and straighten every curve on your road to learning mathematics?

Your instructor should not make learning as difficult as possible; however, he or she must avoid "spoon feeding" you. Your instructor must wean you away from dependence on him or her for learning—to help you to become a lifelong learner. This requires certain actions on his or her part to challenge you in the short term, so that in the long term you become a more self-sufficient learner. An added benefit is that the more you figure out yourself, the greater will be your understanding.

Good instructors will challenge you by: (1) holding you responsible for knowing certain content that will not be discussed in class (but is in your textbook or in a handout), (2) answering your questions by asking you questions, (3) expecting you to read the textbook and begin working the associated exercises before the content is discussed in class, and (4) not taking class time to review for tests.

A good instructor requires you to work diligently to succeed, but supports you. This type of instructor is doing you a big favor, especially if feedback on your efforts is provided in other ways (e.g., by directly answering some questions in class but not others, encouraging small group discussion in and out of class, encouraging you to get help during office hours). Having the belief that *you* are responsible for your learning serves you well, as attested to by Brown and Atkins:

> Research has shown "that the way students perceive themselves and the way they account for their academic successes or failures, have a strong bearing on their motivation and their performance. Students are likely to initiate learning, sustain it, direct it, and actively involve themselves in it when they believe that success or failure is caused by their own effort or lack of it."[1]

False Belief about Getting Behind in the Course

Do you believe it is no problem to consistently get behind in your mathematics course by two or three days because you can catch up by working more diligently on the course when you get around to it?

Chances are you will have to drop the course if you consistently get behind in your course work. The course does not stop to wait for you to catch up. Your classes will be less meaningful to you because you do not have the requisite knowledge to understand the new content that is being developed. This gets you even further behind. The false belief by students that getting behind in mathematics courses does not cause much harm *is a major reason* for their lack of success in these courses.

False Belief on Curving Final Grades

Do you believe that, even though you are not doing well in your mathematics course, your instructor will determine final grades based on a curve because many of your classmates appear to be doing as badly as you?

That may have happened in your high school mathematics courses, but don't bet on it happening in your college mathematics courses. College is a different ballgame. There are no groups of parents complaining to a college mathematics department chairperson about the low grades an instructor gives. High school principals get these complaints, which pressure them to work on rectifying the situation. It is typical these days for secondary instructors with appropriate standards to get considerable grief from students, parents, and administrators. College instructors have a lot of latitude when it comes to grading. The harm that can come from believing your low grades throughout a course will result in a higher course grade because of grading on a curve, include the following: You won't work diligently to make the necessary changes to improve your grades, or you won't drop the course by the stated drop date, even when it is clear that your chances of doing well on remaining examinations is highly unlikely. I saw these behaviors in college mathematics students more than I care to mention.

False Beliefs on Working Assignments

Do you believe that it is okay to leave uncompleted some of the assigned exercises, especially the more difficult ones? That completing the bulk of the assignment is good enough?

I wish you were with me in my office during the times I met with students who did badly on an examination. We diagnosed the reasons why they did badly. There was often more than one reason, but a predominate reason was inadequate completion of the assigned exercises prior to taking the examination. Some students had not started some of them, others had started them but left then unfinished, and others thought they had completed all of them correctly, but some of their solutions were incorrect. How did this impact their examination results? I was able to show them that many of the examination questions they were unable to work were related, in one way or another, to incomplete assignments. There are reasons for a poor test performance, and you need to identify these reasons when you test poorly. Most likely they include the false belief that you can leave your assignments unfinished. Complete your assigned exercises in a timely manner, do them correctly and with understanding, get feedback to see if your solutions are correct, and get help on those that are causing difficulty.

False Belief on Changing Behavior

Do you believe it is impossible to make the necessary changes to be successful in mathematics?

If you believe you can't change, you will despair and continue to study as you always have. This is not a good prescription for success in mathematics. Why might you carry the belief you can't change? In all likelihood, you don't know how to change. It is helpful if you understand the stages of change, and how to progress through these stages. The last two sections of this chapter discuss change.

In summary, false beliefs can lead to actions that are detrimental to learning. List the false beliefs you hold that were discussed in this section, and ask yourself periodically throughout your mathematics course if you still subscribe to them. Add to this list other beliefs you hold as a mathematics student that you are beginning to think are false. Discuss them with your classmates or instructor. For those you view as false, reflect on the changes you will make in your behavior based on your modified beliefs, and get advice if you need it.

3. Basics of Change

There is no doubt that you will need to make some changes in how you study mathematics. The need to make changes in most aspects of our lives is never ending. We are continually confronted with different situations that may demand change. Such situations may involve a place to live, roommate, doctor, child, diet, debt, or job. In college mathematics courses you are confronted with new content, instructors, expectations, classmates, textbooks, assignments, examinations, class times, and methods of grading. The list goes on and on. Positive changes in your study methods will lead to better performance in your mathematics courses.

The issue is not whether there needs to be change in your study of mathematics—that is inevitable, but how you view the change that needs to take place. You have the choice to (1) view yourself as a victim of change, grieve over it, and be terrorized by it; or (2) welcome it, confront it, and view it as a chance for better things to happen. You may not be ready to respond positively to change. It takes time, and the quicker you realize you have to confront it, the quicker you can benefit from the changes you make.

When Is It Time to Change Your Approach to Studying Mathematics?

A poor performance on a mathematics examination seems to motivate most students to take stock of how they study mathematics. For some students a poor performance is devastating, to others it is definitely cause for concern, and to others it hardly raises a blip on their radar screen. Some students who perform badly realize they have to work on making changes and go about the business of doing so. Others will resist doing this for a variety of reasons, not the least of which is not knowing what to change or how to go about it. Some students who do poorly will not question what they can change to do better, but will continue as before, expecting the situation to miraculously change. These students need to ponder the truth of this anonymous statement: "If you always do what you have always done, you will always get what you have always got." How could it be otherwise? Don't treat a poor performance on an examination lightly. If you don't change your study methods, it will happen again.

Other warning signs that it is time to make changes include these: (1) your instructor's concern about specific study behaviors you exhibit, (2) turning in incomplete or late assignments, (3) procrastinating in your study, (4) confusion during most class sessions, and (5) inability to read the textbook or even have the desire to read it. The warning signs are many and *the first signs that you are in trouble come well before a bad examination result.*

There is no doubt that you are using some appropriate study methods in your study of mathematics. What may be problematic is your ability to separate these from what you are doing that is not good. If this is difficult to do, help is available in this book, from your instructor, from your classmates, and from the academic counseling office. Once inappropriate study behaviors are pointed out to you, you may need to be convinced to change them since they are well-entrenched and have become old friends. Habits are security blankets, not suddenly removed.

Change Means Lasting Change

Hopefully, you have decided to change specific study methods after spending time reflecting on the benefits you will receive from making the change. You want to make changes that have far-reaching and far-lasting effects—not just until the end of the course. Change requires making two basic decisions: The decision to *initiate* change and the decision to *sustain* it. Be impatient enough to make changes and patient enough to sustain them until they become habit.

Change Is Difficult

Change is difficult. You need to be patient as you make changes because you can expect one or more of these things to be present:

Assault on your self-esteem. Accepting that some of your methods do not work can be a blow to your ego. Or, you may question your ability to make the necessary changes, to stay the course, or to succeed. These can also negatively impact your self-esteem. There is a wealth of information in chapter 11 that will help you come to grips with self-esteem issues.

Sacrifice/Frustration. Your study sessions may be longer, more difficult, or cut into your recreational pursuits. You may have to overcome a reticent nature, seek more help, or continue to experience less than desirable results until your efforts to change take root. This causes sacrifice and frustration.

Risk. Nothing is guaranteed, so you risk throwing away your old secure habits and incurring sacrifices that might not bring about the desired results.

Fear/Anxiety/Stress. These feelings may be present and you don't know why. You think about the changes you tried to make in the past that did not work, and that invokes fear, anxiety, and stress. You may say, "Why do I want that back in my life?"

Perhaps you can remember when you were about to go from middle school to high school. You knew that you would not know a lot of students and teachers, and that you would have more difficult subjects. You were told many times that middle school and high school were very different, and that you would have to make changes when in high school. Looking back, did you experience an assault on your self-esteem, frustration, sacrifice, risk, fear, anxiety, or stress? Did you experience any of these things when you first entered college? It is not uncommon to experience these feelings when making changes to how you study mathematics. *Changing well-entrenched habits is a dying to yourself*, and that can be painful. What you leave behind is part of you, and you grieve over it. Your unproductive habits have to go; they will be in conflict with the changes you are making. Charles DuBos said, "The important thing is this: to be able at any moment to sacrifice what we are for what we could become."

Change Requires Practice, Patience, Persistence, and Pressure

Minor changes can take root quickly, but significant changes evolve slowly. Patience and persistence are prerequisites. Not only are you changing well-entrenched habits of study, but you are also changing attitudes. You need to know what needs to be changed, and then practice your changes. Practice takes time. Noticeable changes in your mathematics performance may not come as fast as you like. *You will not be patient and persistent unless you believe that change will make things better, and that your old methods of study are of little value to you.* It is this knowledge that keeps you from relapsing into your old ways. Practice, patience, persistence, and pressure are necessary components of change.

Change Is Joy

Even though you see reasons to change, your thoughts may be focused on the trials and tribulations that accompany it. You might ask, "Why bother to change? Who needs that grief?" These are fair questions and ones that I love to answer this way: *If you ask yourself this question, then you don't know enough about joy.* I am not talking about fleeting joys of this world such as an ice cream cone, a great vacation, an exciting movie, or viewing an exciting NCAA basketball playoff game. These joys are here today and gone tomorrow. *The joy I am talking about here is the deep-seated and lasting joy that comes from doing the right thing, from being true to yourself, from knowing that you are on the right path.* This is a different kind of joy, one that is always with you. It comes from knowing it is important to change, and believing that you are headed toward a resurrection. Suffering for suffering's sake is useless. Suffering for redemption's sake is worthwhile suffering and needs to be embraced. Welcome the suffering that may come from pursuing change.

It is beneficial at the end of each day to ask yourself these two questions: (1) What did I fail to do today in my efforts to learn mathematics? (2) What will I do differently tomorrow in my efforts to learn mathematics?

Don't Make Wholesale Changes at Once

A word of caution about making changes: Don't take on more than you can handle, physically and psychologically. Too much overload can cause you to despair and give up. It may be best to begin with smaller changes that are easier to make, and then progress to more major changes, working your way prudently through each one of them. It is a balancing act. You don't want to take on too many changes at one time, but you also don't want to wait too long to improve your performance in mathematics. Being successful in making some changes will spur you on to make other changes. Oliver Wendell Holmes got it right when he said: "The great thing in this world is not so much where we are, but in what direction we are moving."

Your Desires Will Change

The more you change, the more you gain a sense of control, and see what else needs to be changed. When change takes place, your desires will change, and then you know what else needs to be changed to satisfy these desires. Change begets change. You are controlling your fears and becoming a self-perpetuating change agent.

4. The Process of Change

To make changes, it is important to know what your thoughts, feelings, and actions may be as you travel this journey. This knowledge will help you realize that what is happening to you at specific phases of your journey is expected. This realization is motivating and empowering. An overview of the *three phases of change*, as they apply to the study of mathematics, is now given. They are based on a cognitive theory of behavior change appearing in a book by Donald Meichenbaum.[2]

Phase One

In this phase you can expect to experience the following: You begin to feel that there is a problem with one or more ways you approach the study of mathematics. You become more aware of how you approach mathematics; that is, you are looking at your study habits with a clearer sense of what it is you currently do. You are analyzing yourself to see what you do. Your desire to do better is strengthening, but you are not sure what you need to do in order to do better. Your thoughts are very limited and you don't take any real action to change. You want to understand better the feelings that are surfacing. You replay this dialogue with yourself over and over in your mind, maybe even viewing your situation as somewhat futile.

Examples of thoughts you may have on the way you approach your study of mathematics:

1. "I am frequently confused in my mathematics course, yet I do not ask questions, in or out of class, to get clarification."
2. "I have never been one to carefully read my mathematics textbook. It is difficult to do. I know that my understanding of the content in this course is not very good. Perhaps I should rethink this behavior."
3. "When reviewing for an examination, I seem to be spending too much time learning content that I should have learned earlier in the unit. What kept me from learning this earlier?
4. ""I am too easily distracted in class. Perhaps I should sit closer to the front and away from classmates who are distracting me."
5. "I noticed that when I haven't worked on the assignment that was given in a class period, I am more confused during the next class period. This behavior is hurting me. I need to determine why I am not doing these assignments on time."

In this phase it is important to dialogue with yourself and other individuals, including your instructor, classmates, or someone in the Learning Center or Counseling Office. In reading this book you are carrying on a quasi-dialogue with me. It is through the process of dialoguing and asking and answering questions that you develop new ways to approach your study of mathematics. Your dialogue will begin to change in substance and quality, taking on a more positive tone, indicating that you are more in control. Denial is minimized, but you have not completely discarded what is not working, or totally accepted what potentially will work. The illogicalness of portions of your approach to the study of mathematics is clearer to you, and you are aware of the directions you need to take to have a more viable approach to learning mathematics. If you have difficulty getting through this stage, get some help.

Phase Two

At the outset of this phase, your internal dialogue has been changed to the extent that you are now ready to *rid yourself* of what has not been working for you, and take on one or more new study techniques. You are ready to cope with these new techniques. This is not to say that this phase will be problem free since the change in your internal speech may vacillate. However, your conceptualization of your situation has changed to the point that you can organize your practices around these thoughts in a more effective way.

Examples of thoughts you may have on new learning techniques you are adopting follow:

1. "From now on, when something confuses me in class, I am going to raise my hand and ask for a clarification. If I cannot get it answered there, I will see my instructor in her office or talk to one of my classmates, as soon as possible. The same goes for questions that come to me outside of class. I will get my questions answered, one way or another."

2. "I am going to make it a point from now on to understand what my textbook authors are saying. Before I begin working my assignment I will work on understanding the section of the textbook that relates to this assignment. A friend of mine referred me to a guide for reading a textbook and I will consult it. If need be, I will get help from my friend or instructor on content I don't understand."

3. "I am going to listen better in class. To do this, I will do three things: (1) read the content in the textbook, before class, that the instructor will be discussing, (2) find a seat in the first few rows of the classroom so I can see and hear

better, and (3) be more attentive to taking notes. These changes will keep me more focused on what is being said."

4. "I will spend more time getting ready for my mathematics study group by working more diligently on the assignment that we will be discussing. I will make room in my schedule to do this."

Phase Three

In this phase you have adopted new behaviors, and evaluated their outcomes by talking to yourself. You will decide whether the evidence you gather is good enough for you to continue on this path. You may accept or reject it. *Hopefully, your rejection is not based on a faulty interpretation of the evidence or on unwarranted fear.* What you say to yourself about the evidence is critical to determine whether you will maintain this new behavior—this new way of operating. You not only have to believe that you have benefited from your changed behavior, but you have to believe that you are a changed person. You want to know if your specific behavioral changes have generalized to yourself. That is, do you think differently about yourself, including your ability to succeed, to grow, to handle what comes your way, etc. Once again, *it is the evidence you gather and how you view it* that will determine this. This evidence can come from a variety of events and people.

Examples of your thoughts on the evidence collected follow:

1. "Asking questions of my instructor in her office has not helped—my examination score since visiting her was hardly better than the last one." (Is this enough evidence and is it being viewed properly?)

2. "My instructor embarrassed me in class today when I asked a question. I don't think I'll ever do that again. So much for my good intentions!" (Is this enough evidence and is it being viewed properly?)

3. "I have seen my instructor twice in his office and it was helpful; however, I know I need to see him more often for it to have a significant impact on my performance."

4. "I am beginning to feel a sense of excitement over my improved understanding of mathematics. I know that I have a long way to go but I am hopeful that things will continue to improve if I stick with it."

5. "I still don't understand many of the steps presented in the textbook. This is too much work and is taking too much time. I won't be able to do this." (Is this enough evidence and is it being viewed properly?)

6. "With my change in study habits, I seem to be less afraid of the next examination. I seem to look at things differently now, and that excites me. Life is getting better."

7. "My instructor's comment to me today that he noticed a change in my interest in the course makes me want to do better."

8. "In the past when I worked with my classmate, Mary, she was the one answering most of the questions; this now seems to be reversing. More and more I am able to help her. I know that she has noticed this change in me, and I feel good about it."

In summary, the need to make changes in the way you approach mathematics is to be expected. Knowing what to change requires discernment. Whether your changes are long lasting depends on the evidence you collect and how you view that evidence. Will you fight or flee?

11

Increase Confidence and Motivation, and Decrease Procrastination and Anxiety

Confidence, anxiety, motivation, and procrastination are psychological concepts that all students struggle with to varying degrees and at various times. These concepts do not stand apart from each other; they are connected. For example, lack of confidence in your ability to learn mathematics can lessen your motivation to study mathematics, which can increase your anxiety about mathematics, which can cause you to procrastinate in your study of mathematics. Having confidence in your ability to learn mathematics can increase your motivation to study mathematics, which can decrease your anxiety about mathematics, which helps eliminate procrastination in studying mathematics.

It is apparent that a change in one or more of these behaviors can influence changes in one or more of the others. What may appear to be an insignificant and isolated change in a behavior can have far reaching positive or negative consequences. *Much of what you do in your approach to mathematics is connected*—you change one thing and changes automatically take place in other things. In general, negative changes cause other negative changes, and positive changes cause other positive changes. Here are two examples: (1) Procrastinating in doing your homework can make class time and work with other students less meaningful, increase anxiety before and during examinations, lower performance on examinations, and result in a lower course grade. (2) Asking questions when you are confused can increase motivation, enhance understanding, increase confidence, lessen anxiety, help you understand the textbook, and improve work on assignments.

Each of these four psychological concepts plays an important role in your success as a mathematics student, and are addressed in these sections:

1. Increase Confidence to Succeed in Mathematics
2. Increase Motivation to Succeed in Mathematics
3. Decrease Procrastination in Studying Mathematics
4. Decrease Mathematics Anxiety

These sections will help you better understand these psychological concepts, the positive or negative effects they have on you, and the strategies you can use to reduce the negative effects. As you read, apply the information to your personal situation.

1. Increase Confidence to Succeed in Mathematics

Lack of confidence is debilitating. It can prevent you from leaving the starting gate in your attempts to learn. When you have confidence about a situation, many good things can happen. Successful people can lack ability, but are most likely confident people. Talented musicians, athletes, teachers, and students who have confidence in their ability, typically rise above others with comparable or more talent, but less confidence in their ability. Think about the power, exhilaration, motivation, and inspiration you felt when you were confident about a situation you had to face, or an action you had to take. Now think about how you felt when you were not confident about a situation. Do self-esteem and confidence go hand in hand?

The Importance of Self-Esteem

What do you think about yourself? Do you define yourself by your accomplishments or failures? Let's hope not, for failures or accomplishments apply to events, not people. You may consider yourself a failure, but that can only happen with your permission. Unless you come to grips with knowing and accepting who you are, apart from your accomplishments or failures, your actions in making the necessary changes to be successful in your mathematics course will be compromised.

You may have heard the expression, "As a person thinks, he or she is." That is, you operate out of who you think you are. Henry Ford said, "If you think you can or think you can't, you're right." If you have low self-esteem, you don't think much of yourself, and will find ways to sabotage what you need to do to be successful. People with low self-esteem program themselves to fail and people with high self-esteem program themselves to succeed.

You may have weaknesses, but that does not make you weak. People with low self-esteem evaluate themselves by their worst work, not their best. Should you take your work seriously? Yes. Should you take yourself seriously? To a degree, yes, but not to the extent that you beat yourself up—that you become your enemy. The importance of having good self-esteem cannot be over emphasized. According to Nathaniel Branden, a noted author and expert on self-esteem, "Positive self-esteem is the immune system of the spirit, helping an individual face life problems and bounce back from adversity."[1]

Good self-esteem can be developed, but it is never a quick process or an easy one. A necessary condition for it to develop is to become more successful. With success comes confidence, and with confidence comes the willingness to tackle new situations, and as the situations are met, more confidence is gained, and so on. As you confront and conquer more and more challenges, your confidence to take on any challenge grows and builds self-esteem.

It's Up to You

Nobody can learn for you. Others can support your learning but you have to make it happen. No one but you can stop you from learning, and knowing this can build your confidence. Do you act as though you are not responsible for your learning? Do you play the "blame game" by faulting your instructor, textbook, examinations, parents, or your roommates, but never yourself? If you want change in your life and it is not happening, then fault yourself since you are the only one that can make it happen.

Work to be more confident. Think for yourself since no one else can do it for you. Even if your thinking is incorrect, the fact that you did it is significant, not only because you are doing what you need to do to be successful, but also you are more apt to learn from mistakes in your thinking. Your goal is to improve your thought processes and that can only happen if you are engaged in thinking. When I invest my money and lose much of it, I feel so much better than when someone else invests my money with the same result. I do not want big decisions made for me, and, as an adult, neither should you. It is your life and *you* have to live it the best you can.

Before working on assignments with others, put some work into it yourself. I have seen too many students working together on assignments, exercise by exercise, without having first worked on them individually. That's a big mistake. What you accomplish by yourself is definitely yours—you own it. Go to others for help on a topic when you need it, but don't go too quickly and don't expect them to do all the thinking. Get some hints from them and then go off and continue working on the topic by yourself, until it's an integral part of you.

Having the mindset that you are not a victim of circumstances— that you are in the driver's seat—that it is up to you, builds confidence.

Genuine Confidence Versus Pseudo-Confidence

You may confidently strive for some goals that are unattainable because you are not prepared to attain them at that time. Experience teaches you the boundaries of your capabilities as you progress through life. It is valuable to know your capabilities, which includes knowing what you are not prepared to do. Don't confuse genuine confidence with pseudo-confidence (i.e., non-authentic confidence). Your dreams, goals, desires, or aspirations have to be realistic. You have to strike a proper balance between expecting too little of yourself and expecting too much of yourself. Expecting too little of yourself will keep you from reaching your potential. Expecting too much of yourself leads to frustration and despair, which also keeps you from reaching your potential.

Having confidence does not mean that you will succeed according to your timetable—you may have to backtrack. For example, if you lack significant prerequisite knowledge for a course, all the confidence in the world isn't going to get you through the course. But a confident person would drop the course, attain the prerequisite knowledge, then re-enroll in the course, and know that things will be different this time.

Perils of Striving for Perfection

I recall a freshman calculus student who was doing well in my introductory calculus course, working at a high B level. Unfortunately, she was very upset that it was not at the A level. Every now and then, in a very stressed and panicky state, she would say something to me like this, "I have to get an A in this course. I want to go to medical school, and if I don't get an A in calculus I won't be able to compete with those who have better grades than me. The competition is so keen." She had an unrealistic view of what she needed to get into medical school and was unwilling to be a friend to herself. Yes, she should strive for excellence, but not expect perfection. Expecting perfection and not achieving it is demoralizing, depressing, and debilitating. It doesn't build confidence or promote taking care of oneself. To be human is to be less than perfect. Oliver Wendell Holmes, Jr., said, "We expect more of ourselves than we ever have a right to."

We all lack confidence at times; that is the nature of the human condition. The big question is: "What can I do, if anything, to become more confident?" Perhaps you think that nothing can be done about this since you either are, or are not confident, and that cannot change. Change this thinking!

Believe You Can Succeed

There is nothing that bolsters your confidence like success. Until you have successes, you have to *believe* that you can succeed. This belief has to be coupled with taking appropriate action, for without taking action your belief is only wishful thinking. Your sustained belief keeps you working and looking for appropriate avenues to success. Perhaps being confident is nothing more than acting in faith. Eventually, success will come. Doubting that you will succeed is often a self-fulfilling prophecy for you will not try, or only make a feeble attempt, to succeed. You have control over how you deal with the elements of your course such as your instructor, textbook, and assignments. It is my hope that you believe and are motivated by this comment by Arnold Glasow: "The best bet is to bet on yourself."

Trust Your Instincts

It is not uncommon for students who have been unsuccessful to question their instincts. It may make sense to question some of your instincts, for they may have led you in the wrong direction. But doubting most of them does not serve you well. *They are your instincts, nobody else's, and are meant to serve you well.* If you don't trust them, you won't use them, and you then are like a ship without a rudder. Denying your instincts is another problem. You may correctly sense what an appropriate action should be, but suppress it because you would rather do something quicker and easier or less-anxiety provoking. Don't ignore what others have to say, but by all means don't ignore your instincts. If you ignore them, you are eliminating the place where some of your best decisions can be found.

Act with Confidence

You can *act* with confidence even if you lack confidence. I am talking about mind over feelings here. When you act with confidence, you are taking action, and when the action leads to success, confidence comes. For example, you may dread giving speeches, but decide to act as if you don't dread them. You can act confidently when you prepare a speech, when you wait to deliver it, when you walk up to deliver it, and when you deliver it. If the speech goes well it may be because you focused on the actions at the expense of your fear. Similarly, act confident as you go about making changes in your study habits, and see what happens.

Acting confidently requires using your imagination. If you want to be successful in mathematics, imagine what that is like, and then act it out. For example, imagine asking questions in class, taking good notes, working with other students, or taking an examination. If fears arise while you are imagining, keep imagining until they are gone or minimized. You will be surprised how helpful this is. I would bet that many students, maybe even a majority, imagine being unsuccessful rather than successful. Do you?

Confidence Requires Courage

If you want to improve your approach to learning, you need to make changes, and the process of making changes involves taking risks, which invokes fear. Overcoming fear requires courage. Some brave souls say that it is never a risk to do what is necessary, because the consequences of not doing what is necessary can be disastrous. Change is often difficult because it can be inconvenient, create more work, raise doubts, and increase anxiety. Making changes in the way you study mathematics is no exception. You are sacrificing creature comforts that come from the security of entrenched ways of studying mathematics, as ineffective as they may be, to attain honorable goals that are not assured. It takes courage to begin something new and attack it directly, and also to stay with it when difficulties arise. What helps sustain this courage is periodically recalling your goal. Courage ranks high on the list of desirable traits because so many other traits depend on it. Courage is necessary to move forward in your learning, and you don't have to look far to find it. *It is found within yourself.*

Diverse Reactions to Adversity or Success

Students react to adversity or success in different ways. For example, the behaviors of students as a result of performing badly on an examination include—

- Dropping the course
- Giving up in despair, but staying in the course
- Resignation (which leads to poorer study habits)
- Staying with the same ineffective study habits
- Making minimal improvements in study habits
- Doing all that can be done to improve study habits

Behaviors exhibited by students who receive a non-acceptable grade for a required course (i.e., a D or E) include—

- Changing their field of concentration so they don't have to retake the course
- Repeating the course and having poorer study habits than before (poorer study habits may result because they believe they know more of the content than they actually do, or because they are upset over having to repeat it)
- Not retaking the prerequisite course even though it is called for, but re-enrolling in the course and keeping the same study habits
- Not retaking the prerequisite course because they have the prerequisites, but retaking the course, making minimal or major changes in study habits
- Retaking the prerequisite course (so they are better prepared the next time they enroll in the course)

The range of possible student behaviors is considerable. Many students who retake a course do just as badly or worse the second time, because they still don't have the prerequisites or haven't improved their study habits. There are good lessons to learn in adversity—take time to find out what they are for you.

Students also react to success in different ways. A good test performance motivates some students to work harder and smarter on the next unit, whereas it causes others to slack off on the unit. I have seen students go from an A on a test, to an A, B, C, D, or E on the subsequent test that was comparable in difficulty.

It is sometimes better to fail in a big way than in a small way, because a big failure will get your attention. For example, failing an examination will get your attention more than receiving a grade of B- or C+. You might think the B- or C+ grade is not bad, but what caused it to be this grade rather than an A or B, may cause big trouble on the next examination (especially if the content examined is more difficult, or your instructor has higher expectations). Unfortunately, you may not recognize the small problems with your study methods that eventually lead to a big failure. They are often aspects of your study that you thought had no bearing on your failure.

I once heard a business man say that it is important to fail often and fail fast. Tom Monahan, founder of Domino's Pizza, failed twice in his first two pizza stores. When he sold his company he had over 6,000 stores throughout the world, with the distinction of being the world's largest take-out pizza company. He was determined to succeed and worked accordingly. Most importantly, he did not let failure of his short-range goals interrupt plans for his long-range goals. A salesman was told that one out of every 20 customers would buy her product. The first 19 customers said "No." A friend asked if she was depressed. She said, "No, I just happened to hit the 19 who would say, "No." Another person tried 10 times to perform a task and did not succeed. He said, "I didn't fail 10 times, I learned about 10 things that didn't work."

If you are afraid to take risks or to fail, then it will be difficult for you to succeed. Failure or adversity *is temporary* unless you decide to call it quits—to no longer value winning. You may make changes in your study habits, not experience the results you wanted, and then decide to give up. Get additional advice on what else you can do before you call it quits, since you may be on the verge of success.

Do you allow adversity to wreck your confidence? Do you allow success to over-inflate your confidence?

2. Increase Motivation to Succeed in Mathematics

Significant learning doesn't occur unless you study, and study doesn't automatically happen. You have to be motivated to study, so the question to ask yourself is this: "What are the driving forces that motivate me to study?" These driving forces are the reasons you study. There is little you can do if you are not motivated, and there are no boundaries to what you can do if you are highly motivated. It is your duty as a student to study and that should be reason enough to study. As an instructor I always assumed that students knew this was their responsibility, although I did what I could to motivate them to study. Doesn't it make sense that the primary role of being a student is to study diligently in any course? If that is the case, then other driving forces to study are not needed. Unfortunately, being human dictates otherwise.

It is no simple matter to know all the driving forces that motivate you to study your mathematics course. Most likely it is a myriad of forces, and what may motivate you to study in your mathematics course may not motivate other students to study in the course, or at least study to the same degree. If you lack motivation to study in your mathematics course, you need to determine what will motivate you.

Studying mathematics, or the desire to be successful in mathematics, can be intrinsically or extrinsically motivating. *Intrinsic* motivation to study mathematics comes from the subject of mathematics itself. That is, your interest in the ideas and power of mathematics motivates you to learn it. This is different from *extrinsic* motivation to study mathematics, which comes from outside the subject itself. For example, you are extrinsically motivated to study mathematics if you study it to get a grade that qualifies you to take a subsequent mathematics course. Intrinsic motivation affords you more learning benefits than extrinsic motivation, for you are interested in understanding mathematics for its own sake. Nonetheless, the value of extrinsic motivation should not be minimized. It can move you to study and can lead to intrinsically motivated study that is self-perpetuating.

Motivation Is Self-Perpetuating

The reasons that lead you to study mathematics may not include an interest in the ideas and power of mathematics. However, the encouraging news is that an interest in the ideas and power of mathematics can lead you to study mathematics—if you allow it to happen. And it can happen if you study it with understanding and are open to its power. Interest in a subject comes with understanding it—the more you understand it, including its applicability, the more you will be interested in it. This interest increases your desire to study it.

This simple chain of events—understanding leading to interest and interest leading to understanding—will continue to repeat; that is, your motivation to study the subject becomes self-perpetuating. How powerful that is! An added bonus is that your learning of the subject will become easier and more enjoyable. William H. Armstrong is very succinct in capturing the message of this paragraph when he says, "Interest is the fruit quite as much as it is the stimulus of study."

Positive Motivators to Study Mathematics

If your motivation to study mathematics needs to be strengthened, it is good to reflect on a list of positive motivators to study. They are called *positive motivators* because you are being *rewarded* for your efforts. Ask yourself these three questions for each item on the list of positive motivators below:

1. How strongly does this reason motivate me (low, medium, or high)?
2. Is it to my benefit to allow this reason to motivate me more, and why do I believe this?
3. Will I allow this reason to increase my motivation?

List of positive motivators to study mathematics:

1. Anticipating or imagining positive results—the thought of success
2. Confidence that positive results will be realized
3. Recalling success in past similar situations
4. Trusting in a higher power that, no matter what, things will be okay if you do your part
5. Doing what your conscience tells you
6. Accepting responsibility
7. Interest in the subject
8. Desire to learn or become better educated
9. Working with someone who loves the subject
10. Seeing classmates succeed
11. An instructor who motivates
12. Receiving respect or praise for a good performance
13. Wanting peace and happiness in your life
14. Satisfying a passion to be competent and successful
15. Proving to yourself that you can overcome adversity
16. Desire to meet short- and long-term goals that are dependent on your success

17. Using the learning gifts you were given
18. Satisfying your curiosity, which comes from acquiring a new area of knowledge
19. Knowing that you control your success
20. Proving to yourself that you can succeed
21. Trying out new methods of study
22. Knowing that ineffective results were due to a lack of motivation
23. Because it's there (Isn't that a reason people often give for climbing a mountain?)

What are positive motivators for you to study mathematics that are not on this list?

Negative Motivators to Study

If your motivation to study mathematics needs to be strengthened, it is also good to reflect on a list of negative motivators to study. They are called *negative motivators* because you are *avoiding negative consequences* for your efforts. Reflect on these three questions for each item on the list of negative motivators that follows the questions:

1. How strongly does this reason motivate me (low, medium, or high)?
2. Is it to my benefit to allow this reason to motivate me more, and why do I feel this way?
3. Will I allow this reason to increase my motivation?

List of negative motivators to study:

1. Repeating the course, which causes a loss of time and tuition monies
2. Loss of financial aid (because you may have to drop the course and carry less than the minimum number of hours)
3. Receiving a bad grade on your transcript, or a lower grade point average (each of which could affect your ability to stay in college or obtain a desired job)
4. Having to change concentrations
5. Frustration and resentment from sitting through the course again
6. Delay in graduating or retaking the course at an inopportune time (e.g., in a spring or summer term)

7. Delay in taking courses that have this course as a prerequisite
8. Performing inadequately in subsequent courses that have this course as a prerequisite
9. Performing inadequately in future occupation
10. Lack of motivation to work in other courses taken concurrently
11. Humiliation of family, friends, or classmates
12. Anxiety and sadness that comes from confusion
13. Guilt knowing that you could be doing better if you applied yourself
14. Discomfort at being confused or ignorant
15. Fear of being left behind

What are negative motivators for you to study mathematics that are not on this list?

John M. Wilson says, "There are only two stimulants to one's best efforts: the fear of punishment and the hope of reward." Combining the positive and negative motivators in the two lists gives 38 potential reasons to be motivated to study diligently in your mathematics course. Some of these motivators will be more meaningful to you than others. Perhaps one of them is motivation enough for you to be successful in the course. You may have reasons to study that do not appear on this list. These three exercises may be of value to you:

1. Are you motivated more by positive motivators or by negative motivators?
2. Pick the three most important motivators for you from the 38 that are listed, and rank them in order of importance to you.
3. Have a classmate or friend respond to the same questions and then discuss why each of you answered the way you did.

Carry statements of these three motivators with you and reflect on them as soon as you have the urge to avoid studying mathematics. Your motivation to succeed in your mathematics course will most likely fluctuate throughout the term. When it is low, you need to know why it is important to continue working. The consequences of not studying mathematics are not hypothetical—*they are real.*

Taking Examinations Is Motivating

One of the great motivators to work in a course is having to take examinations. How much work would you do in a course if the instructor came up with some scheme for giving you a course grade on the first day of the course? I once took a high level graduate course in educational psychology in which I had the option of having my grade be (1) the average of my examinations taken during the course, or (2) my graduate school grade point average at that time. I took the latter offer, as did almost everyone of my classmates. Thus, I received an A in the course on the first day of class and am not too proud to say that I did not study much in that course. I felt pressure from my other courses that graded in a more traditional way. This maxim was operational for me: "The squeaky wheel gets the oil."

Taking examinations may only motivate you to study what you think will be tested. This is unfortunate because you are then limiting your education. You are more likely to learn more if you are motivated to study for reasons besides receiving good grades on your examinations.

Asking and Answering Questions Is Motivating

Interest in studying a subject can come from asking and answering questions about the subject. Look for questions to ask others or have others ask you questions. Questions arouse natural curiosity. Look forward to answering questions asked by your instructor. Statements of facts are not that motivating, but questions are, since they present a challenge.

Choose an Instructor Who Is Motivating

You may or may not have the opportunity to choose your instructor. If you have the opportunity, there are many things to look for. (Advice on choosing a mathematics instructor is given in section 4 of chapter 13.) What an instructor says, how he or she says it, and the body language used, can be motivating. At the very least you want an instructor who loves, values, and understands mathematics, and *wants the same for you.* Traits to look for in an instructor include expertise, enthusiasm, appropriate expectations, and quality instructional techniques. An instructor's enthusiasm for a subject is contagious and motivating.

3. Decrease Procrastination in Studying Mathematics

The number of college students who come to a mathematics class session unprepared can be staggering. It can be the majority of the class. Far too many students come to class without having read the sections in the textbook or worked on problem sets that were assigned the previous class period. *Procrastination by college students in their mathematics studies is a major reason why they are unsuccessful.* Many of them are always one or more textbook sections behind, due to putting off their out-of-class work. Somehow, they think that mathematics only needs to be studied once or twice a week, and that they can catch up on the weekend. A procrastinator comes to class unprepared for discussion of the assignment from the last class session. He or she is also unable to understand new content presented in class due to lacking prerequisite knowledge developed in recent class sessions and reinforced by assignments. At examination time, procrastinators are trying to learn content when they should be reviewing content. Procrastinating completion of reading and writing assignments causes many students to receive grades of D, E, or W (indicating withdrawal) for their course. Procrastinators receiving a course grade higher than a C, learn less than they otherwise would have if they did not procrastinate. Is putting off your assignments a habitual problem for you?

Describing a Procrastinator

A procrastinator is a person who *regularly* postpones doing something. Each of us procrastinates, now and then, on completing specific tasks. We put things off, but that doesn't necessarily make us a procrastinator; it means we are simply being human. A mild form of procrastination causes minimal harm.

The degree of procrastination can be mild, moderate, severe, or chronic. If you are a mild or moderate procrastinator, you need to overcome some bad habits and that takes effort. However, the causes of severe or chronic procrastination are more complex; hence the solutions are more complex and take more work. These procrastinators will most likely need professional help to alleviate the problem. The content of this section should be of benefit to all types of procrastinators—from mild to chronic.

Who Is Not a Procrastinator?

You do not technically qualify as a procrastinator just because you put things off. It depends on why you put them off. Are your reasons for procrastinating rational or irrational? For example, suppose you are over-extended by carrying more credit hours or working more hours on a paid job than you can handle. Also suppose that you have good reason for being over-extended, and know full well that your studies will suffer. Since your reason is a rational one, delaying your study would not be considered procrastination, although there may be better options for you to pursue. Since there are various causes of procrastination, it is important to know what they are, especially those that apply to many procrastinators. It is difficult, if not impossible, to reduce procrastination if you don't know what causes it. Knowing the causes helps, but you need to go beyond that.

Causes of Procrastination

It is probably safe to say that most students have no idea why they procrastinate. They may think it is because they are lazy or lack self-control. Burka and Yuen say this about procrastination: "Procrastination is not just a bad habit but also a way of expressing internal conflict and protecting a vulnerable sense of self-esteem."[2] They say that procrastinators want to do well on a specific task, but believe that they will not be able to meet this desire of theirs. That is, they doubt their ability to succeed in the task. Rather than finding out if they can succeed by giving it their best efforts, they sabotage their efforts by delaying working on the task. Then, if they don't do well on the task, they have the excuse that they didn't attend to the task in a timely manner. This protects them from an attack on their self-esteem. They avoid thinking that they do not have the ability to perform the task; rather, they attribute their inability to do well on procrastination. They say, "I would have done better if I had attended to the task in a timely manner—if I had allowed more time and started earlier." They have protected their self-worth by not allowing their best effort to be judged. More succinctly, procrastinators fear failure, and they blame their lack of success on many things, such as bad luck, anxiety, or ineffective study methods, but not on their real or perceived limited ability. Fear of failure is the reason some procrastinators are indifferent about their studies, lack goals, or set unrealistic goals. Charles Baudouin comments about the devastating effects of fear of failure: "No matter how hard you work for success, if your thought is saturated with the fear of failure, it will kill your efforts, neutralize your endeavors, and make success impossible."

Do not be quick to say that fear of failure or fear of not doing well is *not* why you procrastinate. Take some time to reflect on this, especially when you find yourself procrastinating. At those times, ask yourself these questions: "Why am I procrastinating?" "Am I worried that I will not do well and find out that there is a limit to my ability?" If you come to the realization that this is the reason for your delay in attending to your studies in a timely manner, then you have taken a first and big step in your efforts to combat procrastination.

It is difficult for procrastinators who are high in ability or high performers to accept that they fear failure. Procrastination can affect students of any ability or performance group, and may affect those of high ability or performance even more. This may be more plausible after you read about the relationship that Burka and Yeun see between procrastination and striving for perfection, regardless of a student's ability or performance level:

> Since most procrastinators equate self-esteem with their performance and doubt their abilities, they must perform extremely well to reassure themselves that they are capable. The need for reassurance creates very high perfectionist standards. However, perfectionism can be a deterrent to working rather than a motivator. Facing very high standards frightens some people who fear they won't measure up. The fear of possible failure may then keep the student from even attempting the task, and the result may be procrastination. Perfectionists expect themselves to accomplish the impossible, without realizing that it's impossible. Any limit to their standards is felt as a compromise and becomes a threat to self-esteem. This all-or nothing philosophy means that less than perfect work is intolerable. If procrastinators can't work perfectly, they won't work at all.[3]

Are there college students with an A or near A overall average in their studies who are procrastinators? You can bet on it. If you suffer from low self-esteem and doubt your abilities, performing well in your studies does not provide a cure. It may help to alleviate this condition, but it can also exacerbate it if you always have to equal or surpass your good performances.

Burka and Yuen state that there are other reasons for procrastination, and they include[4]:

To maintain control. You may resent instructor-imposed deadlines for projects or tests. You want to decide when your work commences.

To minimize the pressure to succeed that is placed on you by various individuals. The individuals who may pressure you include a parent, spouse, sibling, friend, or instructor.

To get back at an instructor you do not like. You may dislike an instructor because of the perceived quality of the instruction, the demands and fairness of the instructor, or the unwillingness of the instructor to accommodate your personal needs. You may think, "I'll show him by not doing my work, turning it in late, or doing it poorly." This is comparable to cutting off your nose to spite your face. You may get your revenge, but your learning and grades will suffer.

Fear of success. You may view the consequences of success to be worse than failure or mediocrity. For example, some of your friends or classmates may resent the grades you get; hence, you have ambivalent feelings about studying hard to get these grades.

Uncertainty about short- and long-term goals. For example, you may not be sure you want to be an engineer or pursue a concentration that requires specific mathematics. Or, you may worry about how well you will perform in your career.

Help Is Available for Procrastinators

In all likelihood, you will need assistance if you are a chronic procrastinator. You can seek assistance at the counseling office and office of academic support at your college or university. Perhaps you can receive one-on-one assistance from a counselor, or participate in a workshop on procrastination that meets over a period of time.

Burka and Yuen have conducted group programs for procrastinators, in which a group would meet once a week for two hours, over an eight-week period. Over three years they observed that "The group experience seems to be comprised of three phases: optimism, pessimism, and realism."[5] The participants' initial optimism, based on some successes, drifted toward pessimism when they found out that not all their strategies to lessen procrastination were successful. More work was done in the last third of the program to look at underlying causes of procrastination, primarily low self-esteem. It was during this stage that the participants came to realize the complexity of their problem. The goal in the last part of the program was to "strike a balance between naïve optimism and despair," and to "accept their own limitations, and still accept themselves." Most group members made progress toward this end. If you need more assistance and a treatment workshop is available, you can benefit by being a participant.

What College Calculus Students Say about Procrastination

I devised a questionnaire for students enrolled in Calculus I, II, or III, which included students who were repeating the course because they received a grade of D, E, or W the first time they enrolled in the course. I told them that I was writing a book for students, and wanted to know the advice they would give students on how to better their chances of succeeding in mathematics. The greatest number of comments I received had to do with the implied or stated misgivings of procrastinating in their study. The fact that these comments were provided by calculus students is not significant here; students enrolled in any mathematics course would have made comparable comments.

Here is a sampling of their comments:

- ❖ *"I was always playing catch-up on the homework."* (Calculus I student)
- ❖ *"No homework was collected. This made it easy to fall behind."* (Calculus I student)
- ❖ *"Never get behind in class because once you do, it becomes increasingly more difficult to catch up."* (Calculus I student)
- ❖ *"Falling a day behind puts you at a disadvantage. Falling a week behind puts you in deep trouble."* (Calculus I student)
- ❖ *"The material in the course is overwhelming. It seems that if you fall behind once, you never catch up."* (Calculus II student)
- ❖ *"The best advice I can give is to never fall behind in your homework. Sure, there are times during the semester when you have back up in work for the many courses you may carry, but never neglect a class like calculus. Even if you*

can't go into as much depth as you'd like, you should al-
ways read through the material and try some examples. The
only time I've ever had trouble in a math class is when I've
tried to cram material in before a test." (Calculus II student)

❖ *"I have noticed that I am more apt to recall the calculus I*
have learned when the calculus was learned one assign-
ment per night. I seem to have a harder time remembering
a few months down the road when I didn't do this." (Calculus II student)

❖ *"Do the homework as it is assigned—don't let it pile up*
and then try to do it all at once—it just doesn't work." (Calculus II student)

❖ *"When I was taking several subjects at the same time I al-*
ways did the easy things first. I would leave the math to
last. Then I would say I didn't do the math because I didn't
have the time." (Calculus III student)

These comments by college calculus students should make it clear
that procrastinating has consequences. Experiencing these consequences is an excellent teacher, but it is a painful and costly way to
learn.

Ways to Procrastinate in Your Study of Mathematics

There are many ways to procrastinate in studying mathematics. They
include:

- Avoid spending time learning all the content of a unit in the
  hope that certain content will not be examined
- Set unrealistically high standards and then avoid failure by
  delaying your study (an example that could be an unrealistic
  high standard is telling yourself that you will get an A on the
  next examination)
- Allow yourself to be easily distracted, including daydreaming, when studying
- Come late to class, or don't come at all
- Put off asking question in class, or don't get help
- Postpone rewriting your class notes after class, or don't
  rewrite them
- Delay receiving help from your classmates, instructor, or a

mathematics tutor until just a day or so before an examination, or not receive help at all

- Avoid completing problems in assignments until just before the examination, or don't bother to complete them
- Begin your more intense study for an examination a day or so before the examination
- Delay analyzing your returned class period examinations, or the answer keys to them, until just before the final examination, or do not analyze them at all
- Spend too much time thinking that you have to study, rather than studying
- Over-socializing and over-indulging in extra-curricular activities
- Believe that there is always time later to get done what needs to be done

Success requires work, and that the work be done in a timely manner. We all procrastinate at various times to some degree, but do you procrastinate to the detriment of your educational health?

Methods for Decreasing Procrastination

There are methods you can employ to minimize procrastination tendencies. Some methods will work better for you than others. You have to discover what works for you. Read the methods presented here and choose those that seem most meaningful to adopt. If you find that a method works, continue to use it as you try other methods. The more methods you have working for you, the more you will find other methods easier to implement.

Methods of decreasing procrastination include:

Believe that your study will not be as bad as you imagined. It is true that at times we imagine the worst about what our study will be like. We anticipate that it will be too difficult, too painful, or take too much time away from other things that need to be done. These thoughts make it easy to procrastinate. Then we find out that living with these thoughts is far worse than doing the task. We spend considerable energy entertaining them. Michel de Montaigne says, "He who fears he shall suffer already suffers what he fears." *Most of your study is never as bad as you think it will be.*

I have had runaway imaginings, not just in my mathematics study but also in other jobs I needed to perform. For example, the banister on the steps from the basement to the first floor of my house had a broken metal bracket. The banister was loose and not very safe to use. It looked like a major job to me since the bracket was fastened to the wall behind the plaster. I wasn't exactly sure what was under there, and knew that I had to break out the plaster, remove the old bracket, find a replacement, replace it, replaster the hole that I made, and then repaint the new plaster. Also, the broken bracket was old, so I wondered if one existed that would match the others. I didn't do the repair work for two years. There were always other things I needed to do. In reality, I avoided a problematic situation that would take more time than I wanted to give to it. Did I expend much energy during those two years? You bet I did—several times a day I would use the steps and often tell myself to fix this banister. It was also on my mind other times, including the many times my wife, family, and friends would mention it. Actually, I am fairly handy when it comes to carpentry skills, but that didn't seem to matter.

One day I bit the bullet and decided to fix the banister. The job had six steps. Excluding the time it took to go to the store for a few items, the entire job took about 25 minutes. This story depicts well why a student might avoid or delay mathematics study. For my banister job, I focused too much on the work that was required to do the job, clearly blowing it out of proportion. Once I began the job, I focused on doing one step at a time, which was not difficult. Success at one step motivated me to do the next step. This comment by Jean de La Fontaine characterizes my banister situation and perhaps your avoidance of study: "From a distance it is something, and nearby it is nothing."

If you procrastinate because you think the job or task will be painful or frustrating, here are steps you can take:

- Break your study into small parts, focusing on one part at a time, not necessarily beginning in a prescribed order. Starting the task is most important.
- If things get difficult as you progress in your study, and you have the urge to procrastinate, think about the positive consequences you experienced when you were successful in your study.
- Periodically repeat this to yourself and believe it: "Focusing on the problems and difficulties I might encounter in my

study will deplete my energy, and absorb my time much more than doing the work itself."

Do not lose sight of the long-term goals and general learning skills you desire. Knowing your long-term goals, including the general skills you will acquire by successfully completing your mathematics course, will motivate you to conquer your tendency to procrastinate. To avoid losing sight of these goals, you have to keep them in front of you. (See section 1 of chapter 9 on benefits of learning mathematics.) With these goals in mind, you will be less apt to sabotage the day-by-day work that needs to be done. These two analogies will help you appreciate this: (1) It is difficult to create new flower gardens, but the vision of them inspires you to create them. (2) There are many obstacles to building your house, but the thought of living in it guides your daily work.

Manage your time wisely. Good time management will help you manage procrastination. Time management issues are involved when you have to decide on the best use of short and long periods of study, choosing the courses to study in a given study session, setting goals for each study session, and rewarding and evaluating your study time. (Chapter 7 provides advice on managing time. Chapter 6 provides advice on scheduling issues related to your courses (scheduling is a time management issue), a paid job, and other non-course activities.)

I want to reiterate the interrelationships among most aspects of your study methods. You change a specific aspect of your study, and it causes other changes; that is, you push in one place and something pops up in another. Positive changes cause other positive changes, and negative changes cause other negative changes. Be aware of changes you make that cause other changes. Our world is interconnected and so is your mathematics study.

Build procrastination into your schedule. This may sound rather strange, but it is a psychological ploy that works for some. Procrastination is not consciously scheduled—it just happens, and that is the problem. However, you can schedule procrastination and then decide to only procrastinate at those times. For example, for a study block of 3 hours on a Tuesday evening, you may decide to schedule two procrastination breaks of 15 minutes each, one occurring after the first hour of study and the second after the second hour of study. In a study session of 2.5 hours on a Wednesday morning you may decide to schedule two

10-minute procrastination breaks, one occurring every 45 minutes. Over the weekend, you may plan to study on Saturday and procrastinate on Sunday. For these three study periods, your schedule looks like this:

Tuesday evening	Wednesday morning	Saturday	Sunday
6:30–7:30 Study mathematics	8:00–8:45 Study mathematics	Study	Procrastinate
7:30–7:45 Procrastinate	8:45–8:55 Procrastinate		
7:45–8:45 Study mathematics	8:55–9:40 Study mathematics		
8:45–9:00 Procrastinate	9:40–9:50 Procrastinate		
9:00–9:30 Study mathematics	9:50–10:30 Study mathematics		

Scheduling in procrastination makes you more aware of it, eliminates any guilt feelings about it, and leaves enough time to complete your study. Eventually, you will be able to wean yourself away from this technique.

Be consoled by small successes. This advice is particularly meaningful for you if you have a tendency to view the glass half empty rather than half full. When you have a lengthy assignment to complete, do you focus on what you have left to complete rather than on what you have successfully completed? If so, this can trigger procrastination. Revel in each small success and reward yourself for it. Even if you can only spend 10 minutes on an assignment on a given day, congratulate yourself on what you accomplished. Success, no matter how small, motivates you to be more successful. String small successes together and you have a big success. It makes no sense to only value a completed assignment or project. That is comparable to only valuing yourself if you do everything well, and that is a prescription for fostering low self-esteem.

Take care of your personal problems. It is difficult to study if you are struggling with a personal problem. Perhaps someone close to you is very sick, your relationship with a close friend is problematic, or you lost your job. You may be anxious and depressed over it. In situations like these, it is difficult to focus on your mathematics study since your problem may be forefront in your mind. Whenever you find yourself distracted by a personal problem as you study, take a brief time out and devote your full attention to the problem. Plan steps you will take later to lessen your worry. Do not go back to your study until you have convinced yourself that any further worry while you study is unnecessary and nonproductive. You may want to set a time after your study to worry some more, but this is an appointment you will probably not keep. Your studies are difficult enough without distractions and your goal should be to minimize turmoil in your life. Plan to tackle the personal problems in your life, and give them your full attention at that time. Combining your study thoughts with those arising from your personal problems is inefficiency at its best. If thoughts on your personal problems arise when you come back to your study, quickly dismiss them.

Increase your confidence and motivation. Confidence and motivation are intimately related to procrastination. A confident and motivated student is less likely to have a procrastination problem. Confidence and motivation are addressed in sections 1 and 2.

Write about your journey to lessen procrastination. Journaling your efforts to lessen procrastination can be very helpful. For starters, it brings this insidious concept out of the shadows into the light. When you write about it you confront it head on. You can evaluate the hold it has on you and your efforts to manage it. In your attempts to minimize the effect procrastination has on your life, you become more aware of your anger, frustrations, joys, hopes, attitudes, excuses, setbacks, successes, and future plans for fighting it. Write about it almost every day, addressing one or more of the items listed above. Write as little or as much as you want. Periodically reread what you have written to see the path of your journey. By writing, you come to grips with your procrastination problem in an overt way and that helps to lessen its hold on you.

Imagine what life would be like without procrastination. If procrastination is a problem for you, you know the effect it has on your life. Imagine your life without it. Here are some examples:

- You did not put off studying some of your subjects for several days or skip some classes because you allowed the necessary time to study for an examination.
- You were able to engage in fun activities because you completed things in a timely manner.

- You do not carry the guilt that comes from procrastinating, but instead have good feelings because you are a responsible learner.
- You do not obsess over an upcoming examination. You are looking forward to it because you didn't procrastinate in your study and are ready for it.

These imaginings are your desires, and eventually become your experiences if you pursue a course of anti-procrastination. Act on your positive imaginings for an hour, half-day, day, or even a week. During this time believe that you are not a procrastinator, and act accordingly. At the end of that time, reflect on how it felt. Think about how good life can be as a student when your procrastination is under control.

Get help from others. One reason we are on earth is to support others. I know individuals who cannot thank others enough for helping them work through a problem. Here are ways to be with others so they can support you in your efforts to minimize procrastination:

- Make appointments to do some studying with one or more diligent and upbeat classmates. You benefit in at least three ways: (1) You won't want to let your classmates down, so you have motivation to honor these appointments. (2) You have someone to help you in case you experience difficulty with the content, so you won't flee from it. (3) You have someone who can encourage you and cheer you on.
- Discuss your procrastination with a good listener—a friend, classmate, instructor, or someone in the counseling office. They may have insights and suggestions that will benefit you. Additionally, you get to know better your thoughts and habits on procrastination.
- Take a friend with you to see your instructor (if you are reluctant to go alone). Additionally, you can ask this friend for advice on how to approach your instructor. See section 1 of chapter 23 on obtaining help from your instructor.

Be realistic about minimizing procrastination. There are a variety of ways to sabotage your efforts to minimize procrastination. Many of them are related to unreasonable expectations on your part, including the following:

- You expect that you can make multiple changes at the same time. Attempting to change too many things at once can cause you to despair and abandon your plan entirely. Make one or two changes at a time.
- You expect to make progress without any setbacks. With almost any change, you will have good and bad days. You have days when you move forward and you will have days when you move backwards. Behavioral change takes time. Well-engrained habits are developed over time and they are changed over time. Be patient with yourself when you slip back into an old habit.
- You expect that your strategies to lessen procrastination do not have to be maintained over the long haul. Since it takes considerable effort to wean yourself away from undesirable and long-standing habits, old habits can resurface quickly without continued vigilance.

A thorough discussion on the basics of change and the process of change are presented in sections 3 and 4 of chapter 10. These sections will help you understand what you will experience as you make changes to minimize procrastination. You are then less likely to be surprised and discouraged by pitfalls you experience that can disrupt your efforts.

4. Decrease Mathematics Anxiety

Some of us suffer from a general form of anxiety that affects much of our life, including our work in mathematics. Others of us may only be anxious about a particular subject such as mathematics, and have no more than the typical worries that come with taking other subjects. This section addresses issues associated with mathematics anxiety. Mathematics anxiety is an unhealthy fear of mathematics that affects your ability to succeed in mathematics. It cannot be discussed without also discussing general anxiety. Some of us have test anxiety. Others may only be anxious about mathematics tests because they suffer from mathematics anxiety. Test anxiety is addressed in section 2 of chapter 27.

Knowledge of issues related to mathematics anxiety can help alleviate undue stress you may experience learning mathematics. Mathematics anxiety is a complex psychological problem; hence, do not be surprised by the concepts, attitudes, myths, circumstances, behaviors, and people that affect it. If you don't understand and address these issues, your efforts to manage mathematics anxiety may be severely compromised. When the phrase, "mathematics anxiety" is used in this section, it refers to an unhealthy, inhibiting, and debilitating anxiety. As with any psychological problem, there are varying degrees of severity.

Anxiety Is Required for Learning to Take Place

Psychologists tell us that learning cannot take place without some degree of anxiety. Imagine how you would work in a course if you didn't have any pressure to learn the course content or to do well on examinations. A degree of anxiety comes with the desire to learn and do well. It gets the adrenalin going, which makes you more alert and ready to respond, and is the impetus for action. On the other hand, too much anxiety can inhibit learning.

How Mathematics Anxiety Manifests Itself

You may suffer from mathematics anxiety if you—

- Hesitate to ask questions in class, make use of your instructor's office hours, or work with classmates for fear of being ridiculed or embarrassed by your lack of knowledge
- Are very apprehensive when trying to understand a lecture, work an assignment, prepare for an examination, or take an examination
- Regularly avoid completing mathematics assignments or preparing for mathematics examinations
- Avoid situations that involve mathematics, in and out of college
- Have an increased heart rate, sweating, trembling, or stomach and intestinal discomfort when confronted with situations involving mathematics
- Obsess day and night with thoughts of being unsuccessful in mathematics
- Are habitually confused about what you know or do not know in mathematics

- Have recurring thoughts of past failures in mathematics, or of statements made by others that you can't do mathematics
- Have repeated thoughts on the real or imagined consequences of doing poorly in mathematics

Everyone Has Some Degree of Mathematics Anxiety

Circumstances can be such that even a person with a doctorate in mathematics can suffer a degree of mathematics anxiety. Perhaps he or she is giving a lecture to colleagues on some research findings and is asked a routine question that he or she cannot answer. Feelings of panic can result if the person feels embarrassed. In many schools, candidates for a doctorate in mathematics have to pass specially constructed written and oral examinations. These are formidable examinations that must be passed. Imagine the pressure on these candidates to pass these exams. They may have spent several years studying post-bachelors degree mathematics and know that their progress is stopped if they cannot pass them. They can freeze on the examinations to the extent that their performance is unacceptable. It is safe to say that everyone has, at some point in life, suffered from mathematics anxiety. It may have been temporary, or to a mild degree, such that the person could continue to function reasonably well in mathematics. You may suffer from it rather severely, but still manage to survive your mathematics courses. Take steps to decrease your mathematics anxiety, not tolerate it.

Mathematics Anxiety May Not Be Obvious to You

Mathematics anxiety can be insidious. It can develop slowly and subtly so that you are not aware it is happening, yet its debilitating effects can happen. It is like living with chronic back or sinus problems. Only when it is gone for a short spell do you realize how badly you were feeling and the negative effects it had on you.

Circumstances That Contribute to Mathematics Anxiety

The good news about mathematics anxiety is that it is learned behavior and learned behavior can be unlearned. According to Dr. Karl Menninger, "Fears are educated into us, and can, if we wish, be educated out." Many circumstances contribute to this learned behavior, and they most likely happen in a student's pre-college years. If you have mathematics anxiety, circumstances that may have contributed to it include:

1. False beliefs on what constitutes mathematics; hence, studying it inappropriately
2. Inappropriate evaluation of your mathematics knowledge (e.g., overemphasis on examinations in determining your course grade; poorly written or graded examinations)
3. Unrealistic expectations you set for yourself (e.g., thinking that a specific problem should not have taken you that long to work, or you must get an A on the next test)
4. Lack of confidence because of past problems with mathematics
5. A parent who didn't do well in mathematics, and believing that you cannot do well because you are like that parent
6. Lack of prerequisites for the mathematics courses you enroll in
7. A mathematics instructor who belittled you for not having prerequisite knowledge
8. A mathematics instructor who was impatient with you when you expressed confusion
9. A mathematics instructor who embarrassed you in front of your classmates by continuing to prod you to come up with the solution to a specific problem, clarify an idea, or work a problem on the blackboard, even though it was obvious that you were not able to do it
10. Cutting remarks from parents, siblings, classmates, friends, or instructors about your inability to do mathematics
11. A parent who frequently compared your mathematics ability or achievement to a sibling's who was a high achiever in mathematics
12. Unrealistic pressure placed on you by your parents to do well in mathematics
13. Not knowing how to study or learn mathematics

Mathematics Anxiety Can Be Alleviated

You are encouraged to gain a better understanding of the symptoms and possible causes of mathematics anxiety, especially as they apply to you. If you have mathematics anxiety, this knowledge will help you control it. Marie Curie says, "Nothing in life is to be feared. It is only to be understood." Suggestions for lessening your mathematics anxiety follow:

Use an appropriate process to learn mathematics. Without question, this is the best advice that I can give to lessen mathematics anxiety. The major cause of mathematics anxiety in most students is not knowing how to study or learn mathematics. Believing false myths about one's ability to learn mathematics, the nature of mathematics, the value of knowing mathematics, and the way mathematics is learned, coupled with psychological issues, contribute to one's use of inappropriate study techniques. These comments were made by Calculus II students in response to my question on advice they would give students to be more successful in mathematics:

❖ *"I've taken Calculus I twice. I had to repeat it because of the anxiety I encountered the first time. I didn't know how to study mathematics. That was two years ago."*

❖ *"I have always been a successful mathematics student and when my grades began to drop in Calculus I, I started to panic. I always kept up with the reading and the homework, even though I didn't comprehend all of the material. So I started practicing former teachers' suggestions."*

A major goal of this book is to present a workable process for learning mathematics, which will enhance your understanding of mathematics and make it less threatening, stressful, and anxiety provoking.

Have well-based perceptions. One of the more important things you can do to alleviate mathematics anxiety is to change your perception of things. Your perception or outlook on what has happened, is happening, and may happen to you in mathematics, is important. Healthy viewpoints and attitudes are based on an awareness of reality, which is based on knowledge of many things, including yourself, mathematics, mathematics anxiety, and how mathematics is learned.

Be encouraged by small successes and changes. In most situations, making significant change evolves slowly. Wholesale change in your study techniques does not happen all at once. Celebrate any positive changes you make, small though they may be. An anxious mathematics student dwells on what still needs to be changed, rather than on what has been changed. Dwelling on the former increases anxiety, and dwelling on the latter increases motivation to sustain changes and make others. Small changes may include asking a question in class, spending more time on an assignment, talking to your instructor, rewriting your notes from a class session, being well-prepared for a class session, or receiving an increase of one letter grade on a graded assignment or examination. Be comforted and pleased with the changes you made, rather than obsessing over the changes yet to be made. Focusing on the latter produces anxiety. There is a time and a place for other changes that you may need to make. Significant changes typically come from a series of sustained small changes, but don't minimize a small change, for its effects can be large, especially if it spurs you on to making other changes.

Don't focus on the time it takes to perform a task. It takes time to learn and apply mathematics concepts and principles. When initially exposed to these ideas and processes, you may be confused about their abstract nature and complexities. This is normal—you should be surprised if you are not confused. You can become anxious if you don't understand that confusion is part of learning something difficult.

Think about a time when you spent many hours learning something difficult. Most likely you were thrilled when you understood, and quickly forgot how long it took. I think of certain house projects, like hanging a door, which took me about two hours to complete. A professional would have completed the job in 15 minutes. When I finished hanging the door I was so proud of my work that my frustration at the amount of time it took paled in comparison. It is encouraging to know that time spent learning will be reduced over time when your improved study techniques become habit.

Set realistic goals. Avoid thinking that the changes you make will completely eliminate mathematics anxiety. That probably will not happen, and believing that it will leads to more anxiety when it doesn't happen. Instead, aim to control your mathematics anxiety to the extent that it is not debilitating. There will be times that it will resurface; a realistic goal is to minimize these times and the severity of the anxiety.

Another important area where you have to set realistic goals or expectations is your performance on assignments and examinations. Improvement in performance comes slowly; therefore, aiming to improve that D or E on the previous examination or assignment, to a grade of C on the upcoming examination or assignment is probably realistic. To strive for a B or A is a worthy goal, but don't expect to receive it. Setting your expectations too high results in increased anxiety as you try to meet them, and when you realize that you have not met them.

Things are often not as bad as they seem. Many of your thoughts on the doom you may experience in your mathematics courses are often imagined or exaggerated. Little fears can be seeds for larger and unfounded fears. Do you often fear something and wonder later why you feared it? I suspect this has happened to you many times. It happened to me more times than I care to admit. Sir Winston Churchill said, "I remember the story of the old man who said on his deathbed that he had had a lot of trouble in his life, most of which never happened."

Write your thoughts and feelings in a journal. It is important to know what causes mathematics anxiety and strategies to control it. It is also important for you to recognize and understand the feelings in you caused by mathematics anxiety. An excellent way to do this is to journal. You will be surprised how revealing and healing this can be. Write about your frustrations, anger, failures, successes, future plans, etc. Write as much or little as necessary, and as often as you have the need. You may want to journal on your thoughts every time you work on a mathematics assignment, before and after every examination, or whenever you feel especially anxious. After awhile, if this doesn't seem to help, stop doing it; but give it a fair chance.

Discuss your thoughts and feelings with trusted individuals. This is a great means for expressing your thoughts and feelings about your anxiety; hence, lessening its effects. It also provides an opportunity to learn more about your thoughts and feelings.

Don't focus on yourself. You have to wean yourself away from preoccupation with your anxiety or thoughts of impending doom. Concentrate on what you need to learn. Work on minimizing a "woe is me attitude" by taking a greater interest in the mathematics content you are studying. How do you do this?

1. Reflect on the rationale for focusing on learning.
2. Make the decision to focus on what you are to learn.
3. Focus on it.
4. If you stop focusing because of anxiety, go back to statement 1.

Regularly exercise and engage in playful pursuits. Of course! Case closed.

Tell yourself it is useless to worry. Needless worry is futile. You have more worthwhile things to do, and worrying about what has happened and what might happen prohibits you from taking action now. Replace pointless worry with thinking and action, as Harold B. Walker so aptly states: "When you worry, you go over the same ground endlessly and come out the same place you started. Thinking, on the other hand, makes progress from one place to another. . . . The problem of life is to change worry into thinking, and anxiety into creative action."

Know the process of change. Knowing the stages you go through in changing unproductive habits will lessen the anxiety that comes from your efforts to change. The basics and process of change are presented in sections 3 and 4 of chapter 10.

12

Student and Instructor Responsibilities

As a mathematics student you have certain responsibilities to yourself, your instructor, and your classmates. Likewise, your instructor has responsibilities to him or herself, you, and your classmates. If you are able to choose your instructor, you will want to know how well he or she meets instructor responsibilities. Unfortunately, an instructor doesn't typically choose his or her students. I can assure you that if an instructor could, he or she would want to know if these students meet their responsibilities as students. Do you know what they are? Do you meet them?

How much better the learning process would be for mathematics students and instructors if each met their responsibilities. There is not complete agreement among students or instructors as to what their responsibilities are. Unfortunately, there are instructors who relieve students of some of their responsibilities (e.g., distributing a solutions key to an examination prior to students having a chance to redo the examination). A major problem with instructors performing tasks that are the responsibility of students, is that students then expect this from other instructors, and they do not easily or willingly give up this expectation. Complete agreement on the responsibilities of instructors and students should not be expected since there can be honest differences of opinion. However, using common sense and sound principles of learning, a solid core of student and instructor responsibilities can be identified, and that is the focus of this chapter. It is *in your best interest* as a mathematics student to know your responsibilities and meet them.

It is unrealistic to expect you to meet your responsibilities all the time. Obstacles present themselves now and then that keep you from meeting one or more of them, but that normally doesn't cause any major problems. It becomes problematic when it happens too often. Why might you have a problem meeting some of your responsibilities? The reasons are many and include these: not knowing your responsibilities, having an instructor who assumes some of your responsibilities, mathematics anxiety, a chronic procrastination problem, lack of goals, lack of prerequisite knowledge, or not knowing what mathematics is and how to learn it.

The responsibilities of a student and instructor do not overlap—they are different. Regrettably, some students confuse their responsibilities with those of their instructors, and vice versa. This confusion often results in hard feelings between students and instructors, and diminished achievement. Here are four examples of confused roles:

1. You may think that it is your instructor's job to conduct a class review for an examination. It is your responsibility.
2. Your instructor may think that it is your job to determine the textbook exercises that comprise the minimum assignment for the day's class session. It is your instructor's responsibility.
3. You may think that it is your instructor's job to let you know the date an assignment is due, *after* it was announced in a class period that you missed. It is your responsibility to find this out.
4. You may think that it is your instructor's job to see to it that you get your graded examination or assignment back, *after* you were absent from class when they were returned. It is your responsibility.

The first section of this chapter presents your responsibilities, prior to enrolling in a mathematics course, and then when you are enrolled in the course. The second section presents your instructor's responsibilities, which *you need to know* since they dovetail with your responsibilities. When your responsibilities *and* those of your instructor are met, you have the best chance of succeeding. You will have to compensate for those responsibilities your instructor does not meet. The responsibilities given for student and instructor are many, are not meant to be exhaustive, and you may not agree with every one of them. The third section of the chapter presents students' answers to the question, "What did you dislike about your instructor?"

1. Student Responsibilities

Along with the student responsibilities listed in this section are comments on the importance of meeting these responsibilities. For each responsibility listed, answer these three questions before moving on:

1. Are you meeting this responsibility?
2. If you believe it is your responsibility and are not meeting it, why not? What do you intend to do about it?

If you believe that a responsibility listed is not your responsibility, you are encouraged to talk to a friend, classmate, or your instructor about it. The discussion may encourage you to change your mind. Your responsibilities in a mathematics course are now presented.

Select an Instructor

I can't think of anything relating to your mathematics course that is as important to you as your instructor. *He or she has a profound effect on the quality of your course.* You will see this when the responsibilities of an instructor are given in section 2. Sadly, choosing a good instructor who is appropriate for you, is not always possible due to scheduling conflicts or because few sections of the course are offered. But if you have a choice in choosing your instructor, a process for choosing one is presented in chapter 13.

Possess the Prerequisite Knowledge

In some mathematics departments, you cannot take a course unless you have the mathematics prerequisites, at least on paper. Many mathematics departments are more flexible, having the unwritten policy that can be described as, "Enter at your own risk." In the final analysis, you have to be responsible for satisfying the prerequisite knowledge, but not just on paper because the grades you receive do not always indicate that you have an appropriate understanding of this knowledge. See chapter 4 if you want to know how to determine if you have the prerequisites.

Know the Course Syllabus

Distribution of a course syllabus is normally done the first day of class. Ordinarily, a syllabus contains (or should contain)—

- A description of the sections covered in the textbook and the order in which they are covered
- Policies on assignments (including those turned in late), examinations (including those missed), grading, and attendance
- Information on contacting your instructor, including office hours
- The instructor's teaching philosophy (sadly, this is often not included)

It is important to *read the syllabus carefully* and *refer to it periodically throughout the term.* Many students, well into the semester, either don't have their syllabus, or have it, but don't know what's in it. They are surprised when something unpleasant happens to them, most likely in the area of evaluation, when it should have been expected since it was alluded to in the syllabus. You can hold your instructor accountable for information in the syllabus, but you have to be reasonable since circumstances may dictate the necessity to make sensible changes during the term. For example, there may be one less examination than the number stated in the syllabus because class was cancelled for legitimate reasons near the end of the term.

Know the Dates for Adding and Dropping Courses

For most mathematics courses you only have a week or two at the outset of a term to add the course. You should add a course as soon as you can within this time period. Missing the first few days of a course is problematic for many students. If I were making the rules, strictly from an academic viewpoint, you would only have one or two class periods to add a course after it begins. After that it could only be added by special permission of the instructor. However, there are reasons for having a more extended time period to enroll once the course has commenced. They include finding out after a week or so of being in a course, that you do not have the prerequisite for that course. Thus, you need to drop the course, and as quick as possible enroll in the prerequisite course; otherwise you would have to wait a whole term to enroll in the course.

It is vital to know, throughout the semester, the course drop date set by your college. You can drop the course on or before this date without academic penalty. I often saw students failing to drop a course by the drop date because they didn't know this date. At many colleges, once the drop date passes, and unless a formal appeal is filed and upheld, you have to suffer the consequences of not dropping the course, which may be receiving a poor grade. There can be other consequences, including raising havoc with your grade point average, perhaps even causing you to be placed on probation or to fail out of college. Additionally, employers do not view as positive low grades appearing on your transcript. There are other drop dates you should know that relate to reimbursement of tuition. This varies from college to college, but if you drop by these specified dates you may get all or some of your tuition reimbursed.

Knowing Whether and When You Should Drop a Course

Knowing drop dates is a simple task. What is more difficult is knowing whether, and when, you should drop a course. No one can make this decision for you, although you can get advice from your instructor or advisor. A thorough discussion on a process for determining whether to drop your mathematics course is given in section 2 of chapter 25.

Attend Class

Attending class is critical to your success. There are times you have to miss class due to illness, a traffic accident, a death in the family, and so on. However, it is almost always unwise to miss or cut class for reasons that include: studying for an examination, working more paid hours during the holiday season, not having an assignment completed, playing cards with classmates or friends, attending doctor appointments (unless it is an emergency, or there are no other times to schedule it), or finding class boring and uninformative.

In my classes, I knew of very few students who did well if they consistently missed class. It is a disastrous course to follow for almost any student. A few times over many years I had several highly talented students who did reasonably well with sporadic class attendance. However, these students could have achieved significantly more in the course by being present more often. You may ask, "What constitutes missing too many classes?" That depends on you, the content of the class periods missed, and what you did about missing them. For a college term, spread over four months, there may only be 40 classes for a three-credit course, each 50 minutes in duration. That's a total of about 33 hours of class time, far less than an average workweek of 40 hours. Suppose you miss as many as eight of these 40 classes. That's missing one-fifth of the classes. You are treading on very dangerous academic ground and are not receiving full benefit from your tuition. If you have to retake the course, based on the consequences of your poor attendance record, you are paying for the course again, to say nothing about the extra time you have to spend redoing it, or the delay in taking a course that has this course as a prerequisite.

In talking with my instructor colleagues about students who miss class, almost to a person they indicated that they hardly ever, if at all, missed one of their undergraduate mathematics classes (and many of them have doctorates in mathematics). If they thought they couldn't miss class, and they were in class with students like you, how can you think that you can afford to miss class? Perhaps you cannot identify with these mathematics instructors. If so, you should be able to identify with classmates who do well in courses. Almost to a person, *they don't miss class.*

The value of what is missed by not attending class varies from instructor to instructor. It has much to do with how the instructor makes use of class time. Nonetheless, no matter the methods of the instructor, you miss much more by not attending class than you may realize. This is not to say that you might not still do well on examinations, but that is not likely. What should you do if your instructor uses class time unwisely? The solution is not to miss class but to discuss your concerns with this instructor, and if you are not satisfied, with the mathematics department chairperson. Hopefully, you can be a catalyst for affecting change that will benefit you and other students.

Reasons to Attend Your Mathematics Class

The reasons to attend you mathematics class include these:

Your professor has a unique way of presenting the content. A significant portion of what you are expected to learn in a mathematics course may not be in your textbook. Your instructor may change a textbook author's definition, bring in a theorem not included in the textbook, prove a theorem differently from the proof in the textbook, or present an application of a theorem not found in the textbook. This is a real problem for you if you miss class. Understanding the textbook authors' development of the content is not enough. The textbook is not your instructor.

The examinations in the course reflect the bias of your instructor. Your instructor will drop many hints and admonitions during class on the content that is especially important to know and what you are expected to do with it. Your instructor may make statements like these:

- *"The most important part of this section has to do with . . ."*
- *"A common mistake in applying this algorithm is . . ."*
- *"You might think that you cannot work this problem using . . . , but . . ."*
- *"This theorem is so important that I will ask you to state and prove it on the next examination."*
- *"Use my definition in working a problem of this type."*
- *"Be sure to know . . . for the next examination."*
- *"When I ask you to solve this type of problem, I want you to follow these steps: . . ."*

It is foolhardy to think that the examinations will not reflect your instructor's class presentations.

The dialogue between your instructor and students, or between students, is often immeasurably valuable. It is through in-class discussion, prompted by your instructor's or classmates' comments and questions, or in-class small group work, that concepts and principles are better understood and problem-solving skills better nurtured—miss class and you miss out.

Your instructor makes important announcements concerning quizzes, examinations, and assignments, and distributes materials such as study guides, solution keys for examinations, or class notes. I had students who didn't turn in assignments, who missed examinations, or didn't have adequate time to prepare for examinations and assignments, because they were not in class to hear due dates of assignments or dates of examinations. They could have asked a classmate or me for these dates, but they even failed to do that. Additionally, not receiving the materials that were distributed in class in a timely manner had negative effects on their performance.

You have a responsibility to your instructor and classmates. There is more to attending class than learning content and finding out about important logistics. You are also there to participate. The morale of a class and the quality of a class session suffer if students are absent from class. You might think this is not your responsibility or that it is not in your best interest to attend class for the benefit of others. Experience the incredible deep-seated joy of helping others—of being outer-centered rather than inner-centered.

Getting to know your classmates. Attending class affords an excellent opportunity to get to know your classmates, which makes it easier to associate with them outside of class. Just recognizing them as your classmates can be the catalyst to talk to them or have a cup of coffee or soda. This can lead to formation of a study partnership or study group. This type of socialization contributes to a quality college life.

Your instructor may refuse to help on class content that you missed. It was not that unusual for a student who missed one of my class sessions to email me or come to my office and ask for help on content that was presented and discussed in the class session. If the student did not have a good reason for being absent, I refused to help on the content of that class session, and mentioned why. I do not help students who don't help themselves. I refused to be an enabler of inappropriate behavior. If they had to miss class, I bent over backwards to help them with what they missed.

Your instructor may have an attendance policy that plays a part in determining your final grade for the course. It is perfectly acceptable for a college instructor to have an attendance policy. He or she knows that you will not learn as much if you are not in class. Even if you received an A on an examination, the examination only *samples* the content of the unit. An attendance policy is there to help you—to impress upon you the importance of attending class.

Some instructors formally take roll each class; I did not. However, I knew the students who routinely missed class. I regularly informed my students of the pitfalls of not attending class, but they were adults and had the freedom to ignore my advice. Those that missed class experienced the reality of these pitfalls.

If you know that you have to miss a class, let your instructor know as soon as you can. In that way, you can get the assignment and the sections of the textbook that will be addressed. Tell your instructor that you regret your absence. Have someone take good notes for you, including due dates for assignments and examination dates. Ask a classmate to get an extra copy of the handouts. Your note taker needs to note more than the mathematics displayed on a board or screen. Good note takers are hard to find. See your instructor if the notes need explanation. If your instructor is teaching another section of the course, you might want to consider attending that session (if it addresses content you missed).

It was not unusual for a student who missed one of my classes to ask if he or she missed anything of importance. This indicates that the student didn't know the value of attending class. Does the student expect me to say, "No. The class period was a total waste for those present. This includes the problems we worked in large or small group, the responses of the students and me to questions that were asked, and my development of new content, including my motivation for it, how it related to previous content, and all of my admonitions concerning it. It would have been useless for you to attend." Perhaps the best way your instructor can answer your question is to say, "No comment." How many times can a student miss class without experiencing some academic loss? I say, "Zero."

If examinations or graded assignments have been passed back in class on a day you were not there, it is your responsibility to pick up your examination later, either in class or in your instructor's office. Do not expect the instructor to let you know that he or she has your papers. I had students who did not pick up their examination for weeks, even though they did not know the grade they received on it. Some may never have picked them up if I did not eventually remind them, but my purpose in reminding them was to have the opportunity to discuss this unusual behavior with them. None of the reasons they gave for being so irresponsible were legitimate.

I asked calculus students for advice they would give students on how to succeed in college mathematics. The responses I received included these:

- ❖ *"Class attendance is very important. Attend or you'll get lost."* (Calculus II student)
- ❖ *"Students should be sure to attend class and do the homework."* (Calculus I student)
- ❖ *"I would recommend that any calculus student attend class. . . ."* (Calculus I student)

How could anyone say something different? Yet the number of mathematics students who miss class these days more than they should is confounding and shocking. My colleagues and I shake our heads in disbelief and say to ourselves, "If only they realized the harm they are causing themselves."

Be on Time for Class

Not much needs to be said about the importance of being in class when it begins. Arriving late disrupts the instructor and students. It is especially annoying if the classroom is small and the only way in is through a door at the front of the room. If I had to pick the most important part of a class session, it would be the first 20 minutes. This is when announcements are made, an overview and motivation of the day's content is given, and some questions are taken on the previous assignment. Also, new development of new content may have begun and missing that can cause you to be confused for the remainder of the class. Imagine having a job and a habit of coming late to work. You may be given a warning or two, but if your tardiness continued you would find yourself looking for another job. Being a student is a job and you have a responsibility to yourself, your classmates, and your instructor to be in class on time.

Unexpected circumstances may be a valid reason to come to class late. By all means don't hesitate to enter the classroom. I knew students who were so reluctant to disrupt class that they stayed in the hallway waiting for class to end so they could get notes from a classmate. Enter the classroom and apologize to your instructor at the end of class, indicating why you were late. Normally, instructors don't mind you coming late if you have a good reason. However, they do mind chronic latecomers, whatever the reason.

Be Prepared for Class

The benefits you accrue from being in class are lessened if you have not read the sections in the textbook that were assigned, worked diligently on the assigned problems from these sections, and read the textbook sections relating to the new content that will be introduced in class. Class preparation not only benefits you, it also benefits your instructor and your classmates, and you have to care about them.

Contribute to Class Discussions

A responsibility that many students shirk is class participation, due to shyness, lethargy, or lack of preparation. The *quality of a class session is highly correlated with the amount of student participation*, and each student has a responsibility to participate. Ask and answer questions posed by your instructor and classmates. You are then engaged in active learning, which will make the class session more valuable to you and to others. Imagine four or five friends visiting with each other and one of them asks the other four this question: "What are your plans when you graduate?" Suppose you cast your eyes downward and say nothing. Rather unusual behavior, wouldn't you say? This is often what an instructor sees when asking a question of the class. View each question in class as a question directed at you. It is too easy to let other students respond. The participants in your educational family are you, your instructor, and your classmates. Each member has to meet his or her responsibilities; otherwise it becomes a dysfunctional family.

Respect Your Instructor and Classmates

The disrespectful actions I saw in college students include—

- Carrying on a private conversation in class that disrupts others
- Making fun of responses of classmates or the instructor, either by laughing or making snide remarks, often heard by the entire class
- Speaking in class without raising your hand and being acknowledged by the instructor
- Asking a question at an inappropriate time (You can sense when a time is inappropriate by the body language of your instructor. It is important for instructors to finish a point before being interrupted by a question; otherwise the impact of the development is lessened. Don't put pressure on your instructor to acknowledge you by continually thrusting your hand upward until you are called on. This is very disruptive to everyone. Do it gently and at selective times.)
- Not turning off cell phones, text messaging in class, and leaving class to make phone calls
- Doing homework, playing games, and Internet-surfing

Help Maintain Discipline in Class

You think I am kidding, right? Not so. If you are frequently distracted in class by another student's rude behavior, and the instructor allows it to continue, there are two things you can do: (1) In a quiet and respectful manner tell the student that you are annoyed, or use body language to express your annoyance. (2) Talk to the instructor about the problem after class and ask him or her to please take steps to stop it. If this last option doesn't stop the problem, talk to the department chairperson who should then talk to the instructor. Is it courageous of you to pursue these options? I suppose so. Is it your responsibility? Yes.

Take Notes in Class

Attending class is not enough. You have to take good notes; otherwise what you learned in class will be forgotten shortly after you leave class. I was flabbergasted and disheartened when students did not take notes, even after I reminded them of its importance. Taking notes is discussed in detail in section 5 of chapter 14.

Turn in Assignments on Time

An instructor does not have to accept an assignment that is turned in after the due date if there is no acceptable excuse for the tardiness. Some students who have not completed an assignment will deliberately miss class the day it is due. They turn it in later, often waiting until the next time the class meets. Some instructors see through this ploy and do not take it kindly. If you have to miss class the day an assignment is due, get the completed assignment to your instructor as soon as you can. Perhaps you can turn it in before it is due, or give it to a classmate to take to class the day it is due. If need be, call or email your instructor explaining your situation. For logistical reasons, it is difficult for your instructor to grade an assignment that is turned in after the other assignments have been graded. Some instructors will accept a late assignment regardless of the reason, but may grade you down on it.

Don't Miss Examinations

It is your responsibility to take your examinations when they are scheduled. You have to do all that you can to avoid missing examinations. There are instructors who will say that you cannot make up an examination that you have missed, regardless of the reason for missing. If you have a legitimate reason for missing an examination, I cannot see how an instructor can stick to this position. You have the right to see how well you know the content covered by the examination, and to have your grade on an examination covering this content figure in your final course grade. Rather than give you an equivalent examination as a make-up examination, your instructor might have the policy that you can delete one examination grade in the course. Consequently, if you miss an examination, that examination is viewed as the one deleted. I think this is unfair for the reasons I gave earlier. If you had a legitimate reason for missing an examination, you may want to appeal this policy. I cannot understand how you can be penalized for something that was out of your control.

If your instructor will give you a make-up examination for an examination that you legitimately had to miss, be aware that it most likely will be different from the one given to your class. It may be shorter, longer, more difficult, etc. Some instructors will give, as a make-up examination, the same examination that was given to your classmates if you take the examination near the time they did (perhaps within a few hours). If you know you have to miss an examination, talk to your instructor as soon as you can, and make arrangements to take the examination or a make-up examination.

Support Your Instructor When Warranted

Some students like to gang up on the instructor because misery loves company. For example, if you thought the length of an examination was reasonable, don't support students who complain that it was too long. Some students blurt this out to the whole class during an examination. Their complaint has to be legitimate in your mind before you support it, and that is not the time or place to do it. It is common practice for many students to fault the instructor for their shortcomings. Students who have shortchanged their preparation discover that an examination is too long or the questions too difficult for them, yet many of them blame the instructor by complaining about the examination. Even outside of class, if a classmate is unjustifiably criticizing your instructor, defend the instructor.

Be Fair Evaluating Your Instructor

Many colleges require their faculty to pass out student evaluation forms. This gives students the opportunity to evaluate their mathematics courses and instructors. It is your responsibility to be fair in your evaluation. It is not uncommon for a student to have an axe to grind. Perhaps the instructor upset the student and the instructor's actions were not justified in the student's mind. The student then gets back at the instructor by judging the instructor badly in almost every category on the student evaluation form. This is despicable behavior. You have a responsibility to be fair even if your instructor has some shortcomings. And please be aware of your shortcomings and own them so you don't make them your instructor's shortcomings. Be critical, but truthful in your evaluation.

Realize You Are Not the Instructor

Be mindful that you are the student, not the instructor. Your instructor, not you, is in charge of your mathematics course. You can respectfully disagree with him or her in many areas, ranging from how the class period is run, to the questions on the examinations, to how your examination was graded. You may have a valid point in some of your disagreements; however, any unresolved disagreement has to be resolved through proper channels. But *you do not make the decisions on how the course is run*, and this should be evidenced by the way you disagree. Instructors have the credentials to make the decisions, and have been employed because of them. They have to take into consideration many different factors when making a decision. Do you have the knowledge to consider all these factors? The answer is "No." They will make poor decisions at times, and some of those you have to live with. There are students in mathematics classes who act as though they are the instructor—that their opinions should carry as much weight as their instructor's.

Pursue Proper Channels When You Have a Complaint Against Your Instructor or Classmate

If you have a problem with the way the course is being taught, talk to your instructor. An instructor cannot read your mind. It doesn't do much good to continually complain about your instructor to your classmates or other students. Approaching your instructor might not resolve the issue, but it is the place to start. If there are others who think that your problem with the instructor is legitimate, ask them to accompany you. Your instructor may agree with your complaint, but doesn't want to openly admit culpability. That does not mean that he or she will ignore your concern. Allow time for change to take place. If the problem is serious and remains unresolved, go to the chairperson of the mathematics department. If it still remains unresolved, go to the dean of the college or to the ombudsperson. Your college has student grievance procedures that give proper channels to pursue.

Assume Responsibility for Your Education

In the final analysis, you have to take responsibility for your education. You cannot delegate it to your instructors, your classmates, your friends, or your parents. Know your responsibilities and meet them.

Don't settle for a mediocre education. If you experience mediocre instruction, compensate to overcome it. Education is a cooperative process and you have the biggest role to play. If you think your responsibilities as a student are too much, or your instructor has to accept more responsibility for your learning, I urge you to reflect on this position again after you read the next section on instructor responsibilities.

You have a myriad of responsibilities as a student, employee, family member, etc. It is not possible to meet all of them; consequently, you have to determine those that are most important to meet, and do the best you can with the rest. You are an imperfect person in an imperfect world, and all that can be asked of you is to do your best—and rest easy with that.

2. Instructor Responsibilities

Your instructor has responsibilities to you and your classmates. Learning mathematics has to be a team effort and your instructor is a significant member of the team. If you are able to choose your instructor, choose carefully. (See chapter 13 for detailed information on choosing a mathematics instructor.) A major consideration should be how well the instructor meets his or her responsibilities. The maxim that an ounce of prevention is worth a pound of cure, applies here. Being selective in choosing your instructor helps ensure that problems you encounter in the course are minimal. You have the right to expect certain things of your instructor. There are remedies you can pursue if there are serious violations of important responsibilities. However, you have to be fair and reasonable and not expect your instructor to be perfect. He or she is human like you and will not carry out all of his or her responsibilities to perfection. What you can expect from your instructor is now presented.

Provide a Written Course Syllabus

You have the right at the outset of the course to know certain things about your instructor and course. A syllabus should minimally include—

- The sections covered in the textbook, and the order in which they will be covered

- Policies on class attendance, examinations (including those missed), and assignments (including those turned in late)
- The factors that will be used to evaluate your performance, and the relative weights assigned to them
- Information about contacting the instructor, including office hours, phone number, and email address
- The instructor's philosophy of teaching (e.g., what he or she values as an instructor, and how that relates to his or her expectations of you)

  It is reasonable to expect this, but don't be surprised if you don't receive it. If a philosophy statement is not included in the syllabus, it should be discussed in class at the outset of the course. You have the right to ask your instructor for this philosophy statement, but don't be surprised if he or she doesn't know how to respond.

Record all relevant assignment and examination dates in your planner. Some instructors will not have this information in the syllabus, but for those who do, they might forget to remind you of it; nonetheless, if stated in the syllabus it is normally binding.

Follow the Syllabus

It's one thing to have a syllabus and another for the instructor to follow it. In most *multi*-section mathematics courses, the textbook sections to be covered are mandated by the mathematics department. There may be some instructor leeway in selecting content from specific sections, but an instructor is not to delete sections unless they are considered optional by the mathematics department. Multi-section courses that are part of the calculus track are prerequisites for other courses in the track. Hence, the sections in the syllabus for a course need to be covered so you are properly prepared for subsequent courses. Instructors who delete sections near the end of the course, usually because of poor pacing during the course, are doing you a disservice. Perhaps the poor pacing was due to demands made by students who were struggling with certain sections of the course. Regardless, students who are able to handle the syllabus should not to be short-changed by an instructor's deletion of sections whose content is prerequisite for subsequent courses. The sections of the textbook listed in the syllabus should be the ones mandated by the mathematics department.

Sometimes unforeseen circumstances may cause the instructor to vary slightly from the sections of the textbook identified in the syllabus. Many college mathematics courses have overly ambitious course outlines that move an instructor to delete certain sections. If that is the case, the instructor needs to work diligently to convince the mathematics department to revise the syllabus, which most likely necessitates revisions of syllabi for some courses that have this course as a prerequisite. An instructor's unilateral decision to delete one or more sections of the course, especially those near the end of the course, is generally not the way to go. When this happens, some instructors tell their students to study those sections on their own after the course ends. An instructor has to use his or her professional judgement as to what is best for the students, and you have to be fair in evaluating any decision he or she makes. At the very least, the instructor owes the class an explanation if certain sections of the textbook are deleted or shortchanged.

Another example of how an instructor may deviate from a department's approved syllabus for a course is if he or she does not honor a department directive such as, "Graphics calculators are to be integrated into the course." If your instructor doesn't abide by a stated policy of the mathematics department, you have legitimate grounds to file a complaint.

Post Office Hours

Expect your instructor or a teaching assistant to provide a reasonable number of posted office hours. Unless they are posted somewhere for you and others to see, they do not exist. It is common to post two to five office hours weekly; but, part-time faculty, especially if they teach in the evening, may have no hours posted or at most one hour. Office hours may be posted in segments as short as a half-hour. Many instructors will also meet at other times by appointment. No appointment is typically necessary for scheduled office hours; however, it is recommended that you let your instructor know, if possible, when you are coming. The instructor may know of other students who are coming at that time and suggest you come at another time. If there is no way you can meet with your instructor during scheduled office hours, *the instructor has an obligation to work out a time when you can meet.*

Not only must an instructor post office hours, but he or she has an obligation to let you know, if possible, when specific office hours are cancelled or will not start on time. Usually the instructor leaves a note on the office door. If the instructor does not show up on time, wait a short time before leaving. Sometimes a meeting will last longer than it was scheduled, which means your instructor may arrive a little late for an office hour.

Start and End Class on Time

If your instructor is *consistently* late for class by more than a few minutes, there is cause for concern. Both students and instructors have an obligation to be on time. The instructor should apologize for being late. Habitual lateness to class is inexcusable.

Classes should end on time. Once in awhile your instructor may go over the stated time by a few minutes and this should not be cause for concern. Clinching a key point in a lecture or giving a few extra minutes on an examination are often reasons a class might run long. Regardless, you have the right to leave the classroom at the stated time. You should not be forced to arrive late to your next appointment.

Cancel Class Infrequently

You should be concerned if your class is cancelled more than once or twice in a term. Suppose a three-credit course meets for about forty one-hour class sessions. If the class is canceled five times, then you receive, in terms of class time, about 88 percent of what you paid for. There would have to be some unusual circumstances for this to be acceptable, such as a combination of severe illness of the instructor with no time to plan for substitutes, and inclement weather with no opportunity to reschedule the cancelled classes. Tuition is expensive and you should expect to get what you pay for.

Maintain a Respectful Classroom

Your instructor has to treat you and your classmates with respect, demand respect from each student in the classroom, and expect students to respect each other. You shouldn't tolerate any disrespect shown to you or to others in the classroom. The same remark applies to interactions during office hours. An instructor who consistently makes or allows disparaging remarks is not meeting his or her responsibility. You have to be careful here because what constitutes a disparaging remark is often in the mind of the beholder. An overly sensitive student, or one with a low self-concept, can view a perfectly innocent and appropriate remark as a put-down. Please do not infer that expressing disagreement with someone is disrespectful. However, it may be expressed in a disrespectful way. Slips of the tongue do occur, and if you walked in the shoes that day of the person making one, you may be more tolerant of the faux pas.

Maintain Discipline

Can college instructors have discipline problems? Yes. I was a student in a doctorate level mathematics class where the instructor allowed despicable behavior to go on, at his expense. Discipline can be easily maintained in college but the instructor has to do what needs to be done to have it. It's as simple as making this statement and then backing it up: "Please behave or you will have to leave." Morale and learning is lessened in undisciplined classes. You have the right to learn in a disciplined environment.

Use the Class Session Wisely

There are a variety of ways in which the class session can be structured, some good and some not so good. Good use of the class session enhances learning. Poor structure results if there is an inappropriate division among the various class activities. There should be an appropriate division between developing new content and giving feedback on the previous assignment, and between students speaking and the instructor speaking. Are the same few students involved in class discussion or does the instructor spread the discussion among all the students? Does the instructor wait for students to think about what was asked before calling on someone? Is there an appropriate division between small group and whole class discussion, and between in-class use and non-use of technology?

It is good for the instructor to begin class by raising a current topic of interest to students such as the result of a sporting event, a recent political event, or a university decision or concern. It can be motivating and draws the students together as a learning group. A minute or less of class time is all that is needed for this, unless it is a critical and urgent social concern. The instructor should not discuss non-course related matters for considerable periods of time.

Another misuse of class time is when an instructor spends considerable time with one student who doesn't understand something when it is clear that the confusion is unique to this student. After an attempt or two with this student, the instructor should meet with the student immediately after class, or encourage the student to come to office hours for further conversation. If there are 30 students in class and the instructor spends five minutes with one student, there could be five minutes of down time for *each* of 29 students. This can also negatively affect students during the remaining class time, due to loss of focus and motivation.

Accepting Late Assignments and Giving Make-Up Examinations

Instructor responsibilities in these two areas are given in consecutive subsections of section 1, titled, "Turn in Assignments on Time," and "Don't Miss Examinations."

Grade Problem Sets, Examinations, and Projects in a Timely, Careful, and Fair Way

It is ill advised for an instructor to return a graded mathematics examination more than a week or so after it is taken. Circumstances may dictate that the return of a graded examination has to be delayed, but that should be the exception, not the rule. Motivation dissipates quickly and learning suffers when graded examinations are not returned in a timely manner. You may have to make some important and timely decisions based on the graded results, and you cannot do this if you don't have them.

It is very difficult for a college instructor to get an examination graded and back to you by the next class session, especially if it is the next day. It is not atypical, for example, to administer an examination on a Wednesday and get it back on Monday of the next week. Don't expect any better than this even though many instructors would have it back by Friday, two days after it is taken. Realize that your instructor has a life beyond your course. He or she attends conferences, teaches other courses, and has service and research responsibilities, along with personal responsibilities. Your instructor will need more time in returning extensive problem sets and projects, unless he or she has a grading assistant. If an instructor is *habitually* late returning examinations and papers, he or she is acting irresponsibly.

If your instructor makes an assignment to be turned in and graded, then it needs to be graded in a manner that reflects the quality of your work. Receiving only a check mark or a letter grade for *completing* the assignment is not careful grading. That does not mean that your in-structor is obligated to grade every item on your paper. Grading a representative sampling of the same items for all students can reflect the quality of the work. Your work should be returned to you so you can see how it was graded.

It is not always easy for you to determine if your examination is graded fairly, either because of your bias, or lack of understanding. I had students who would come to my office and say they were graded unfairly on a particular problem or two. They might say, "There is nothing wrong with my solution, yet I did not receive full credit," or "There is a classmate who has the same solution, but received more credit." In most instances, after reviewing the solution, I could not change the points I gave. (See section 3 of chapter 27 for more on this topic.)

Allow Students to See Their Graded Final Examinations or Project Papers Within a Specified Period of Time

You have the right to see how your work is graded, including your final examination and papers turned in at the end of the term. This becomes problematic if you took the course in the last semester of the academic year or during a spring or summer term. Many students and faculty scatter for an extended period when these terms are over. Fortunately, many colleges have a policy that instructors must keep final examinations or papers for the duration of the next regular semester. You may not be allowed to remove your final examination from your instructor's office, but you have the right to look at it in his or her office. If you are allowed to keep your final examination after it is graded, but cannot pick it up, leave a stamped, self-addressed envelope with your instructor and he or she will most likely mail it to you, along with other papers that have not been returned.

Speak in Class with a Voice That Can Be Heard and with Clear Pronunciation

Your instructor's voice has to project to all corners of the classroom, and you have to know what words are spoken. This can be more problematic with foreign-born instructors, but you have to do your part in trying to get by an accent. Don't hesitate to ask your instructor to repeat something. If it is a serious and persistent problem for you and others, take steps, pursuing appropriate channels, to rectify the situation by beginning with the mathematics department chairperson.

Use Handwriting That Is Readable

It is not enough for an instructor to speak clearly in a mathematics class. The instructor must also write clearly. Learning is enhanced when these two communication tools are good, since students learn through auditory and visual means. You cannot take good notes if you cannot read what is on a board or screen because of poor handwriting. Ask your instructor to rewrite something that you cannot read. I have been on the receiving end of such requests, and applaud students who make them.

Announce Class Period Examinations in a Timely Manner

You need a reasonable amount of time to do a final preparation for an examination. For example, an instructor announcing in Monday's class that an examination will be given this coming Wednesday, does not allow adequate preparation time. However, announcing to the class on a Friday that an examination will be given the following Wednesday is reasonable.

Use a Variety of Teaching Modes That Accommodate the Learning Styles of as Many Students as Possible

This is an important instructor responsibility. Not all students learn the same way. Some are more visual learners (who need to see), others more auditory (who need to hear), and others more tactile or kinetic (who need to move, do, and touch). Consequently, to meet the needs of all students in the course, instructors have to have a proper balance of (1) using written or printed words, drawings, graphs, and diagrams; (2) articulating ideas, relationships, and solutions to exercises, and encouraging the same in students; and (3) promoting laboratory or hands-on activities (when appropriate), including conduction of experiments. Choosing a mathematics instructor based on your learning style and instructors' teaching styles is discussed in chapter 13.

Stay Current with Innovative Methodologies for Teaching Mathematics

This is another important responsibility of instructors. Methods of teaching mathematics are *not* an integral part of the formal training of most doctoral students in mathematics who plan on teaching at the college level. Mathematics departments need to in-service their instructors, and instructors need to welcome the training. It is the responsibility of mathematics instructors to be familiar with the mathematics education standards and recommendations of the Mathematical Association of America or of the National Council of Teachers of Mathematics. It is inexcusable for college instructors to remain ignorant of these standards, or resist employing them without good reason. You receive mixed messages as to what you should be doing as a mathematics student when different instructors use *widely* divergent instructional techniques. Should there be some differences? Yes, no two instructors are alike and they will have their different strengths and weaknesses. What works well for one instructor may not work as well for another. However, an instructor must work on using teaching techniques that best help students attain important educational objectives. Some would say that this instructor responsibility is infringing on the instructor's academic freedom. Better arguments can be given to the contrary.

Avoid Being an Oral Textbook

Your instructor's work in class should complement the textbook, not repeat it. This is not to say that your instructor should never work a textbook example in class. This example may have been confusing to you and others and requires more explanation. But your instructor needs to avoid consistently working problems in class that were worked in the textbook. It lessens your motivation to read and learn from the textbook, and limits the number of problems that you have available to study. Similarly, your instructor should not repeat the proof of every theorem that is proven in the textbook. If a proof is repeated, the instructor needs to avoid using the exact language of the textbook. Your ability to learn from your reading is an essential educational objective and has to be supported by your instructor's actions. The instructor has to know the textbook well, and let you know when, and how, he or she is deviating from it, and what is considered optional reading in it.

Teach Students a Viable Process of Learning Mathematics

Comments have to be made throughout the course by the instructor, but especially early on, about what to study and how to study it. Some instructors incorrectly think that all they need to do is present the subject matter as clearly as they can. They need to make you aware of viable processes of learning mathematics, and support them with their teaching methods. *Teaching students a viable process of learning mathematics is as much a responsibility of instructors as teaching them mathematics.* Students will not achieve a good understanding of mathematics unless they employ a viable process to learn it. It would be helpful if all students had such a process when they enter college, but that is not the case for many of them—perhaps a large majority of them. Success in high school mathematics does not automatically translate into success in college mathematics since there are many differences between taking mathematics in high school and taking it in college. (Many of these differences are given in section 1 of chapter 2). Before an instructor can help students gain a viable process of learning mathematics in college, he or she must have a viable process of teaching mathematics. Similarities abound in these two processes.

Teach for Understanding

A primary focus of your mathematics course is to *understand* content. Mathematics relationships have to be examined using reasoning; hence, considerable time has to be spent on this by instructors, and by you. A process of reasoning is supported in the course by the instructor's use of it in class and the types of assignments and examinations given. Rote teaching and learning are unacceptable. (Understanding and thinking in mathematics are discussed in sections 2 and 3 of chapter 9.)

Address Student Tardiness

Instructors need to talk to students who are chronically late for class. The instructor and the students are disrupted when students come late to class. Allowing these disruptions is a misuse of class time, and so is backtracking to redo content that the latecomers missed. Backtracking enables the latecomers and bores those who arrived in class on time.

Assign Specific Problem Sets

Your instructor knows best the basic set of problems from the textbook that should be assigned. An instructor who leaves it up to the students to decide the number and types of problems to work is not doing his or her job. How can you be expected to do this if you have just been introduced to the content that these problems address? To discern what is accomplished by working a specific problem, and the appropriateness of doing so, requires an experienced person, one who has the necessary mathematics background and experience. Most sections of the textbook contain an almost unlimited number of problems to work, and the instructor needs to choose a viable set of problems that comprise a minimal assignment. More capable students should choose problems to work beyond those assigned by the instructor.

Respond to Telephone Calls and Emails in a Timely Manner

Instructors are busy people. It is not always easy for them to get back to you as quickly as you desire. Nonetheless, they have an obligation to respond as quickly as they can to your telephone calls and emails. But don't expect this if you are not meeting your course responsibilities.

Inform You of Your Status in the Course

Your instructor, upon request, is expected to tell you how you are performing in the course. Included in this assessment may be your instructor's advice as to whether you should drop the course. Students often approach their instructor for a current evaluation of their performance. Many wait until close to the last day they can drop the course without penalty. There are good reasons not to wait that long for an appraisal of your work, and to see what your instructor says about your chances of receiving a specific grade for the course.

Provide Academic Counseling or Referrals Upon Request

Most instructors can provide academic counseling for their students. Your instructor has a responsibility to directly help you, or refer you to the appropriate personnel or offices for assistance. You may need to hire a tutor or obtain help with a writing problem or personal problem. Details for obtaining assistance *in mathematics* from a variety of sources are given in chapter 23.

3. Students' Complaints about Their Instructors

I gathered evaluative comments of students on their calculus instructors. The comments came from student evaluation forms that were completed in Calculus I, II, and III courses. Many different classes and instructors are represented. The comments were in response to the question, "What did you dislike about your instructor?"

Exercise A. Based on each of the student comments that will be given shortly, do you believe the instructor met his or her responsibility? Explain your thought process.

Exercise A would be more valuable if you complete it with at least one other student, discussing your opinions. If you believe information is needed before you can adequately give an opinion, state the additional information you need. A word of caution is in order before you proceed: What at first glance may appear to be a negative aspect of the instructor's teaching, may be a positive aspect—if you think beyond the superficial. To help you with this exercise, here are remarks from two students, and my comments about each one:

❖ *"The instructor did not cover material in class that was on the test. However, it did make me think and study harder."* (Calculus I student)

The first sentence of the remark does not necessarily mean that the instructor was irresponsible, perhaps far from it. Specific examination questions may have been based on the assigned problem sets or on a reading in the textbook, and the instructor might have deliberately decided not to discuss the content in class and for good reason. As long as he or she prepared the student, in some fashion, for these examination questions, they are fair game. Perhaps the instructor wanted the student to put some ideas together based on his or her reading and assigned problem sets. The second sentence in the remark above indicates that this student benefited from this instructor's actions. Instruction is not good when all your instructor does is solve specific problems in class, has you practice them over and over, and then has you spit back the solutions on an examination. This type of instruction is educating you to be a robot—it is deadly instruction.

❖ *"Some of the explanations went fast and made comprehension difficult. However, having read the section before it was discussed in class alleviated the problem."* (Calculus II student)

The first sentence is not necessarily a negative comment about the instructor. Mathematics in college is, of necessity, fast-paced. You are expected to use additional methods in your efforts to understand an explanation by your instructor. The second sentence indicates that the student compensated for the pace by reading the section beforehand. This, as a matter of course, is an appropriate thing to do. If a student said that an instructor's explanations were *always* carefully and slowly developed, I would wonder if the instructor was avoiding difficult content, able to complete the syllabus, or spoon-feeding students (not leaving anything for them to grapple with).

Work *Exercise A* for these students' remarks about what they didn't like about their instructors:

1. *"Tests were very hard in comparison to assignments."* (Calculus I student)
2. *"Didn't always lecture on the material before assigning the homework."* (Calculus I student)
3. *"Did not assign any odd problems so I could not look up the answer in the back of the textbook."* (Calculus III student)
4. *"She waits too long to go over homework."* (Calculus I student)
5. *"Not enough tests. Because there weren't enough tests, each test couldn't cover everything, so I studied for the wrong stuff."* (Calculus I student)
6. *"I think she should have reviewed before exams and tried to leave more time for questions."* (Calculus I student)
7. *"There wasn't sufficient time to check my work on the test."* (Calculus II student)
8. *"I disliked the method of teaching directly out of the book. I would have felt more comfortable if he would have had some outside material (applications, different methods of solving problems) that would have aided our reading of the text."* (Calculus III student)
9. *"We were not expected in other math classes to do a whole lot of reading on our own."* (Calculus I student)
10. *"No homework collected. This made it easy to fall behind."* (Calculus I student)

11. *"He seemed impatient when it came to answering questions in class—made you feel stupid."* (Calculus III student)

12. *"Not enough application to practical problems. Too much theory."* (Calculus II student)

13. *"I want more explanations of topics even if fewer examples are presented."* (Calculus III student)

14. *"Not enough proofs shown."* (Calculus I student)

15. *"He didn't tell us where in the book he was. It was hard to follow along."* (Calculus I student)

16. *"I sometimes was confused by the different formulas or methods the instructor used compared to those used in the textbook."* (Calculus II student)

17. *"I wanted the instructor to grade on a curve."* (Calculus II student)

18. *"He lectured on three sections from the next subject matter when being tested on current subject matter."* (Calculus II student)

19. *"He gets mad when you go for help and don't understand something that he said."* (Calculus I student)

Did you find it difficult to complete *Exercise A*? You should have.

This quotation by Meiland provides a summary of the essence of student and instructor responsibilities:

It is true that there are bad teachers—teachers who do not prepare for class, who are arbitrary, who subject their students to sarcasm, who won't tolerate (let alone encourage) questions and criticism, or who have thought little about education. But there are also bad students. By "bad students" I do not mean students who get low grades. Instead, I mean students who do not participate sufficiently in their own education and who do not actively demand from their teachers. They do not ask questions in class or after class. They do not discuss the purposes of assignments with their teachers. They do not make the teacher explain the importance and significance of the subject being studied. They do not go to office hours. They do not make the teacher explain comments on papers handed back to the student. They do not take notes on the readings. They just sit in class, letting the teacher do all the work. Education can only be a cooperative effort between student and teacher. By actively participating in these and other ways, you play your proper and necessary part in the cooperative process.[1]

Choosing a Mathematics Instructor Based on Learning and Teaching Styles

You come to a mathematics course as a learner, bringing with you your personality, thought processes, emotions, mathematics background, and learning experiences. You use all this in your efforts to learn, and this comprises your *learning style*. Your learning style is unique, although there are other learners who will possess some attributes of your learning style.

Your instructor comes to your mathematics course as a teacher, having been influenced by many things, including his or her own personal learning style, mathematics instructors he or she has had, formal education courses he or she has taken, conversations with mathematics colleagues, attendance at educational conferences, readings of professional mathematics or mathematics education journals, teaching experiences, and personal beliefs on what constitutes good mathematics instruction. All of this is used in your instructor's efforts to teach the course, and this comprises his or her *teaching style*. Your instructor's teaching style is unique, although there are other teachers who will possess some attributes of his or her teaching style.

As a student enrolled in a mathematics course, you want these two questions answered: (1) how will you interact and respond to the learning environment created by your instructor's teaching style, and (2) what can you do to make this interaction and response better? That is, how well will your learning style mesh with your instructor's teaching style, and how can you compensate for what may be lacking in either style? If you had the opportunity to select your mathematics instructor, you would want these questions answered: What information do I need

215

to characterize an instructor's teaching style, how well does the teaching style dovetail with my learning style, and how well does he or she meet the basic responsibilities of a college mathematics instructor? The three questions just posed are addressed throughout these sections:

Section 1. Learning Styles
Section 2. Teaching Styles
Section 3. Complementing Your Instructor's Teaching Style
Section 4. Choosing Your Mathematics Instructor

1. Learning Styles

Since each person is predisposed to learn in a way that works best for him or her, it makes no sense to say that one learning style is better than another. The style that helps you process knowledge most effectively is the style you should primarily use. Any attempt you make to *significantly* change your natural propensity to learn is problematic. In a mathematics course you want to look for opportunities to use your learning style; that is, to enhance the natural tendencies you have for learning. This doesn't mean that you should ignore aspects of other learning styles. You can take them on, although at the outset it may be unnatural or difficult to do so. But with desire and experience, they can become part of you, and give you a more comprehensive learning style. Certain educational objectives are better achieved by using features of a variety of learning styles. Thus, to maximize your learning, you need to use a more integrated approach to learning, which necessitates going beyond, but not dispensing with, your preferred way of learning.

Know Your Learning Style

Like many students you may not know your learning style. I suspect that you have not given much thought to this, and most likely no instructor, counselor, or learning style expert has broached the subject with you. You need to understand your learning style before you can determine (1) the teaching style that serves you best, (2) the learning activities that will complement your learning style, and (3) how to compensate for differences in your learning style and in your instructor's teaching style.

Formally Determining Your Learning Style

The information that will be given shortly on learning style models and the type of learning activities that are in sync with them, may be enough for you to identify your basic learning style. You know yourself better than you think; thus, you have a good grasp of the activities that do or do not work well for you in your attempts to learn mathematics. If you feel the need to have a more formal or systematic means to determine your learning style, there are survey tools available that will help. Information on these tools can be found by accessing Internet sites on learning styles. The counseling office at your college may be able to help you assess your learning style, or suggest credible survey tools for doing this. One appropriate survey tool is the Myers-Briggs Type Indicator (MBTI), which is discussed in the following subsection.

Learning Style Models

Your learning style preference can be formally determined by completing an inventory based on a specific learning style model. Two major learning style models, one based on a multiple intelligences viewpoint and the other on a personality viewpoint, are described:

Multiple intelligences viewpoint. Howard Gardner has identified seven intelligences[1]:

- Visual/Spatial Intelligence (ability to perceive the visual)
- Verbal/Linguistic Intelligence (ability to use words and language)
- Logical/Mathematical Intelligence (ability to use reason, logic and numbers)
- Bodily/Kinesthetic Intelligence (ability to control body movements and handle objects skillfully)
- Musical/Rhythmic Intelligence (ability to produce and appreciate music)
- Interpersonal Intelligence (ability to relate to and understand others)
- Intrapersonal Intelligence (ability to self-reflect and be aware of one's inner state of being)

You can go online if you want more information on multiple intelligences identified by Howard Gardner. Gardner's model will not be pursued in this book; instead, our focus is on the Myers-Briggs Type Indicator (MBTI), which is now discussed.

Personality viewpoint. The Myers-Briggs Type Indicator describes 16 personality types, determined by preferences on four dichotomies [2]:

Extroversion (E)—Introversion (I). This dichotomy describes how people prefer to focus their attention and get their energy: from the outer world of people and activity (E), or from their inner world of ideas and experiences (I).

Sensing (S)—Intuition (N). This dichotomy describes how people prefer to take in information: focused on what is real and actual (S), or on patterns and meanings in data (N).

Thinking (T)—Feeling (F). This dichotomy describes how people prefer to make decisions: based on logical analysis (T), or guided by concern for their impact on others (F).

Judging (J)—Perceiving (P). This dichotomy describes how people prefer to deal with the outer world: in a planned orderly way (J), or in a flexible spontaneous way (P).

If you want to *formally* determine your personality type, take the Myers-Briggs Type Indicator. Go online for more information on how you can do this. Your scores reveal the strength of your preferences. Your personality type will be one of these 16 distinct types, indicating your preferences, with one preference selected from each of these four pairs of preferences, (E, I), (S, N), (T, F), and (J.P):

ISTJ	ISFJ	INFJ	INTJ
ISTP	ISFP	INFP	INTP
ESTP	ESFP	ENFP	ENTP
ESTJ	ESFJ	ENFJ	ENTJ

Two examples are given:

If your personality type is ENTJ, you prefer to—

- Focus your attention and get your energy from the outer world of people and activity (E)
- Take in information focused on patterns and meanings in data (N)

- Make decisions based on logical analysis (T)
- Deal with the outer world in a planned orderly way (J)

If your personality type is ISFP, you prefer to—

- Focus your attention and get your energy from your inner world of ideas and experiences (I)
- Take in information focused on what is real and actual (S)
- Make decisions guided by concern for their impact on others (F)
- Deal with the outer world in a flexible spontaneous way (P)

These two personality types could not be more dissimilar—their preferences are different on *each* of the four dichotomies. Thus, if one student has personality type ENTJ and another student personality type ISFP, they will not respond the same to an instructor's teaching style.

Informally Determining Your Learning Preference

You will now be shown an *informal* approach to determining your personality type or learning preference. This will be done by giving you descriptions of students who favor specific learning preferences. A specific learning preference of yours is not an all or nothing thing, or set in stone. You have a more balanced learning style when you assimilate other learning preferences. If your learning style is far from being balanced, you goal should be to work to include aspects of other learning styles. You can make this happen if you desire it, and work diligently to achieve it. This may not be easy to achieve since you may be going against your natural inclination, but it is necessary to do if you want a more complete understanding of mathematics.

Following are descriptions of students with specific learning preferences that are based on these three pairs of dichotomous learning preferences: (extroversion, introversion), (sensory, intuitive), (visual, verbal). These six learning preferences were chosen because they give a sufficient description of the learning preferences of students. As you read these descriptions, you should be able to recognize your learning preferences, at least to the degree that you can benefit from the descriptions. More is said on this later.

Extroversion or Introversion Preference

Extroversion preference. A student with an extroversion preference for learning—

- Is energized by being with others, and interacting with them and the external world
- Bounces ideas off others to understand better
- Thinks spontaneously
- Is quick to offer opinions
- Learns by explaining things to others
- Values and enjoys working in groups
- Expresses thoughts and feelings as they are felt
- Prefers an open and tolerant view of things
- Is more sensitive to the rewards inherent in most social situations such as warmth, affection, and close emotional bonds

Introversion preference. A student with an introversion preference for learning—

- Gets energy from within
- Can be sociable, but needs to have time alone to get re-energized
- Chooses confidants carefully
- Is private in his or her thoughts when developing ideas
- Listens to others
- Likes private places
- Thinks he or she should be more sociable
- Wants his or her need for privacy accepted

Sensory or Intuitive Preference
Sensory preference. A student with a sensory preference for learning—

- Finds security in well-structured development of content
- Desires an instructor who is well organized, and presents content in a clear, logical, detailed, sequential, and straightforward way
- Places more confidence in the instructor than himself or herself
- Struggles with complex ideas, preferring the practical to the abstract, and the more mechanical over the more intellectual
- Prefers beginning a theoretical development with applications from which the theory will evolve
- Wants to see a need for what is learned, which motivates

him or her to learn. In this sense, bringing in the real world is important

- Is prone to struggle with basic tools of learning, including reading and writing
- Does not like confusion and wants to be told precisely what to do and what to plan, including the types of questions on examinations, the length of an examination, how much needs to be written for an assignment or paper, and what to include in lecture notes
- Does not like uncertainty ("Can I learn this content?" "What grade will I receive?" "How long will this take to understand?")
- Likes working problems whose solutions call for the use of well-understood mechanical or algorithmic procedures
- Has difficulty working with symbolic form
- Is primarily interested in learning for the purpose of satisfying practical needs such as passing a course, satisfying a concentration requirement, making more money, or helping in a future career

Intuitive preference. A student with an intuitive preference for learning is—

- Comfortable with instruction that is not rigidly organized
- Focused on the big ideas and how they relate to each other, and ignores or misses details
- Very interested in the structure of mathematics (that is, how ideas are related)
- Not very interested and adept at working with details and concrete applications
- Secure in relying on himself or herself for learning, and favors independence
- Nourished by ideas, concepts, abstractions, theories, and the potential possibilities emanating from them
- Motivated by theory and wants to begin with the theory and progress to applications
- Inclined to gain knowledge internally; that is, knowledge that comes from inside through thought and reflection
- Creative and innovative
- Interested in learning for learning's sake

- Tolerant of ambiguities
- Open to differing viewpoints, ideas, and approaches to learning
- Absent-minded at times because of focus on the world of ideas; hence, oblivious to the real world

Visual or Verbal Preference

Visual preference. A student with a visual preference for learning—

- Perceives information most effectively through seeing
- Learns best through the use of pictures, diagrams, illustrations, videos, charts, drawings, graphs, and schematics
- Translates an explanation into a mental picture to better understand it
- Creates mental pictures to remember specific information (You have often heard someone say, "In my mind's eye I see. . . .")
- Interprets visual images
- Makes sketches to improve understanding and solve problems
- Makes good use of computer and calculator graphics
- Creates visual metaphors and analogies
- Is attuned to a speaker's body language to better understand what is being said

Verbal preference. A student with a verbal preference for learning—

- Perceives information most effectively through written and spoken words
- Learns best through lecture and discussion
- Works well with symbolic form, including formulas
- Has acute hearing skills
- Is good at remembering information
- Is adept in using words and language, whether written or spoken
- Is convincing when explaining a point of view
- Is good at teaching
- Is a good note taker
- Uses verbal analogies to make a point
- Benefits greatly from reading the textbook and instructor handouts

All six learning preferences are normal preferences. No value judgments should be placed on any of them. For each of the three pairs of learning preferences, most students fall somewhere between the two extremes. Take time to ponder these questions:

1. Based on your reflections on these six learning preferences, what seems to be your learning style (i.e., select one preference from each of these pairs: (introversion, extroversion), (sensory, intuitive), (visual, verbal)? For example, are you a sensory—visual—extroversion learner?

2. Attach one of the following statements to each of your preferences, indicating how certain you are of your preference in comparison to the other preference that was part of the pair: very sure, moderately sure, marginally sure. For example, your result might read this way:

> sensory (very sure)—visual (marginally sure)—
> extroversion (moderately sure)

How does understanding your preferences for learning (i.e., your learning style), benefit you in learning mathematics? For starters, if you are choosing an instructor you will want one who teaches in a manner that supports your learning style. If you have an instructor whose teaching style is not in sync with your learning style, then you need to take steps to ensure that you are engaged in activities that lend themselves to your learning preferences—for that is how you learn best. However, it is not enough for you to have an instructor who supports your learning style, for this instructor may not use teaching methods that support other learning styles, yet you can benefit from being exposed to methods that do support other learning styles. You will have to compensate for what you are not getting from your instructor. And your knowledge of various learning preferences that are discussed in this section will help you know what you should include in your study.

Learning Styles of Students Versus Learning Styles of Instructors

There is a fundamental mismatch these days between the learning styles of most students and the learning styles of most college instructors. It appears that a majority of current college students prefer the *sensory* mode of perceiving knowledge, whereas a majority of college instructors prefer the *intuitive* mode. It also appears that a majority of college students have an *extroversion* preference for learning, whereas a majority of college instructors have an *introversion* preference for learning. How aware instructors are of this mismatch is not known—it most likely ranges from no awareness to a strong awareness. Those that are aware understand that their learning preferences need to be tempered or played down in their teaching style, so as to better match their students' learning style. But even for these instructors, it is not easy to make these accommodations, and they may not be made to the necessary degree.

It is axiomatic that this mismatch in learning preferences can cause discontent and despair in students since an instructor's teaching style reflects his or her learning style. Members of either group are quick to reach the conclusion that something is wrong with the members of the other group. Is something wrong with either group? No. Is there a mismatch between the groups? Yes. Can you do something about this? Yes. Read on.

2. Teaching Styles

It would be gratifying if every college mathematics instructor met all the teaching responsibilities outlined in chapter 12, but this is an unrealistic expectation. Also, those responsibilities that are met are not necessarily met in exactly the same way across instructors. Instructors have different innate strengths, weaknesses, and experiences that influence their teaching preferences or teaching styles. There are as many mathematics-teaching styles as mathematics instructors. However, most mathematics teaching styles are probably variations of one of two general teaching styles that can be characterized as *lecture*-oriented or *discussion*-oriented (to be described, shortly). When you choose a mathematics instructor, you have to decide which of these two general teaching styles best describes that instructor's teaching style, and whether that style meets your needs. A teaching style that may best serve your needs is a blend of these two styles, one that gives a more appropriate balance.

Mathematics instructors have an obligation to modify their teaching styles if they are not meeting key instructor responsibilities such as encouraging class discussion, requiring students to relate ideas in a meaningful way, or grading examinations on time. They need to determine what to add or delete from their teaching styles that will address these responsibilities. You can be a catalyst for this change, but don't get your hopes too high. Resistance to change is in all of us. Similarly, you will have to step out of your comfort zone to choose a mathematics instructor whose teaching style is not exactly in sync with your learning style, but who may provide you with a more varied approach to learning. Making such a selection means you will have to make adaptations to your learning style, which requires considerable effort. In summary, you and your instructor have mutual responsibilities: your instructor needs to grow in his or her teaching style, and you need to grow in your learning style.

You will now be introduced to the basic teaching styles of two hypothetical mathematics instructors, Dr. Lector and Dr. Dialogue. (I was assisted in the ensuing remarks involving Drs. Lector and Dialogue by my university colleague, Professor Ronald Morash.) Dr. Lector's teaching style is lecture-oriented, and Dr. Dialogue's is discussion-oriented. The characterizations of these two general teaching styles place them at the *extreme* ends of styles similar to them. I mentioned before that most mathematics teaching styles are variations of one of these, and to lesser extremes. (*Note.* The masculine gender for both Dr. Lector and Dr. Dialogue is used to avoid having a male or female being identified with one of these general teaching styles.)

Dr. Lector—A Lecture-Oriented Teaching Style

Dr. Lector is known among his students and colleagues for his orderly approach to classroom activities, running each class session in pretty much the same way regardless of the level of the course. Specifically, he spends most of his time with his back to the class, filling the board systematically with a succession of definitions, axioms, theorems, proofs, remarks, and examples. To his students, he always seems to be proceeding rapidly and, as they take hurried notes, they feel the pressure of a fast paced course. He clearly conveys to the students that the road to success in his class is to copy or digest the textbook he is writing before their eyes. He is not hostile to questions from the students, but responds most comfortably to questions about problems from recent homework assignments. His answers are, like his lectures, carefully prepared finished products (even when he talks to students outside of class). His testing calls mainly for student reproduction of material presented in class and mimicking problem solving techniques covered explicitly in homework assignments. Before each test, he conducts a systematic review of content they need to know and his examinations are faithful to that review. In summary, his courses totally revolve around his lectures and his students regard notes from his class as among their most valued possessions, at least until after the final examination.

Dr. Dialogue—a Discussion-Oriented Teaching Style

Dr. Dialogue is known among his students for his whirlwind style in the classroom. His enthusiasm for mathematics and for discussing any aspect of it with students at any time, including class time, knows no bound. Depending on the level of the course he's teaching, he either speaks extemporaneously about some aspect of the day's material, which he finds interesting, or else simply allows the students' questions and comments to determine his agenda. In either case, whether the course is elementary or advanced, he doesn't habitually allow himself to become bogged down in the messy details and treats covering the material as something that the students, not he, must do. In class, he is completely open to questions and at any time will let one student's question determine a significant portion of the class period, even though the question may not be directly related to the material immediately at hand. This is not to say, however, that he always gives the students the types of answers they want, since he sometimes answers questions with a question of his own, or with a suggestion to help them get the answer on their own (e.g., further reading, working a related problem, etc.).

He frequently carries on an audible conversation with himself in class, wondering whether there is a relationship among certain ideas or if a way can be found to justify a result. He continually asks his students to commit themselves to guessing what an answer might be, and he likes to point out disagreements that may have surfaced from the class. If there is no disagreement, he will create it by playing the role of devil's advocate. He encourages students to disagree with him. His students, to say the least, are required to be actively involved in the course, since their own reading is usually their first introduction to new ideas and, in particular, since he does not take them over the rough spots in advance of their reading.

He lets the students know how a mathematician thinks, and asks questions and poses problems that help them think like one. He periodically forms small groups in class to discuss an idea, to work a problem that surfaced in class, or discuss the previous class assignment. He frequently mentions the value of students being part of a mathematics study group outside of class.

Additional Details about the Two Teaching Philosophies, Including Justifications for Their Use

When Dr. Lector was asked about his teaching style and the philosophy on which it was based, he responded that his main goal in teaching the course is to convey the specific mathematical information contained in that course (which the students should systematically absorb) and to illustrate certain problem solving techniques (which the students could imitate and eventually use). He believes most real learning (i.e., the careful digestion of the content of the class notes) takes place as students study the notes and perhaps the text and do problems, all after seeing a careful presentation in class. Thus, he believes that he is using class time in the most efficient way by giving as much specific information in the notes as time will allow. He agreed that most of his notes are a rehash of the textbook, but said that he had yet to find a textbook whose organization and presentation he couldn't improve upon in his class notes. He also claimed that his students, in their notes, have a better-integrated treatment of the material than in the textbook and one whose emphasis is tailored to their own interests and abilities, as he perceives them, as well as the type of testing he intends to do. In addition, he said that students seem not unhappy with his approach in that most pay attention in class and some have even told him that they appreciate the security which his systematic teaching style gives them. Finally, he mentioned that he had learned his method of teaching by observing his favorite teachers and since this style is so ingrained and comes so naturally, it enables him to have more time for his many other professional commitments because he has to spend less time on teaching.

On the other hand, Dr. Dialogue told us that he regards each course he teaches as his opportunity to convey what mathematics is about, how a mathematician thinks, and how mathematics actually develops. In his opinion, students should be able to learn the basic course material from reading assigned sections of a text, and should come to class with a definite point of view, which in turn largely determines the day's agenda. Inevitably, he said, many questions would be requests for solutions to homework problems, and he said he hardly ever attempts to work such problems in advance. He believes that observing his (often dramatized) thought process as he works out a problem in class, as well as students having the opportunity to contribute themselves to solving it, is more valuable to developing the students' problem solving skills and understanding than the finished product. He thinks the best way to get students to be curious and adventuresome about mathematics is for him to display these traits in class. He creates controversy to pique student interest. He mentioned that his students are less bored in class because they are active participants, not passive observers. He stated that the pleasure and satisfaction of mathematics lies in doing it rather than watching it done; hence, a change from a strict lecture approach promotes a more positive attitude toward mathematics as an interesting subject.

He also believes that his teaching style conveys a point of view as to what is important in mathematics, amplifies connections between different parts of mathematics, and projects an enthusiasm for mathematics. It also emphasizes the importance of understanding the great ideas, rather than memorizing the trivia. What a student learns in class, he stated, will be retained long after the course is over since the student truly understands mathematics. For this reason too, he or she will be able to apply this knowledge to a wide variety of situations; that is, to transfer the knowledge. Also, through the continuing give and take of his class, students are better able to develop facility in speaking about mathematics. Finally, he pointed out that his approach enables him to be aware of students' opinions, concerns, and struggles.

All the teaching characteristics displayed by Dr. Lector and Dr. Dialogue are desirable in one way or another. But *each teaching style could use some of what the other has to offer.* Unfortunately, many teachers are uncomfortably close to one of these two teaching styles. *My goal in this section is to point out the necessity of choosing a mathematics instructor who has a more balanced teaching style, one who has the best attributes of both styles.* These attributes don't need to be equally balanced in a teaching style; a tilting toward one or the other may be better for you depending on your needs. This tilting could favor your learning style and the goals of mathematics instruction that are most important to you. The amount of this tilting also is a function of your instructor's particular strengths and weaknesses. It is unreasonable to expect your instructor to develop a teaching style that addresses well the learning styles of all his or her students. That is impossible. A reasonable expectation is that he or she works toward providing more learning experiences that accommodate more learning styles.

3. Complementing Your Instructor's Teaching Style

If you don't have the luxury of choosing your instructor, then you need to compensate for what is missing in the teaching style of the instructor you have. This section addresses how you can complement an overly used lecture or discussion teaching style. The rationale just presented by Drs. Lector and Dialogue for employing their respective teaching styles allows you to see what is missing in each of them (i.e., what does not enhance or complement your learning style), and therefore, what you have to provide.

Compensating for an Overly Used Lecture-Oriented Teaching Style

Suppose you are in a course with an instructor whose teaching style is too closely aligned with Dr. Lector's. How do you compensate for what is lacking? The general answer is that you have to create opportunities to make up for the deficiencies. You may get a good grade in the course if you do nothing about this, but there are more important reasons for being in the course. You don't want to shortchange your knowledge of mathematics because you didn't compensate. At some stage in your life, and that may be as soon as you begin your next mathematics course, or as late as when you are on a future job, it is likely that you will suffer consequences because of this.

The main thing you have to do when you have an instructor like Dr. Lector is to compensate for the lack of discussion. That is, you have to plan to be more engaged in discussion activities, in and out of class. Dr. Dialogue commented on what can be accomplished through discussion and those are the goals you are seeking. In class you can create more discussion by asking more questions of the instructor, and encourage your classmates to do the same. These questions can be on your textbook readings, assignments, and class lectures. If your instructor is not willing to answer many questions, express your concern about this to him or her, preferably out of class. Perhaps you can affect some change with healthy pressure applied by a group of your classmates. You can engage in discussion out of class by having a study partner, forming a study group, making use of your instructor's office hours, or seeing a tutor. Your discussions can include anything associated with the course.

You can compensate for your instructor lecturing so closely to the textbook that you don't see a need to read it, by reading the textbook before specific sections are lectured on in class. In this way, your ability and willingness to read mathematics is fostered. Additionally, you will gain a better understanding of mathematics, and be equipped to ask questions in class when these sections are discussed. This gives you more reason to politely interrupt your lecture-oriented instructor's agenda. Also, working exercises before class relating to the textbook sections you read in advance of class, provides more opportunity to ask questions on similar exercises that your instructor will work in class. Your efforts here to get ahead of the game by this advance work is a way to compensate for the mismatch between your learning style and your instructor's teaching style. To be successful in doing this, you have to know the value of doing it, and be diligent about doing it. This is accepting responsibility for your education.

Compensating for an Overly Used Discussion-Oriented Teaching Style

In a discussion-oriented teaching style, class time is not used to provide you with your primary access to the mathematical content of the course (which is done in Dr. Lector's class through taking comprehensive class notes). Yet you must have access, in some form, to a careful presentation of the content. Your textbook primarily provides this access. You can get the structure and organization of the content by carefully reading the textbook and taking appropriate notes, as well as outlining, organizing, integrating, and reviewing the content. (See section 1 of chapter 15 on the basics of reading a mathematics textbook.) However, you need support to do this, for many questions will arise in your reading. You may need further elaboration of a difficult concept, proof of a theorem, solution to an example, or an illustration. This can come from a study partner, a study group, your instructor, or a tutor, but you have to take the initiative. Since discussion-oriented instructors like students to be actively involved in class, this is your opportunity to help him or her set the agenda for the class meeting. That is, ask questions on your reading and encourage some of your classmates to do the same. This helps your instructor stay more on track. Your questions will also help refamiliarize your instructor with the textbook, for he or she will most likely have to look at the textbook before giving a response. There is a tendency for Dr. Dialogue types, and even some Dr. Lector types, to not have looked carefully at the textbook for a long time, perhaps even years (a sad situation, indeed). It does not play an important role for them as they prepare for class; hence, they have forgotten the important details that you will be exposed to in your reading, which includes how they vary from the textbook. This can confuse you; for example, you may be thinking, "Which definition am I to use—the one given in class or the one in the textbook?"

Some sections of a textbook are not well written, or they may be well written for some students but not for you. Each student's reading and comprehension skills are unique. Thus, you are encouraged to read other mathematics textbooks on the same topic. I have read books that seem to be written for me. A particular concept or principle that I was struggling with became understandable when I consulted other textbooks and through them was exposed to different wordings, organizations, and examples.

I always made it a point to have one or two different mathematics textbooks around when I was in an undergraduate mathematics course. I borrowed them from an instructor, checked them out of the library, or purchased them at used book sales. In reading other textbooks you will be exposed to mathematical notation or theorems that may be different from those in your textbook for the course, or what you are given in class. These are added benefits of reading other textbooks, but know what you can or cannot directly use from them in your mathematics course. For sure you can use the added *understanding* that comes from reading alternative textbooks. For more details on using alternative textbooks, see section 2 of chapter 18.

4. Choosing Your Mathematics Instructor

In choosing a mathematics instructor, it is wise to take into consideration the instructor's teaching style, and how well it—

- Favors your learning style, yet also complements it
- Helps you understand the nature and evolution of mathematics (see chapter 8)
- Helps you attain the benefits and goals of learning the structure of mathematics (see chapter 9)

Additionally, you want to determine how well the instructor meets instructor responsibilities besides those discussed in this chapter, such as maintaining class discipline, returning examinations in a timely manner, or availability for office hours. (See section 2 of chapter 12.)

If your learning style is too closely aligned with that of Dr. Lector or Dr. Dialogue, you will not want to select an instructor whose teaching style hardly accommodates it. That may be highly problematic for you since your propensity to learn will not be supported. Likewise, because of this high degree of alignment with one of these instructors, you also do not want to select an instructor who teaches wholly to your learning style since you will find it very difficult to attain other learning objectives that better lend themselves to other teaching and learning styles.

Once you know what to look for in an instructor's teaching style, your next task is to get this information, which will take time. The more complete your investigation, the more accurate your information and the wiser your decision.

Ways to Obtain Information on an Instructor's Teaching Style

There are a variety of ways to get information on an instructor's teaching style. It is only when you see the same or similar information surface from a variety of sources that you can feel comfortable in describing an instructor's teaching style. Methods of obtaining this information follow.

Visit one or more classes that the instructor is teaching

Ask the instructor for permission to attend one of his or her classes. If you know someone who is taking the course, have that person find out if a guest can be brought to class. Generally, this should not be a problem if space is available. Be aware that some instructors change aspects of their teaching style depending on the content of the course they are teaching or the types of students enrolled. An example would be an upper-division mathematics course versus a lower-division mathematics course. When you visit, don't ignore your general feelings as you observe, but go beyond those to focus on the particulars of the instructor's teaching style. Unless you come to the class with a *checklist of particulars* you want to see in an instructor, you will forget to check for many of them. You can create a checklist from the topics that were discussed in this chapter (e.g., see descriptions of Drs. Lector and Dialogue), and those in section 2 of chapter 12 on instructor responsibilities. If you need more information on the instructor you visit in class, visit more than one class.

Talk to students who are or have been enrolled in the instructor's course

Be careful to interpret students' remarks correctly. Some of them may not be keen enough observers of an instructor's teaching style to give a well-formed opinion, or they may be unfairly biased toward the instructor (e.g., making their shortcomings the shortcomings of their instructors). Talk to students whose opinions and outlook you respect. Also, be aware that the instructor may not be that suitable for some of the students you are interviewing, but may be for you. If you are a very capable mathematics student, you may relish having a challenging instructor who stimulates you to reach your potential. Some capable and non-capable students don't want this challenge, and their comments about the instructor will reflect this. Have a checklist of specific items you want to cover in your discussion, as was suggested when you visit a potential instructor's class. This will help you make a decision based on facts and not emotions. You want to know if the instructor has appropriate expectations, is concerned that you meet them, and can help you meet them by using an appropriately balanced teaching style. The more students you talk to, and the better you know them, the better informed you will be to make an appropriate decision. Interview at least four students.

Be wary of choosing an instructor when you hear *too* many comments such as these:

- *"The instructor is easy."*
- *"I didn't learn anything in this course."*
- *"The tests were easy."*
- *"Almost everyone in the course got a good grade."*
- *"On the examinations you are not asked to write about what things mean, to make connections, or to apply the mathe-*

matics to real world problems that are not carbon copies of what was practiced in the homework."

- *"I know students who are taking courses out of order just to get this instructor again, even though they don't have requisites for these courses."*
- *"Many students miss class a lot but they seem to be doing well on the examinations."*
- *"The wording of the test questions is exactly like the wording of problems in the textbook."*
- *"I know of many students who get good grades from this instructor yet struggle greatly in subsequent courses taught by other instructors."*
- *"The instructor seems to want to please all students."*
- *"There was almost no class discussion."*
- *"There were a lot of students in this course and hardly anyone dropped."*
- *"There were many opportunities to improve my grade on course examinations such as redoing the test outside of class, taking another test, or handing in other work to be graded."*
- *"Most students completed the examinations early."*
- *"There was a lot of important content that was never tested."*
- *"The lectures were always so clearly presented."*
- *"Much of the time class ended early."*
- *"The instructor hardly ever seemed prepared for class."*
- *"There are students in the course who are very confused, yet do well on the examinations."*

Interview the instructor

I had students interview me as a potential instructor. It was always an enjoyable experience for me since I know myself well as an instructor and could articulate this to these students. Feel free to ask the instructor any question that gets at his or her teaching style. The instructor may ask you to clarify what it is you are asking, and help you ask relevant questions. Perhaps the most important information gleaned from the interview is your comfort level with the instructor. How did you relate one-on-one? How was the body language of the instructor? Was your exchange warm, genuine, and meaningful? If the instructor is defensive over your questions or comments, that may be a mark against him or her when it comes to making your final selection.

Read the instructor's syllabus for the course

Ask the instructor for his or her course syllabus, or obtain it from a student who was or is currently enrolled in the course. The information in the syllabus will help you formulate questions for the instructor.

Consult the mathematics department chairperson and other mathematics instructors, especially the instructor designated as the mathematics department advisor

Most likely the information you receive is credible, but it could be biased or based only on rumors. Instructors you ask must uphold professional ethics; however, there are no violations if they respond to specific questions you ask on the instructor's teaching style. Don't expect to hear the response, "This is a bad instructor," but you may hear the response, "This is a good instructor." In some situations, you have to read between the lines to gain the information you need.

Look at student evaluation forms that the office of student government makes available in its office, or are online

You have to proceed with extreme caution when reading student evaluation forms that have been constructed by student government. They are problematic. Not all faculty members participate in this endeavor; hence, the results are automatically biased. It is possible that the forms used are poorly constructed—not assessing the instructor's teaching style in a meaningful way. It is difficult enough for mathematics departments to evaluate the teaching effectiveness of its instructors. I have seen completed student evaluation forms *misinterpreted* by faculty evaluation committees; hence, there is good reason to believe that you can also misinterpret them. Sometimes student evaluation forms, whether constructed by the mathematics department or by student government, give no more information than who has an engaging personality or who gives the highest grades.

Student evaluation forms, in and of themselves, can border on being meaningless. However, *they can be meaningful* if they are well-constructed and are coupled with other means of analyzing an instructor's teaching style, including:

- The instructor's stated teaching philosophy
- An analysis of the instructor's examinations and assignments
- A look at sample examinations graded A through E
- The instructor's distribution of final course grades

With this added information student evaluation forms can be properly interpreted. However, you do not have access to most of this information. Some of the worst instructors in mathematics departments have the best student evaluation forms, and some of the best instructors in mathematics departments have very mixed student evaluation forms. A colleague of mine said this to me: "Tell me if an instructor brushes his teeth and I will tell you what his student evaluation forms are like." Clearly, this is an exaggeration, but an important message lies within: Be wary of selecting an instructor based strictly on his or her student evaluation forms.

Be especially vigilant not to choose an instructor because he or she uses a teaching style that you are accustomed to. This style may not dovetail well with your learning style. Most teaching styles are very similar to that used by Dr. Lector as described in the last section. This style has its strengths, but also its weaknesses, as discussed. You will have difficulty selecting an instructor who deviates much from this style if you believe this is the way mathematics should be taught, you are used to it, or other teaching styles require you to accept more responsibility for your learning and you don't want this responsibility.

In asking students for advice that I could give to readers of this book, this is what three students said on selecting an instructor:

- ❖ *"Get a good teacher."* (Calculus II student)
- ❖ *"Most importantly, select a "reputable" instructor. Just ask fellow students who've been through it."* (Calculus I student)

These students' comments reveal their feelings on the importance of choosing a good instructor, but they do not reveal what they believe are the attributes of such instructors.

The most important decisions you can make in college, save for your decision to work hard and smartly on your studies, are choosing good instructors for your courses. When all is said and done, you have to choose your instructor—if this option is available to you. This decision cannot be left to your classmates, instructors, department chairperson, or friends, although they can help you. Do your homework, make your decisions, live with them, and plan on complementing the teaching styles of the instructors you choose.

Learning Mathematics Through Reading, Writing, Listening, Discussion, and Taking Notes

It may seem unusual to include information on using the basic skills of reading, writing, listening, discussion, and taking notes, to learn mathematics. After all, you most likely have used these skills in your mathematics courses for as long as you have been in school. You may ask, "What more needs to be said about them?" My response is, "More than you realize, and you probably have not used them to the extent you should have." If you can answer these four questions with a resounding, "Yes," then perhaps you can bypass this chapter:

1. Do you know the many valuable educational objectives that can be achieved by an appropriate use of these basic skills?
2. Do you know how to use these skills to attain maximum benefit from them?
3. Do you know the many opportunities that exist in college mathematics courses to use these skills?
4. Do you consistently and appropriately use these skills to learn mathematics?

During my college teaching career, I knew very few students who could answer, "Yes" to all four questions. Many students would not be able to answer "Yes" to any of them. If you are aware of the educational objectives in mathematics that can be accomplished by an appropriate use of these four skills, you will be motivated to find out how to use them and be more apt to use them. Knowledge is power, and knowing the benefits of these skills can empower you to learn mathematics. They are discussed in these sections:

1. Reading Mathematics

The key message of this section is that the most important skill you need to learn mathematics is the ability to read mathematics. Many mathematics educators, including myself, believe that the root cause of many students' difficulties in learning mathematics is their inability, coupled with perhaps a lack of desire, to read mathematics. Your instructor cannot give you the knowledge that you can and must attain from reading mathematics. The pace of college mathematics is too fast for the instructor to give all the necessary details of the subject, and even if that were possible, it would be poor teaching to do so. Possessing the ability to read mathematics makes you an independent and life-long learner, and that ability has to be cultivated by all your mathematics instructors.

It might be true that in high school you did not get into the habit of reading your high school mathematics textbooks. In college you have to learn from reading your mathematics textbooks if you want to grow mathematically, and be more successful in your mathematics courses. Unfortunately, there are college mathematics instructors who are oral textbooks—basically repeating the textbook before their students' very eyes. Regrettably, far too many college students use this criterion to determine the quality of college mathematics instructors. You may get a good grade in such a course, but won't achieve many important learning objectives. For example, you will be impaired in subsequent mathematics courses where the instructors do not do this, to say nothing about the difficulties you may face in your career by not being able to learn new material from your reading. When you graduate from college you do not receive, along with your diploma, an instructor you can place in your hip pocket and pull out whenever you are faced with a mathematics difficulty. You may be able to get assistance from individuals in learning mathematics when on a job, but it will be time consuming and limited. You need to be an independent learner and understanding what you read is a necessary component of that. Mark Twain said, "The man who does not read books has no advantage over the man who can't read them."

Mathematics Reading Materials

Your mathematics textbook will be your major reading source. In almost all mathematics courses, it is your instructor's planning guide. Reading a mathematics textbook means much more than just reading the exercises you have been assigned, or paging back in the textbook from an exercise, looking for procedures that will help you work it. More than anything, your textbook addresses the structure of mathematics (i.e., the relating of ideas), which is critical to understanding mathematics. It is structure you get when you read the material between two consecutive sets of exercises.

Your reading and problem assignments will primarily be out of your textbook. The course syllabus will most likely identify the sections of your textbook that comprise your course, and the order in which they will be covered. Thus, you know well in advance what you have to read in the textbook, and when to read it. Hopefully your textbook is well written, but that doesn't necessarily mean that you can read it with understanding. Information will be given on how to enhance your reading of a mathematics textbook.

Some sections of the textbook may not be as well written as others; hence, alternative textbooks may do a better job clarifying the content of these sections. Similarly, alternative textbooks may do a better job of motivating content, and have more appropriate explanations, illustrations, solved examples, or exercises. A revealing sentence now and then, or a particular example, may make a big difference in your understanding. An alternative textbook may do a better job giving the big picture. Not only might it do a better job than your textbook in developing a specific section, but it may do a better job overall. You may find an alternative textbook that seems to be written for you. That is, you and the authors of this textbook are in sync. Because of the significant benefits you can accrue from reading alternative textbooks, section 2 of chapter 18 is devoted to this topic.

You may be skilled in reading a mathematics textbook, but in case you are not, or want to determine if there are better ways to read it, see chapters 15 and 16. Chapter 15 covers the basics of reading a mathematics textbook. The focus of chapter 16 is on reading specific types of mathematics content, including definitions, statements of theorems, logical arguments (or proofs), graphs, tables, and solved examples. Other course reading materials include your instructor's handouts, solutions manual, laboratory or technology manual, and compact disks.

Instructor Assistance in Reading

One of your instructor's responsibilities is to assist you in your reading, not attempt to replace it with lectures. He or she assists your reading by motivating it, giving an overview of what you will read (i.e., helping you see the big picture), indicating what is optional reading, and pointing out unusually difficult content that you may run across. Your instructor will leave many of the details of the textbook's development to you, including many proofs of theorems. The exercises that your instructor works in class should not be the examples in the textbook, save a few now and then that need further explanation. You need to learn from reading the examples in the textbook, and you miss out on this if the instructor works them in class (thus, not motivating you to read them before they are discussed in class). Furthermore, when the instructor works other exercises in class, you then have more exercises from which you can learn—your instructor's and those in the textbook.

An instructor needs to be familiar with the content in the textbook and how it relates to his or her development of the content. Many instructors operate strictly from their class notes, perhaps written long ago. They may have lost sight of what is in the textbook and that can be problematic for you. To impress upon my students the importance of reading their textbook, I often referred them to a specific place in the textbook to answer a question they asked me on an assigned exercise. I found that many of them had not read the textbook, or did not understand what they had read. If it was the latter, I told them that their questions to me should have been on what they read, not on the exercise. If they had not read the textbook, I told them to read it, and then come back to me with questions if they still needed help.

Reading mathematics is not like reading a novel. You have to move at a much slower rate since the content is dense. In a small space many symbols appear representing many ideas and relationships. To understand what is being said, your reading rate will of necessity be at a snail's pace. When first exposed to new concepts and relationships, whether in your reading or in class, they often seem incomprehensible. However, with more time and further work, you wonder why you were confused in the first place. Improving your ability to read mathematics is a gradual process. Stay with it since the benefits are great. At times you may spend a half hour or more trying to decipher one page and leave it still confused. As you get better at reading, your tempo will pick up. *If you are able to get a good grade in a college mathematics course without reading much of your textbook, you should be concerned about your instructor's expectations.*

Other Reasons for Reading

You accrue many benefits from reading your textbook, in addition to becoming an independent learner. They include the following:

You attain learning objectives associated with discussion. When you read you are carrying on a silent discussion or conversation with yourself and the textbook authors. Hence, some of the learning benefits of discussion are realized (which you will learn about in section 4). To read well you have to think, so reading promotes thinking. Reading well is not a passive activity. You are (1) asking the authors to tell you what they are saying, (2) disagreeing perhaps with the way it is being said, and (3) perhaps telling the authors that you are going to consult another source to attain more information. Your goal is to be in sync with the minds of the authors—to understand their thoughts.

You can learn at your own pace. No one is going to tell you that you have to understand a specific section or chapter of the textbook in the next 50 minutes or so. You can learn it at your pace. This is not true of class lectures. If the pace of a lecture is too fast for you to comprehend, there isn't much you can do about it at the time.

You are better prepared for class lecture. If you read content before class that will be developed in class, you will gain far more when in class. You can focus on what you couldn't understand from your reading by paying closer attention when it is presented in class, and by being prepared to ask questions. You will be able to follow the class presentation more closely because you already have some knowledge of it, and it will reinforce what you already know. Finally, your interest in the class session will be heightened because you know what is coming. It frequently happens that if you are confused at the outset of a lecture, the rest of the lecture is incomprehensible to you. That is less likely to happen if you have read the content of the lecture beforehand. *Reading your textbook before class on content that will be introduced* *in class is one of the more important things you can do to increase your learning.*

It helps you write mathematics. There is little doubt that reading mathematics can help you write mathematics or write about mathematics. When you read you are exposed to good writing that includes good organization, appropriate use of symbols, and good grammar. Why wouldn't some of this influence carry over to your writing, especially if you are cognizant of these things while reading?

Reading the Textbook Cannot Replace the Class Session

It is important to be in class even if you understand what you read in the textbook. More strongly stated, you should attend every class, regardless of how poorly you think your instructor uses class time. *All* instructors say some valuable things in class, and you never know when this will happen. These things are often not recorded in a classmate's notes that you might consult. Included among the many unique things your instructor may do in class are these: lecturing on content that does not appear in the textbook; wording a definition that is different, although equivalent, to what is in the textbook; giving a summary of the lecture, including how it relates to what came before and what comes later; cautioning you about errors that students typically make; and making remarks that will assist you in your reading. Miss class and you miss these things, among others. In my nine years of being a college mathematics student, I missed only a handful of mathematics classes, and I only missed them because I was too sick to attend.

Read the Textbook's Development Before Working the Problem Assignment

Does this describe you as a mathematics student: When beginning your mathematics study, you start with the assigned exercises, and then search through the appropriate textbook section to find information to work them, and that basically is the extent of your reading of the textbook's development? That is not a good approach to use; it is putting the cart before the horse. You will miss important ideas, the flow of ideas, and not be adequately prepared to work the exercises. Reading mathematics gives you a foundation to work exercises. When you read mathematics you are learning *about* mathematics, and you are assigned exercises to test your understanding of the mathematics needed to work them. To understand mathematics—to learn mathematics—you have to *do* mathematics. You find out, when working problems, that you do not understand as you should. As a college student, *you begin mathematics study with your reading*, not with the assigned exercises. In the work world, the process is typically reversed—you begin with a problem that needs to be solved and look for the mathematics that will solve it. The focus for a college student is to learn mathematics, and that is a two-step process that takes place *in this order*: Read mathematics and then do mathematics by working the exercises.

I have argued that learning to read mathematics is a must for everyone, but the students who can benefit the most are those who struggle the most with mathematics. The pace of their instructor's development of the content is too fast for them, yet they do not compensate for this by reading in the textbook, which can be done at their pace. They may struggle more than others as they begin to focus on reading, but they will improve in this area, and eventually their struggles with mathematics will lessen.

What Students Say about Reading Mathematics

What follows are annotated comments on reading mathematics, taken from students' evaluations of their calculus instructors. The comments represent their responses to the question, "What did you like or dislike about your instructor."

❖ *"I disliked the method of teaching directly out of the book. I would have felt more comfortable if he would have had some outside material (applications, different methods of solving problems) that would have aided our reading of the text."* (Calculus III student)

I could not have said this better. This instructor needs to make some changes; his or her students are being shortchanged.

❖ *"He didn't explain things well. Many times I had to learn something on my own."* (Calculus I student)

Perhaps this student is implying that the instructor did not dot every "i" and cross every "t." If so, then this instructor should be congratulated. I would be concerned if this student did not have to learn some things on his or her own.

❖ *"Some of the explanations went fast and made comprehen-sion difficult. However, reading the section before discus-sion alleviated this problem."* (Calculus I student)

College mathematics classes are fast-paced. This student knew how to compensate for that.

❖ *"We were not expected, in other mathematics classes, to do a whole lot of reading on our own."* (Calculus I student)

This lack of expectation by previous instructors doesn't surprise me. Unfortunately, this makes it more difficult for this student to succeed in his or her current course, where reading the textbook was expected. Nonetheless, the stu-dent should be thankful to finally have an instructor with this expectation.

The following comments on reading mathematics are from stu-dents who told me what they would say to students reading this book:

❖ *"Perhaps read in another textbook. Regardless of how good the instructor and lectures are, a student depends on the text for learning. I got through this course with the aid of a library book."* (Calculus I student)

❖ *"Some texts describe or analyze a certain concept better than others. Don't limit yourself to one book. You may spend hours trying to figure out what one book is trying to tell you, when you may understand in just ten minutes the same concept discussed in another book."* (Calculus III stu-dent)

❖ *"Read the textbook and try to get as much as possible out of it."* (Calculus II student)

❖ *"If time was available, I found reading the text before the lecture on that topic was helpful."* (Calculus II student)

❖ *"Read, read, read everything relating to the lectures."* (Cal-culus II student retaking the course)

2. Writing about Mathematics

This section is on writing *about* mathematics, which is using writing to learn mathematics. Writing about mathematics focuses on giving explanations using words and sentences, and minimizes the use of symbols. Your writing process becomes a vital part of your thought process. The subject matter may be any aspect of mathematics. In contrast, writing mathematics has as its focus the use of symbols and minimizes the number of words. Examples are now given that help point out the differences between writing about mathematics and writing mathematics, although at times the distinctions blur.

Example Exercises on Writing *about* Mathematics

1. How are the zeros of the equation $x^2 - 7x + 10 = 0$ related to the graph of $y = x^2 - 7x + 10$?

   Solution. The zeros of the equation $x^2 - 7x + 10 = 0$ are the x-intercepts of the graph of the equation $y = x^2 - 7x + 10$.

2. Describe how the graph of a function $y = f(x)$ can be used to find the domain and range of the function.

   Solution. To find the domain, project the points of the graph onto the x-axis. The x-values of these points constitute the domain. To find the range, project the points of the graph onto the y-axis. The y-values of these points constitute the range.

3. Argue that 2 is the only even prime number.

   Solution. By definition, a prime number has exactly two divisors. Note that 2 is prime since its only divisors are 1 and 2. By definition, a number is even if and only if it is divisible by 2. Any even number greater than 2 has at least three divisors, namely, 1, 2, and the number itself. Hence, it is not prime. Therefore 2 is the only even prime number.

Example Exercises on Writing Mathematics

1. Solve the equation $x^2 + 4x - 9 = 0$ by completing the square.

   Solution. $x^2 + 4x - 9 = 0 \Rightarrow x^2 + 4x = 9 \Rightarrow x^2 + 4x + 4 = 9 + 4 \Rightarrow (x + 2)^2 = 13 \Rightarrow x + 2 = \pm\sqrt{13} \Rightarrow x = -2 \pm \sqrt{13}$.

2. Rewrite $\sin x + \tan^2 x$ in terms of the sine function.

   Solution. $\sin x + \tan^2 x = \sin x + \dfrac{\sin^2 x}{\cos^2 x} = \sin x + \dfrac{\sin^2 x}{1 - \sin^2 x}$.

Early in my teaching career, the phrase, "writing about mathematics" was not in the mathematics educators' lexicon. The examinations, homework assignments, and projects I assigned consisted almost exclusively of writing mathematics. Once I became aware of this method of learning, I used it extensively in all my mathematics courses. The result was increased understanding of mathematics by my students, but this did not come without them paying a price: The course was more difficult for them than it would have been without this requirement. Not only did they have to do more preparation, but writing about mathematics required them to focus more on understanding mathematics at a higher cognitive level.

How wonderful it would be if more mathematics instructors required their students to write about mathematics. If your instructor does not require this, you are still urged to do it. It will be more problematic for you without his or her support, but you can do it. Once you realize the benefit you will be hard pressed not to do it. Likewise, if instructors knew how writing about mathematics benefits their students, they would be hard pressed not to require it. The reasons to write about mathematics are many. *Nothing I did in my teaching has equipped my students to understand mathematics as much as requiring them to write about mathematics.*

Benefits of Writing about Mathematics

The benefits of writing about mathematics include these:

It helps the thought process. We all know how difficult it is to write. We think we know our thoughts until we put them on paper. In our efforts to do this, we discover that our thoughts are not as well formed as we believed; hence, we reflect more on them, which improves them.

It helps gather, formulate, organize, relate, analyze, integrate, internalize, and evaluate thoughts. We write to communicate our thoughts to others and to ourselves. In your efforts to be understood in your written communications, you have no choice but to gather, formulate, organize, relate, integrate, internalize, and evaluate the thoughts you acquire from your class sessions, readings, and discussions. Active involvement in discussion can promote some of these intellectual skills, but there is something unique in writing about mathematics to achieve them.

It assists in attaining an overview or summary of the content. An important skill is to be able to pick out the important ideas from a body of content, how they relate to each other, how they relate to ideas that came earlier, and what their uses are. You have to prepare to write about mathematics, and in the preparation you acquire an overview of the content. Once again, if your instructor does not expect this of you, then you have to expect it of yourself.

It exposes confusion and misunderstanding. You become aware of this as you write, or when getting feedback from others on what you have written.

It improves writing skills. What a wonderful by-product this is. In addition to learning and communicating mathematics better, you become a better writer. Tying your thoughts together using mathematics vocabulary, common words, and the prepositions and conjunctions—"to," "of," "if," "or," "and," "but," is not a simple task. Writing with clarity has to be a goal, and in order to do that your thoughts have to be clear, and you need to be skilled in word usage. All mathematics instructors have an obligation to help their students become better writers, and requiring them to write about mathematics is an excellent way to accomplish this.

It promotes intellectual development. If writing helps clarify your thoughts, it stands to reason that, with sustained and appropriate writing exercises, your intellectual development will advance.

It promotes development of higher order thought processes. Achieving this benefit depends on the type of writing exercises you do. Exercises on mathematics can be constructed whose solutions require the use of a variety of higher order thinking skills, including application, analysis, synthesis, and evaluation.

It enhances your attentiveness in class, helps direct your discussions, and motivates you to critically read your textbook and other related mathematics materials. If you are not asked to write about mathematics you may have a tendency to tune out your instructor in class when he or she is lecturing on content that requires higher order thinking skills. You may also avoid carefully reading the theoretical parts of your textbook, as well as avoid engaging in meaningful conversations about mathematics.

It prepares you for the world of work and increases your employment opportunities. Good communication skills, whether general or more specific—such as communicating highly technical information to those less knowledgeable about the subject, are of utmost importance in our increasingly complex world. It could be the deciding factor in whether or not you get the job you are seeking. This benefit alone should motivate you to focus on writing about mathematics in your college mathematics courses.

Critiquing Your Writing about Mathematics Reveals What You Don't Know about Mathematics

When an instructor looks at the initial efforts of students who are writing *about* mathematics, he or she realizes, *in a big way*, how little they understand mathematics, or how challenging it is for them to communicate what they do understand. It is difficult to assess the level at which you are understanding mathematics by *only* observing how you write mathematics (as opposed to writing *about* mathematics). If all that your instructor sees from you is computational, algorithmic, or memorized knowledge (i.e., low-level knowledge, not requiring much more than mimicking what your instructor might have done), then it is difficult for him or her to know what you understand about the mathematics you are manipulating. The more symbols your instructor sees in your responses, the less he or she knows about your understanding of what is behind these symbols and how they are related. You need to be asked questions about the concepts and principles embedded in your computations. The point is this: you are in your mathematics course to learn mathematics, and you are helped greatly in this endeavor by making sure you write about mathematics, and have this writing critiqued.

A goal of any mathematics instructor should be to employ teaching methodologies that do the best job of increasing students' understanding and communication of mathematics. Knowing the benefits of writing about mathematics should be all that is needed to convince instructors to use this technique with their students. If your instructor does not use this method of instruction, perhaps you and some of your classmates can *gently encourage him or her to read this section and to employ this teaching technique.* Instructor support for your efforts to write about mathematics can be a huge boost to your efforts.

Include Writing about Mathematics Exercises in Study Guides for Examinations

Many mathematics textbooks do not include many writing about mathematics exercises. A key means I used to give assignments on writing about mathematics was a study guide I constructed for impending examinations. Students' written responses to this type of exercise in the guide afforded them an excellent opportunity to reap the benefits that come from writing about mathematics. Whether or not any of these exercises appeared on the examinations was beside the point. Students frequently have difficulty responding to these exercises, which motivates them to ask their classmates or me about them, resolve their issues, and then formulate meaningful responses. Hence, in addition to the benefits derived from their writing, they derived the benefits of discussion. My students' completed examinations changed greatly after I incorporated writing about mathematics in my courses. From a distance, they looked like completed essay examinations, even though the essays they wrote were short and the examination also contained exercises on writing mathematics. *You are encouraged to construct, and then work, study guides for your examinations that contain "writing about mathematics" exercises.* You are supported in writing these exercises later on in the section.

Opportunities to Write about Mathematics

Assuming that your instructor does not require you to write about mathematics, or does not require it to the extent he or she should, here are opportunities for you to engage in this activity.

Respond to writing about mathematics exercises that are in your textbook or in alternative textbooks. Some textbooks have writing about mathematics exercises that are dispersed among the exercises for a specific section or chapter review. Look for them and respond to them, whether they are assigned or not. Get feedback on the quality of your response, if needed, by consulting your instructor, classmates, mathematics department tutor, etc.

Ask at least one classmate to create some of these exercises for you. This can be done anytime, including when you are studying for an examination. Repay the favor by creating some for your classmate, and evaluate each other's responses. Writing a response for a classmate to read requires you to include more in your writing since your classmate is not as expert as your instructor, who can easily read between the lines. Your writing has to have fewer implied assumptions and more details, which can enhance your understanding and improve your writing skill.

Create exercises for yourself, respond to them, and then get feedback. You can do this as you progress through a unit. The very act of writing these exercises is beneficial since you have to discern what content is important to write on.

Construct writing exercises, often in the form of questions, based on the following:

Class notes. A good technique is to write your questions on one side of a page and the class notes on the other side of the page, with a vertical line dividing them.

Textbook reading. Examples of questions to create when you read your textbook are discussed in section 2 of chapter 15.

Review for an examination. Ask questions on what you need to learn (i.e., make a "writing about mathematics" study guide for the examination).

Examination results. After reflecting on your examination performance, write questions that will help you better understand the content of the examination.

Weekly review. Reflect on the week's work and ask yourself questions that relate to it. The generic types of questions you can use include these: How is . . . related to . . . ? What is . . . used for? Where did . . . come from? How do I know when to use . . . ? How do I solve a . . . type of problem?

Problem assignments. Write exercises on what you have learned, or on what you need to learn from completing a problem assignment.

Feelings about course-related matters. Exposing and dealing with your course-related feelings is often accomplished through journaling. You may want to write on questions such as these: What am I enjoying in mathematics? What am I not enjoying in mathematics? What do I fear in this course? What am I, my classmates, or my instructor doing or not doing that troubles me? Expressing your feelings in words helps relieve anxiety, and is a catalyst for finding ways to handle some of your concerns, many of which will involve specific aspects of mathematics. Overtly exposing your feelings to yourself allows learning to occur.

The opportunities to create meaningful "writing about mathematics" exercises are many. Getting feedback on your writing is important, especially when you are new at using this learning technique. Eventually you will need little feedback beyond your notes and textbook. If you have difficulty writing adequate responses, you will need to discern if the problem is the result of poor writing skills, or your difficulty understanding mathematics content. If it is your writing skills, you might want to consult someone in your campus writing center.

Types of Exercises for Writing about Mathematics

Types of exercises for writing about mathematics are listed. All that is needed as a response to some of the exercises is a few sentences. They are not only here as examples of what you can create, but also for you to work (if you have the mathematics background). You don't need to write much to gain many of the benefits of writing about mathematics. The content of the exercises is chosen from calculus, precalculus, and mathematics for preservice elementary teachers. When you create exercises, don't be restricted by the *types* appearing here.

Justification

1. Convince me that if I toss a coin 20 times and get heads each time, I am not *due* to get tails on the next toss.
2. Argue that there are an infinite number of fractions between 1/2 and 2/3. Now, argue that there is no fraction that is greater than 1/2 and closest to 1/2.
3. Explain why the slope of a line can be determined by using the coordinates of *any* two points on the line.
4. Argue that if a graph is symmetric to at least two of the *x*-axis, *y*-axis, and origin, then it is symmetric to the third.
5. Convince me that if a function is not one-to-one, it cannot have an inverse function.

Describe a process, procedure, or algorithm

6. Describe a process for determining the maximum value of a function defined on a closed interval.
7. State the procedure for finding the inverse of a one-to-one function.

Characterization

8. Characterize the fractions that do not have finite decimal representations.
9. Characterize the functions that have an inverse function.

Identify the uses or applications of a concept or theorem

10. What major uses are there in this chapter for The Mean-Value Theorem?

11. How does knowing the symmetry of the graph of an equation help to graph it?

12. What uses are there for linearization of a function?

13. Give several applications of $\dfrac{d^2 f}{dx^2}$, detailing for each one how it can be applied.

Good or bad reasoning

14. Gary and Hugh are working with a specific function. Gary says that the maximum value of the function occurs for the independent variable having a value of 3 and Hugh says that it occurs for the independent variable having a value of 4. Can they both be right? Explain. Can the maximum value of a function occur at every point in the domain of the function? Explain.

15. Is the solution of the following problem incorrect? If incorrect, why is that so?
    Problem. Solve for $\dfrac{x+4}{x-1} > 0$, $x \neq 1$.   *Solution.* Multiplying both sides of the inequality by $x - 1$, gives $x + 4 > 0$, which implies $x > -4$.

16. An eighth grader says, "If I multiply 7 by itself enough times, I will get a number that is divisible by 11." Do you agree with this? Explain.

17. Rosie says that $|-x| = x$. Do you agree with her? Explain.

18. In simplifying $\sqrt{\dfrac{4}{9}}$, Tom wrote $\pm\dfrac{2}{3}$, and Karen wrote $\dfrac{2}{3}$. Who is correct and why? In solving $y^2 = \dfrac{4}{9}$, Tom wrote $\dfrac{2}{3}$ and Karen wrote $\pm\dfrac{2}{3}$. Who is correct and why?

19. In graphing the equation $y = \dfrac{4x^2 + 7x - 2}{x + 2}$, Tom graphed $y = 4x - 1$ (which is a simplification of the first equation). Is Tom's graph the graph of the initial equation? Explain.

20. Mary said that $c(x) = 3x + 4$ and $d(x) = \dfrac{1}{3x + 4}$ are inverse functions. Is Mary confused? Explain.

Identify graphs of functions

21. Suppose the graphs of a function and its first and second derivatives are given, but not identified. Explain how to identify the graph that goes with each function.

22. Suppose the graphs of functions $f(x)$, $f(x + 3)$, and $2f(x + 3)$ are given. Explain how you can determine the graph that goes with each function.

Compare and contrast

23. Compare and contrast the first and second derivative test for determining relative maximum and minimum values.
24. Compare and contrast the Trapezoidal Rule and Simpson's Rule for finding the area under a curve.
25. Compare and contrast the graphs of sin x and cos x.
26. Compare and contrast definite and indefinite integrals.

Summarize a topic

27. In a half page or so, summarize in story form all that you learned about the inverse of a function.
28. In a page or less, give in story form the highlights of what you learned about decimals.
29. In a page or less, summarize what you learned this week about definite integrals.

Discuss concepts

30. What is the meaning of equivalent equations? (No examples allowed.)
31. What is meant by an extraneous root of an equation? Give at least two ways in which they arise in solving an equation. (No examples allowed.)
32. What is meant by the graph of an equation? (No examples allowed.)
33. Discuss various discontinuities a function may have.
34. What is implicit differentiation and what use is there for it?

Develop concepts

35. Describe activities you would give fourth graders that would help them understand the concept of the fraction number 1/3.
36. Using discrete sets, illustrate and describe what it means to say that $\dfrac{1}{2} < \dfrac{3}{5}$.

37. Show how approximating the length of a curve $y = f(x)$ by finding lengths of various polygonal paths, leads to the expression $\int_a^b \sqrt{1 + (\frac{dy}{dx})^2}\, dx$ for the length of the curve.

Relate concepts

38. How are average and instantaneous rates of change related to each other?

39. How are differentiability and continuity related to each other?

40. Discuss the relationships between fractions and these decimals: finite, infinite repeating, and infinite non-repeating.

41. How do area and definite integrals relate?

Give the geometric interpretation of a concept or theorem

42. Given the graph of a specific antiderivative of a function f, draw a picture showing the graphs of several more antiderivatives of f, and explain what you did to get the graphs.

43. Using words and an illustration, give the geometric interpretation of the partial derivative of $f(x, y)$ with respect to x at (a, b).

44. Give the geometrical interpretation of The Mean-Value Theorem.

45. With respect to tangent lines, give the geometrical interpretation of the following: (a) $f'(x) > 0$ on an interval I. (b) $f''(x) > 0$ on an interval I.

46. Give geometrical interpretations of (a) $h(x) = h(x + 5)$, and (b) $g(-x) = -g(x)$.

Help avoid the incorrect use of formulas or procedures

47. Some of your classmates mix up the formula for the circumference of a circle with the formula for the area of a circle. Both formulas use the numbers π, 2, and r (the radius). What meaningful thing can you say to these students so that they will not continue to make this mistake?

48. Some of your classmates often think that $\sqrt{a+b} = \sqrt{a} + \sqrt{b}$. What would you have a classmate do so that he or she will not continue to make this mistake?

49. Carla likes to mechanically memorize rules for performing certain actions, rather than memorize with understanding (i.e., learn why the rule works). Consequently, when her memory partially fails her, she has no way to determine which is the correct rule to use. For example, she is not sure whether, in graphing $y = h(x + 3)$, she should translate the graph of $y = h(x)$ three units to the left or three units to the right. What meaningful advice can you give her so she can determine with understanding which one to use?

Describe a process for performing a specific procedure or algorithm

50. In a brief paragraph, write a story of the basic steps in solving an applied max-min problem.

51. Before irrational numbers were invented, it was thought that every line segment had rational number length. Accepting the fact that the square root of 2 is an irrational number, explain how to construct a line segment with this length, using only line segments with rational number lengths.

52. Given a one-to-one function, describe the procedure for finding its inverse function.

53. Describe a process for graphing a rational function.

54. Describe a method for solving related rates problems.

Describe how to find something, only using words (no symbols allowed)

55. State in words how to find the derivative of the quotient of two functions (no symbols allowed).

56. Given a function, how would you explain to a classmate how to evaluate the function at a specific number (no symbols allowed)?

Read, write, and interpret symbolic form

57. What is the meaning of $\int f(x)dx$?

58. Informally speaking, what is your understanding of $\lim_{x \to \infty} f(x) = L$?

Questions to ask yourself at the end of a class period

59. What was the main point of today's class?
60. What is the main unanswered question I leave class with today?

Questions to ask yourself after finishing an assignment

61. Which concepts, theorems, exercises, or problems gave me the most trouble, and why? What do I intend to do about them?
62. What do I need to change about the way I go about learning mathematics?

In some mathematics textbooks, review questions appear in the chapter reviews, and these are often "writing about mathematics" exercises. Look for them.

3. Listening in Mathematics

Do you believe that you are listening if you can hear someone? Listening is much more than hearing someone's spoken words. You may hear sounds, but are you digesting their meaning? Do you need to become a better listener? My wife frequently tells me that I could be a better listener, and I agree with her. Listening is one of the more difficult educational processes to learn, but it is necessary for learning. We all listen to a degree and perhaps we take more solace in that than we should. Some people say that not listening well is equivalent to not listening at all. It can even be worse than that. When you don't listen well (i.e., when you half-listen), there is the potential to misinterpret what you hear.

You listen to learn, and it may be the fastest, but not necessarily the best way to learn. Most mathematics students have to gain most of their knowledge from reading, but *listening is indispensable to learning*. Opportunities abound to listen to your instructor and classmates, in- and out-of-class. Before you can listen to learn, you have to *learn to listen*, and advice on learning to listen is the focus of this section.

Listening Has a Complexity of Its Own

When you listen, another person is involved. In most listening situations in college, your instructor, classmate, or another helper is physically present when you are listening to them. This is not true when you read or write. Hence, you have to pay attention to more than just the words that are being spoken. You have to be aware of other aspects of the speaker as you listen. This demands that you listen, not only with your ears, but also with your eyes and mind. You have to watch the speaker and put yourself into his or her mind. The speaker's body language (i.e., the non-verbal cues) is revealing and you have to pay attention to it. It includes the waving of arms, facial expressions, rate of pacing, or body deportment. For example, the relative importance of a specific theorem or the difficulty understanding a particular concept can be conveyed by the speaker's body language, words spoken, or voice inflection.

Before you read suggestions for becoming a better listener, pause for a moment to rate the quality of your listening: Is it poor? Is it good? Is it great? What is your rating based on? If you had to give four suggestions to a classmate on how to be a better listener, what would they be?

Suggestions for Becoming a Better Listener

There is a body of knowledge on improving listening skills, including the following suggestions:

Have the mindset to listen. You have to approach a lecture or presentation with the mindset that you will listen attentively because the speaker has valuable information that you need to learn. Instructors have to carefully select what they talk about in class because of limited class time; that alone should give special status to what they say. Whenever you find yourself drifting away during a lecture, remind yourself to listen. Tell yourself to focus on the speaker and what he or she is saying because of its importance to you, and then do it.

Read on the content of the lecture before class. You listen better if you have some familiarity with the content being presented. Confusion can cause you to tune out your instructor. Thus, reading content in your textbook that will be developed in class helps you stay focused on the class presentation. You will anticipate your instructor's next remarks, and look forward to clarification of what you didn't understand in your reading.

Take notes as you listen. There are several important reasons to take notes as you listen, whether or not your instructor is writing down what is being said:

- **It helps you focus on what is being said.** Knowing that you will be taking notes on what you hear motivates you to focus on what is being said. You know that you can't write down what you don't hear. Taking notes trains you to listen and is probably the best vehicle for doing this. Some people say that note taking gets in the way of listening. That is, if you focus on writing notes you won't hear what is said as you write. This sounds plausible, but it is only true if you close your mind to what is being said as you are writing. The time you spend writing good notes should be minimal compared to the time spent listening. To write less and listen more, you need to train yourself to separate the important from the unimportant and only record the important.

- **It helps structure what you hear**. You write the important aspects of what is being said, and this is the content that gives structure and organization to your instructor's words. As you hear, and then write, glancing down at your notes quickly displays the structure that has been developed up to that point. This will help you insert additional remarks. Without taking notes, this structure can be difficult to detect, or is quickly forgotten. Your instructor will make important statements that do not appear elsewhere.

- **There are important categories of statements made by instructors, often prefaced with often-used phrases that many students don't write down, but should. They include these:**

  - ◆ **Relating the content at hand to past and future content.** ("This comes from . . . ," "This will be used or applied to . . . ," or, "The key ideas used in this development are . . . ")

  - ◆ **Indicating the relative importance of content.** ("It is important that you know this: . . . ," or, "Pay special attention to what I am about to say . . .")

  - ◆ **Cautioning you on how to work with the content, including pitfalls to avoid.** ("A common error made is...," "Make it a point not to . . . ," or, "Many students incorrectly apply . . .")

  - ◆ **Indicating what you may be expected to do with the content on assignments or examinations.** ("If you are given a problem like this on an examination be sure to . . . ," or, "I don't want to see you do . . . on an examination.")

Many of these statements convey your instructor's content bias, which will show up on assignments and examinations. Instructors make these statements but typically do not write them. What a grievous mistake it is to record only what your instructor writes.

Minimize distractions. In your efforts to listen, try to minimize distractions. Since listening involves the ears, eyes, and mind, a distraction interferes with one or more of these. What you should hear, see, or think is compromised by other noises, visual images, or thoughts. Here are ways to minimize distractions:

- **Stay focused when energy levels are low.** It is not easy to listen to your instructor for an extended period of time. This is especially true if he or she is not an effective speaker, or if the content is difficult or uninteresting. Listening attentively consumes energy—it is tiring and that is a distraction. It takes extra effort to concentrate, but it can be done.

- **Take care of your physical condition**. Hunger pains are a distraction. If you are not in a position to eat when you are hungry, your ability to listen is compromised. Eating a quick snack to tide you over until class is over is wise. Similarly, you are fighting a losing battle if you are fighting sleep when trying to listen. Get a sufficient amount of sleep each night, and if that is not possible, take a catnap before class.

- **Sit in the front of the room or as near to the front as you can.** If you don't sit in the front of the room, you are more likely to hear distracting conversations or snide remarks of classmates sitting close to you and out of earshot of the instructor. You will also notice movements of those who sit in front of you, and your instructor's remarks will be more difficult to hear. It is easy to hear other students' questions and responses to questions if they sit behind you, and that is beneficial. If for some reason you can't sit in front, sit as near to the front and to the center of the room as you can. Stay away from the door to the hallway because of potential hallway distractions.

Sitting in front also allows you to see the instructor better and make eye contact. Eye contact is important since instructors often look at students in the front row to discern if they are being understood. When the instructor meets your eyes you can indicate your understanding with a "yes" or "no" movement of your head. Eye contact with your instructor will help you stay focused on him or her. Sitting in front also makes it easier for you to see your instructor's writing. It is then easier to take notes, which makes it easier to listen.

- **Don't daydream or worry about your personal problems**. Once again, you need to have the mindset to avoid daydreaming or worrying. Make this part of your preparation to listen. Set a time before class to worry about a personal problem, after class. Knowing that you will address it at a later time can help you avoid worrying about it when you are trying to listen. If you start to worry about it in class, remind yourself that this is not the time. This simple technique does work. An added bonus is that you may discover that postponing your worrying makes it less problematic later.

- **Ignore idiosyncrasies of your instructor**. A favorite pastime of some college students is to talk about the idiosyncrasies of their instructors. Your instructor may display some annoying habits as he or she is speaking. It may be a facial tic, speech problem, unkempt appearance, voice inflections, nervousness, accent, etc. All of this will be distracting unless you focus on the content rather than the delivery.

Pay attention to key words or phrases that place into context what is to follow. You will want to write words or phrases in your notes that indicate content that is to come. They include these: "This follows from . . . ," "This is used for . . . ," "There is a problem with . . . ," "Be careful . . . ," "For instance . . . ," "Don't forget . . . ," "Reasons why. . . ," "This is related . . . ," "Finally . . . ," "There is a better way . . . ," "In summary . . . ," "In the same way . . ."

Listen actively and critically. There are many moments during a presentation when you have time to mentally contribute to what is being said. Your mind travels much faster than your instructor can speak so you will have time to fit in your thoughts. You can:

- Question why something is true
- Speculate on what is coming next

- Relate what is said to past knowledge
- Paraphrase what has been said
- Review or reflect on what has been said up to this point
- Wonder about the importance of what is said

It is a balancing act for you have to relate your thoughts to those of your instructor. There is time to do this and this activity will make you a much more effective listener.

Listen to your classmates. In most mathematics classes students ask instructors questions, and vice versa. These questions and responses to them are important. Listen carefully to your classmates as well as your instructor. Don't pass up an opportunity to ask or respond to a question. This too will make you a better listener.

Listen at crucial times. You should be a keen listener throughout a presentation or discussion. However, it is especially important to stay focused at the beginning and end of a presentation. If you don't listen at the outset you can be confused for the whole presentation. An instructor will often provide an overview of the whole presentation at the beginning—topics of the discussion are introduced, and requisite knowledge is briefly reviewed. Just imagine walking into class after all of this has been done. How much would you understand from that point onward? There isn't much difference between missing class, and being there—but listening poorly.

Valuable class time also takes place near the end of a presentation. It is then that important conclusions are revealed, as well as an overall picture of what was presented, why it was presented, and how it relates to future content. An added bonus to listening well during the early part of the presentation is that you will be more apt to listen well to the remaining portion.

Avoid bad habits of poor listeners. Perhaps the most destructive habit that interferes with listening is to focus too early on what you want to say to the instructor before the instructor has finished his or her thoughts. We all do this to varying degrees. We only hear part of what the speaker says and ignore the remainder because we are in a hurry to give an opinion or ask a question. You have to sense when it is a good time to formulate what you want to say, and when to say it. Speakers have continuity to their presentations and there are good and bad times to interrupt. It is better to err on the side of waiting too long as opposed to not waiting long enough. And don't just speak to hear yourself talk.

Have something meaningful to say. Do you think you learn more from listening than from talking?

Practice listening skills. Anyone can improve his or her listening skills by practicing. There are several ways to practice. One way is to ask a classmate to work with you on this activity: Ask him or her to speak on any subject. It may be on what they did yesterday, their plans for the weekend, people they know, or their last mathematics examination. Your role as listener is to paraphrase what was said with understanding. You have to decide when to interrupt the speaker to paraphrase what was said up to that point, but don't make it too early or too late. Paraphrasing requires you to listen carefully. When you finish have your partner evaluate the completeness and accuracy of your paraphrasing. Then reverse roles so your partner can also learn from the experience. Practice this activity until you are satisfied with your ability to listen.

Another way to practice listening is to listen to a taped presentation and take brief notes. Your library should have taped presentations. Look at the notes you took as you were listening and review in your mind what you heard. Then replay the tape to determine what you missed, noting how you need to improve your listening skills and note taking. Do this for a variety of presentations until you are satisfied.

Excellent language translators have excellent listening skills. They may only need to jot down a word or phrase as a reminder of certain content being presented. I am always impressed by their translations and critical listening skills. They cannot translate well without understanding what was said. Your role as a listener is to be like that of a good translator.

In summary, you listen to learn, and to be a good listener you have to work to improve. The benefits of listening well are summed up by Wilson Mizner: "A good listener is not only popular everywhere, but after a while he gets to know something."

4. Discussing Mathematics

If you think mathematics is a spectator sport, consisting mainly of watching and listening to your instructor, then your actions in a mathematics course will reflect this. You will be a passive learner rather than an active one. Learning mathematics has to be an active pursuit, and engaging in discussion is a key ingredient. It is interesting to note the number of students who think that disciplines like sociology and psychology must engage learners in discussion, yet do not think this for mathematics and science. This makes no sense whatsoever when you consider that, in mathematics, you have a myriad of complex ideas, complex relationships among these ideas, a variety of problem solving skills that need to be learned and employed, and a new language to be learned. The new language includes the use and understanding of symbolic form. Discussion is essential to accomplish this. It doesn't play much of a role if learning mathematics is incorrectly viewed as nothing more than performing computations and memorizing facts.

If you are an introvert, you will not gravitate naturally toward discussing mathematics. On the other hand, if you are an extrovert, you are attracted to discussing mathematics with others. Regardless of your learning preferences, learning mathematics well means that you have to include an appropriate amount of discussion in your mathematics study. It may be easy or difficult for you to do this, but it is necessary for more comprehensive learning to take place. For students who struggle with mathematics, it is absolutely indispensable to surviving their mathematics course.

Learning Objectives Accomplished by Discussing Mathematics

The learning objectives accomplished through discussion or conversation in mathematics are many and are vitally important. Some of them can be achieved through other means (e.g., lecture), but often not as well. Your learning of mathematics suffers if discussion is stifled, including your ability to verbally communicate mathematics to others. Research shows that discussion nurtures most important learning objectives better *than any other means of learning*. An ample amount of discussion, consistently practiced in various settings that are available, should be the rule rather than the exception in your mathematics study.

Learning objectives accomplished through discussing mathematics include:

Enhanced understanding and application of concepts and generalizations. Discussion is an excellent means to promote more complex thinking and the ability to apply knowledge. It is often difficult to know that you understand something unless you present your thoughts to others for their evaluation. The exchange of comments and questions indicative of a discussion will help clarify your thoughts and understanding. We know the value of the give-and-take of verbal feedback between two people on issues of everyday life. It increases our understanding and helps us lead better lives. Why wouldn't discussion of mathematics be just as valuable?

Clarification of thinking and promotion of insight. Your thoughts can be clarified by rejecting, revising, rephrasing, or reinterpreting them based upon reactions from your classmates. There may be differences of opinion, and you will have to differentiate among them and make a choice. These opinions may pertain to the interpretation or application of a concept or generalization, as well as to the techniques or approaches to use in justifying principles or theorems. When asked to give your opinion, you are forced to evaluate it in the broader perspective of the discussion that has already taken place; hence, in response to this situation, new insights are acquired. How often we have seen someone pause in the middle of a sentence and change his thinking because of a new discovery. Your discussion partner is analyzing and evaluating your thinking strategies, and you are doing the same for your partner.

Promotion of the development of problem solving skills. It is clear that the ability to problem solve is cultivated though discussion. Watching your instructor solve problems is meaningful, but that will not take you very far unless the problems you are asked to solve are very much like those. What you have to be able to do is solve problems that you have not seen before. This is a horse of a different color. In conversing with other students you will see that there are many different ways to approach and solve a problem, and your attempts to work a specific problem will be analyzed to see if they work. If they do not work, you have to revise them until you are able to solve the problem.

Increased ability to communicate, and more specifically to communicate mathematics. You learn to speak coherently through practice. If you are not understood, your listeners will let you know; hence, you try again using other thoughts and formulations of words. Also, thinking clearly is a requisite for speaking clearly, and discussion clarifies thinking. With practice, your skill to participate in a discussion will be enhanced. The wonderful thing about this is that it is self-perpetuating: The better communicator you become, the more you want to do it.

Increased confidence through helping others. You will take pride when your classmates understand something because you helped them. This will boost your morale and increase your confidence. An additional benefit of teaching others is you gain a better understanding of the content. Approach your study of content as though you have to teach it. Ask any teachers about the value of teaching and they will tell you that they never learned as much about a topic until they had to teach it. They know that they will have to respond to a variety of questions on a topic, so they study it until they thoroughly understand it. Students' questions sometimes reveal the need for the instructor to learn a topic better, or at least come up with a better way to present it.

Changed behaviors, improved attitudes, and increased motivation. All of these changes come about due to the positive influence that students can have on each other. You see participants in a discussion struggle, succeed, agree, and disagree. You see many different approaches to a problem, and the opportunity for recognition and praise. Discussions evolve into issues that are alive in the group, including course policies. Participants have the freedom to express their ideas, attitudes, interests, and needs. Discussion with your classmates, instructor, and others, contributes to a healthy climate for learning.

Promotion of active learning, thus increasing retention of knowledge. Discussion is active learning, which is more permanent (i.e., retention is longer). Intellectual content is not only involved in discussion, but it is integrated with your beliefs and attitudes—it becomes a part of you, not to be soon forgotten. Active learning is available throughout your life, and preparing to use it is a vital educational objective.

Development of leadership qualities. To engage in discussion you have to take the initiative to ask questions and make comments. You have to work for the welfare of yourself and others, and this requires management and leadership skills. You can plan for it, but it is also the outgrowth of discussion.

You may be reluctant to engage in discussion with your classmates or instructor, but reflecting on the objectives that can be accomplished by discussion should motivate you to do so. The next section presents the opportunities that are available to engage in discussion.

Opportunities to Engage in Discussion

The opportunities to engage in discussion in mathematics are many and include the following:

- During class with your instructor and classmates
- Out-of-class study group
- Out-of-class study partner
- One-on-one help from the mathematics learning center
- One-on-one help from on-campus student professional societies (e.g., engineering, education, business)
- A private mathematics tutor that you hire
- A private mathematics tutor provided by your learning center

These opportunities are elaborated on in chapter 23, including preparation for discussion and pitfalls to avoid when employing it.

5. Taking Notes in Class

At the risk of sounding too traditional, taking notes in class is highly recommended. This is recommended in the same vein that it is recommended you wear a seat belt if you are a racer in the Indianapolis Memorial Day 500 mile race. Taking first-rate class notes is indispensable to your success. It can make the difference between a mediocre or superior performance. I was always surprised when students didn't take notes in my classes. I always stressed the importance of note taking, yet there was always a small minority who didn't. This is comparable to ignoring a sign on a beach that says: "Proceed at your own risk—there is a strong undertow and swimmers drown here every year." Students who have excellent note taking skills benefit in many ways, including success on examinations.

Difficulty of Taking Notes

Taking good notes requires knowing what to record and how to record it, while maintaining an awareness of the big picture and subtle aspects of the presentation. It is not easy to take good notes and at the same time listen to what the instructor continues to say. Instructors do not typically pause for you to take your notes. You have to take notes because you won't remember what was said and the order in which it was said. On the other hand, you can't record all that is said or written since you will lose significant parts of the presentation due to inadequate listening. This is the dilemma you face and it makes note taking difficult. However, practicing good note taking techniques will help you handle this balancing act.

Benefits of Taking Notes

You are more apt to follow advice if you know the benefits of doing so. The benefits of note taking include the following:

You have a record of the content your instructor deems most important for you to learn. Just as you have to be selective in the notes you take, your instructor has to be selective in the content presented. There is much more content in your textbook than what can possibly be presented in class. Your instructor has to determine the content that is most important, including the types of problems you should be able to work. You need to have a record of this for no other reason, and there are other reasons, than it most likely will appear on examinations.

You have a record of the manner in which the content is presented. There are a variety of ways to present content and your instructor will present it based on his or her biases, including what he or she deems most meaningful considering the backgrounds of the students. It has been said that a beginning instructor teaches what he or she knows and a little more, an experienced instructor teaches what he or she knows, and a master instructor selects from what he or she knows, and teaches accordingly. Hopefully you have a master instructor.

You have a record of the more general and evaluative remarks of your instructor related to class content and textbook content. The importance of these remarks can be quickly ascertained by reflecting on these examples of instructors' remarks:

- *"Consider pages 210–215 in the textbook optional reading."*
- *"This next concept is very elusive so you will have to double your efforts to understand it."*
- *"On the examination I will expect you to do this when working this type of problem."*
- *"You are responsible for knowing the proofs of theorems 2, 6 and 7 that are in section 3 of chapter 10."*

- *"I have placed a book on reserve that gives a greater variety of examples on applying the definite integral."*
- *"Pay special attention to example 6 in section 4.5 of chapter 10. There is a special technique developed there that I want you to learn."*
- *"The significance of this theorem is that it is used to . . ."*
- *"These are pitfalls to avoid when using . . . and they are . . ."*

If you are not in class you miss this important information. You might say, "I can get notes from someone who was in class." Perhaps, but realize that many students only record what the instructor writes in class, and that is often recorded inaccurately.

You have a record of what your instructor is eliminating, changing, or adding to the content of the textbook. Your instructor should not be equated with the textbook. He or she should not repeat almost verbatim the content of the textbook. The roles of the textbook and the instructor have to be different. Your class notes should supplement the textbook, not repeat it. View them as separate but related, and be mindful that one of your jobs is to integrate them.

A lot of effort goes into good teaching. How could it be otherwise considering that instructors need to determine the motivation for a particular topic, requisite knowledge to review, examples to provide, exercises to work in class based on students' requests, changes in language for a definition or theorem appearing in the textbook, theorems to prove and the changes to make in the textbook's proofs, and identifying textbook content that students should ignore. This latter task implies that instructors have to know what their students are exposed to in the textbook, and then structure their lectures accordingly. They also need to expose you to content in class not covered in the textbook. *It is prudent for you to take good notes on all of this!*

Many instructors do not let their students know the relationship between their class presentations and content of the textbook. This means you have to know the textbook well and compare textbook content to class content. You won't be able to make this comparison if you don't take class notes. It is difficult enough to remember, near the end of the class session, content covered at the beginning, to say nothing of remembering the class content a day or two later. You need your notes to help recall class work. You need this knowledge to further your work outside of class, including reviewing content.

You can clarify the content of the class session. It is in the process of reading and reworking your class notes that you realize what you don't understand. Your class notes prompt you to seek the assistance you need to clarify content you don't understand. This is not possible if you don't have good notes.

It helps you listen. You cannot take good notes if you are a poor listener. Knowing that you have to take good notes motivates you to pay closer attention to what is being said. Contrast this with how easy it is to be distracted when you don't take notes since you are less active and can more easily entertain thoughts that are not germane to the presentation.

Determining the Notes to Take

You don't want to be a compulsive note taker; hence, be selective about the notes you take. One of my classmates in graduate school wrote down everything the instructor wrote on the board. I have no doubt that if the instructor dropped a piece of chalk on the floor and it made a mark, this student would have put a mark near the bottom of his notepad. Some instructors will help you take notes and that is good. They will indicate when something should or should not be written in your notes, although you have to be the final judge. Other instructors give no overt guidance at all—it is completely up to you. If your instructor is basically copying the book on the board right before your eyes, then you don't need to take notes. The course content appears in your textbook and most likely in a more thorough way than what your instructor is imitating. However, to enhance understanding, a first-rate instructor will reword and reorder some of the textbook's definitions and theorems, add clarifying statements, bring in different concepts and applications, work different examples, and give summaries that are uniquely his or hers.

The instructor's changes and additions to the textbook typically take precedence over what is in the textbook. Hence, you should record these changes and additions in your notes. Also, the responses to the questions and comments made in class by your classmates may be worth recording. You need to seriously consider recording your instructor's examples, solutions to exercises based on questions from the class, graphs, charts, derivations, and formulas. You will lose the flow of the presentation if your notes are too sketchy.

In summary, class notes will help you learn the content of the course. Take notes on what will supplement, strengthen, and illuminate textbook content, and on what conveys your instructor's bias, admonitions, and logistical guidance on deadlines you are expected to meet. Good note taking is learned behavior and must be coupled with good listening so you don't lose sight of the big ideas being presented and discussed. (See section 3 on listening as it relates to note taking.)

Help Your Instructor Support Your Note Taking

There are things your instructor can do to help you take notes. If they are not being done, you can respectfully ask that they be done. Your classmates will appreciate this and you can ask them to support you by doing likewise. Here are several reasonable expectations you can request of your instructor:

To pause at times to allow you to record lengthy definitions, statements of theorems, or extensive and complex developments. You could ask, "Would you please pause while I write this down?"

To keep content on the board or overhead for a reasonable amount of time. Instructors are prone to remove content quickly to free up space. If your instructor is ready to remove a portion of the work, say, "Would you please leave that on a little longer so I can copy it?" Be reasonable in making this request.

To rewrite or restate written statements or words that are illegible. If your instructor's writing is generally difficult to read, frequently asking him or her to rewrite or restate portions of it may motivate him or her to write more clearly.

To repeat oral statements that you do not understand or missed during note taking. You have to be careful that this request is not abused due to shortcomings on your part. Most instructors will accommodate your requests if they are reasonable and not overdone. You help the instructor and classmates by doing this. You have rights as a student and it is up to you to stand up for them. Muster the courage to do so.

Note-Taking Techniques

There are note taking techniques that will help you maintain the proper balance between (1) taking too many notes and missing some of the presentation and ensuing discussion, and (2) taking too few notes. Note-taking techniques include the following:

Decide on the basic format of your notes. You need to identify a basic note-taking format that works best for you. Your work is minimized if your instructor is organized and uses an outline format since many recommend that format for note taking. In an outline format the headings and subheadings are typically short, the content of the presentation is organized and brief, yet contains the main ideas. The exact nature of your outline is left to you. You may want to use roman numerals for major headings and upper-case letters for the first level of subheadings, with a mixture of lower-case letters, numerals, and bullets to identify related content, including examples, supporting arguments, details, definitions, and theorems. Indentation of headings, subheadings, etc. is your decision. You need to be brief without leaving out major ideas.

Note taking becomes more difficult if your instructor is disorganized. In that situation you may need to take notes in a format that is more closely related to the instructor's format, realizing that you will reorganize them later. Regardless of your format, leave spaces to fill in later what you may have missed due to inattention, confusion, or lack of time to write.

Use abbreviations. In your attempt to be brief, use of abbreviations is recommended for words that appear often. For example, you can use *thm* for theorem, *cor* for corollary, *defn* for definition, *eg* for example, *pf* for proof, *fct* for function, *der* for derivative, *cont* for continuous, *int* for integral, *inv* for inverse, *det* for determinant, and # for number. Similarly, use *p* for page, *ch* for chapter, *b/c* for because, and so on. You know abbreviations for many common words so don't hesitate to use them. Be consistent in your use of an abbreviation. Their use will expedite note taking.

Develop your own system to denote emphasis. You want to avoid taking time to write phrases or sentences to show emphasis or your instructor's bias such as, "This is very important," "Use this definition rather than the one in the textbook," or "Pay special attention to the third line of this proof." Emphasis can be noted using these and other symbols: underline, double underline, stars, asterisks, exclamation points, and arrows.

Use your own words. You may be able to write faster if you use some or all of your own words. The idea is to avoid using complex or flowery language but still maintain accuracy and meaning. How much you use your own words depends on the importance of using the precise language of your instructor.

Save time by noting page numbers in the textbook. If your instructor writes items on the chalkboard that are taken directly from the textbook, indicate the page number of these items in your notes (but don't write out these items). If you always bring your textbook to class, you can quickly locate these numbers. You may want to write some of this textbook content in your notes after class so you have a more complete development.

Denote what is review. Instructors often say or write "aside" comments. These are remarks, often very abbreviated, about a review of some requisite material for the concept being presented, or an interesting sidelight that the instructor wants to point out. You have to decide whether you need to record these comments. For example, if they constitute a review, you probably won't write them in your notes if you don't need the review. If you take notes on these asides, then either write "review" or "aside" by them, box them off, or write them in the margins of the pages in your notebook. Then they won't interfere with the flow of the major ideas of the development.

Take Accurate and Complete Notes

You would be shocked at the number of students whose notes are incomplete or contain errors. Students would come to my office with questions, either on their notes or on something related to their notes. I would ask to look at their notes and found that important ideas were not noted or there were transcription errors. They would omit key sentences, parts of equations and formulas, and write a lower-case letter when it should have been an upper-case letter. This indicated that it is challenging to take complete and accurate notes; hence, as you take notes, double check to see that they are accurate and complete. You cannot rely on a classmate's notes, unless you know the classmate is an expert note taker.

Use a Notebook

Using a notebook for class notes and assignments is important for organizational reasons. It was frustrating when students came to my office for help and then spent considerable time trying to sort through a stack of unbound and poorly arranged papers, either looking for their work on an assigned problem, or trying to find their class notes. What came to mind is the adage that "a cluttered desk is a sign of a cluttered mind." There are many occasions when you have to consult your notes; hence, using a notebook makes sense.

Your notebook can be bound or loose-leaf. A divider can separate your class notes from your work on assignments. The advantage of a loose-leaf notebook (e.g., a three-ring binder) is that you can add or delete pages. For example, you can replace pages of your notes with notes that you have rewritten or corrected. The same is true for problems that you have to correct—it is no more than a snap of the binder to reinsert them. You will also be asked to hand in specific problem assignments, and using a three-ring binder makes it easier to remove and reinsert assignments. Instructor handouts can be hole-punched and inserted into your notebook in appropriate locations.

There are various ways to divide a notebook page, and how you do it depends on your preference for placement of specific entries. You could have a margin on one side of the page; say 2.5 inches, leaving a 6-inch wide space for your main notes. In the margin you can record a variety of things such as the date of the next unit examination, a short reflection on some notes, a question or concern, a reminder of what you need to do, admonitions from the instructor, and text you want to insert in the notes that was an afterthought. Most likely you will have lots of white space remaining in the margin at the end of a class period, which allows you to add content later on if need be.

Write the date and day of the week at the beginning of your notes for that class day. This allows you to quickly find your notes for any given day.

The well-known mathematician Paul Halmos conveys some of the key reasons for taking notes, as well as some techniques for doing so *that worked for him* (your notes will be far less skimpy):

> It's all true, the arguments both for and against taking notes. My own solution is a compromise. I take very skimpy notes, and then, whenever possible, I transcribe them, in much greater detail, as soon afterward as possible. By very skimpy notes I mean something like one or two words a minute, plus possibly a crucial formula or two and a crucial picture or two—just enough to fix the order of events, and, incidentally, to keep me awake and on my toes. By transcribe I mean enough detail to show a friend who wasn't there, with some hope that he'll understand what he missed.[1]

Other Note-Taking Issues

Your work with your notes is not finished once you leave the class-room. You have to clarify them, integrate them, review them, take care of them, and evaluate your note-taking ability. These issues are now discussed.

Clarify and integrate your notes. As soon after the presentation as you can, expand or clarify your notes while the material is fresh in your mind. If you have 20 to 30 minutes right after the session ends, use that time to work on your notes. Either stay in the classroom or find an empty classroom. You will be surprised how helpful it is to do this activity soon after class. Waiting as long as a half or full day to do this can be problematic because you will have forgotten things. Fill in the parts that you did not have time to record when in class, including missing details. You will be more aware of what you need to fill in after seeing all of your notes for the class period. If they are not organized because your instructor was not organized, now is the time to organize them in a logical way. Place questions in the margins that arise as you rewrite your notes. Asking yourself this question prompts other questions: "What questions are answered in the notes?" Write the questions in the margin opposite the answer. These questions can be reviewed when you study for examinations and don't be surprised if they turn out to be on the examination.

If some of your notes are unclear or confusing, talk to your instructor or a classmate, preferably before the next class period. It is better to visit your instructor during office hours if your confusion cannot be cleared quickly. Instructors like it when students ask them questions on notes taken in class. It is an indication that you are taking notes and want to understand them.

It is a good idea to compare and integrate your class notes with notes you have taken on the related textbook reading. Notice where they differ and where they are alike. (See section 2 of chapter 15 on a method for reading a mathematics textbook, which includes asking yourself questions that are answered in your reading.)

Take care of your notes. Be careful not to lose or misplace your notebook. Students leave their notebooks in classrooms, restaurants, buses, instructors' offices, etc. This can play havoc with your academic life. Students sometimes know where they left their notebooks, but logistics often keep them from reclaiming them when they are most needed, and many never get them back. In the front of your notebook write the name of the course, the section you are in, the room where the class meets, the name and office phone number of your instructor, and a statement that says, "If found, please call or take this notebook to (insert instructor's name) office." Another method is to have two sets of notebooks. Take one notebook to class and use it to record your abbreviated notes. Use the second notebook for your completely rewritten notes, and leave it at your residence.

Review your rewritten notes. Your class notes are not much use to you if you don't review them periodically. It is not too early to review them within a day after you have rewritten them. Frequent and short reviews are far superior for remembering than infrequent and long reviews. Early review of your class notes will help you read the textbook, complete assignments, and reap more benefits when in class. Some students have a fixed time each week to review notes during the previous week. (See chapter 24 on remembering, reviewing, and summarizing mathematics.)

Evaluate your note taking. Are you benefiting from your notes? Perhaps you need to get better at taking notes. Reworking them is one way to get better at taking them. Have your instructor look at your notes and make evaluative comments. Ask some of your classmates who are doing well in the course to look at your notes, and ask if you can look at their notes. A student who is performing well in a course does not necessarily take good notes, but is probably taking better notes than someone who is doing poorly in the course. Ask yourself these questions about your note taking: "Am I writing too much and not listening enough?" "Am I writing too little, leaving gaps in my understanding?" "Am I writing my instructor's evaluative remarks about specific content?" "Am I reviewing my notes often enough?"

15

Reading a Mathematics Textbook

If you are like most mathematics students, you have not had any formal training on how to read a mathematics textbook. That is not to say you don't read them, but the question becomes, "How do you read them?" Any two students will read mathematics differently, whether or not they have been formally trained to read mathematics. There are guidelines for reading a mathematics textbook, which, if implemented, will significantly increase your understanding of mathematics and ability to apply it. Awareness of the specific benefits of reading will increase your motivation to read. These benefits were addressed in section 1 of chapter 14.

The two main components of your mathematics study are learning the explanatory material in the textbook, mainly through your reading, and applying it to work assigned exercises that largely come from the textbook. Both of these are important to pursue since (a) learning the explanatory material does not necessarily mean that you can apply it to work the exercises, and (b) being able to work exercises, especially if they are a mechanical repetition of what you have practiced, does not necessarily mean that you understand the explanatory material used to work them.

General comments about reading a mathematics textbook are given in section 1. A specific method of reading a mathematics textbook is given in section 2. There are various methods of reading a mathematics textbook, most of which are variations of each other. You have to practice a reading methodology before it becomes second nature to you. The time you spend doing this will pay huge dividends, not only in your mathematics courses, but also in related courses and in your career. The reading method presented in section 2 teaches you how to learn actively, which is motivating and increases your understanding and retention.

1. Basics of Reading a Mathematics Textbook

General information on reading a textbook is presented here and in the next section. Reading specific *types* of mathematics content is presented in chapter 16.

Reading Is Thinking and Dialoguing

When you read a textbook you are engaged in a silent conversation with the authors. It is not a mindless and relaxing activity. It requires you to think about what is being said, question its validity, relate it to what you already know, and anticipate where it is going. You are challenging the authors to convince you that what they are saying is meaningful. You may or may not like the way something is presented, and certain feelings will arise. When you read, you engage your heart, mind, and body, and it takes energy to do that.

What You Should Know about What You Read

When you read a section of a chapter in your mathematics textbook, you will be exposed to a logical development of some mathematical results or outcomes. To understand any one of these developments, you have to understand the result, the reasonableness of it, the purpose for it (that is, how it is used in and out of mathematics), and, for many of them, its formal justification or proof. Chapter 16 specifically addresses how you can accomplish this.

Reading Mathematics Is Different from Reading a Novel

Mathematics makes heavy use of technical words and symbols that make written mathematics compact and dense. This is one of its advantages since substantial information can be conveyed in a small space. This allows you to see it, hold it in your mind, reflect on it, and infer from it. Without this shorthand, it would be difficult if not impossible to carry out mathematics. However, there is a downside to this compactness and that is the difficulty of quickly digesting the information in each statement. It takes time to grasp the technical terms and symbolic forms. As you read, have paper and pencil available, construct drawings that will help you understand, ask yourself questions (and work to answer them), devise examples that clarify, and perhaps consult your class notes or previous pages of the textbook. This is a slow process. You most likely will have to come back at later times to portions of what you have read to understand them better. It is unusual to gain an excellent understanding in a first, or even a second, reading. Compare that to the ease of reading a novel where you can comfortably read and understand 20 to 30 pages in a significantly short time period.

The time required to read mathematics can discourage you. You may think that the speed of your reading will never increase, or worse yet, when confused, that you will never be able to understand what you are reading. This will not happen if you persevere. With repeated attempts over a span of time, your reading speed will increase and you wonder why you were ever confused on a topic you were reading about. Unfortunately, too many students think that you either understand something rather quickly or you will never understand it. If you are one of these, think about this: New ideas come from combining previous ideas in a particular way. It is not a magical process and that should be reassuring to you. If you struggle to understand a particular topic, then you may need to spend more time with it, see more steps of the development, attain a better understanding of the prerequisite knowledge, or get help from someone. It is no more complicated than that. There is nothing wrong with having to spend more time in your efforts to understand. I still have to spend more time learning something than I care to, but *I know* I will eventually understand it, and that makes the time spent on it tolerable.

Relate Reading to Class Work and Vice Versa

What happens in the mathematics classroom enhances what you do in the course outside the classroom, and vice versa. When you read your textbook, you will reflect on how that content fits with class content. Likewise, when in class you will relate what is being done there to what you learned in your textbook reading. When you read your textbook, you can consult your class notes, and when in class, you can consult your textbook (always bring it to class), and the notes you took on it. To be a successful mathematics student you have to integrate your textbook work with your class work. Slighting either of these, and keeping them separated, is not a viable option.

Underline, Outline, and Question

Many students underline key information in their textbooks. There are better options to emphasize key information. However, a *sensible* use of underlining has its role. Underlining in a textbook is often poorly done, especially when done too soon. As a passage is read for the first time, too much underlining does not distinguish the important from the unimportant. It is not recommended that sentences, to say nothing of whole paragraphs, be underlined. *Your underlining should consist mainly of key phrases.* More will be said on this in the next section.

It is wise to delay underlining until you have carefully read a section of the textbook and can identify its main ideas. If you underline, you most likely read the underlined portions of the relevant textbook sections as you review for an examination. It is your expectation that this will help you recall needed content when taking an examination. The problem with this technique is that *you delay your learning.* You rely too much on your memory to recall what you may need. You may underline a lot of content, perhaps way too much, and review all of it in a shortened time in preparation for an examination. This is a pre-scription for confusion when confronted with an examination. Most likely you are hoping for mechanical or rote examination questions that can be answered by writing what you have underlined and memo-rized. Many examination questions are not this type. They are more global in nature, and require more of you. Responding to them appro-priately requires adaptability on your part, which means selecting the appropriate content and piecing it together.

There are two alternatives to underlining: (1) constructing an out-line of the content, and (2) constructing questions on the content. Using these two alternatives to underlining will improve your ability to understand the content and to respond better to examination questions.

Construct an outline of the relevant sections of the textbook. This technique does something that underlining doesn't do. It displays the key ideas of the relevant content and how they relate or flow from one another. You are more active when outlining because you have to decide (1) what to put in the outline (e.g., important concepts and prin-ciples), (2) where to put it (the progression of ideas has to be meaning-ful to you), and (3) the words you will use (your own words makes it more meaningful to you). You cannot construct a good outline if you don't understand the content and the flow of its development. An outcome of outlining is increased understanding. Just the act of creating an outline increases learning. You can reflect on an incredible amount of knowledge when you reread or review your outline. Outlines need to be brief, containing only the essentials.

There is a way to underline that also qualifies as outlining. It is underlining in your textbook and then marking this up in a meaningful way. This is not difficult to do since most content of a section is developed in a logical way. You can use numbers to mark the major ideas and letters to mark their subparts. Additionally, you can "star," "asterisk," or double (or triple) underline important words or phrases.

Construct questions on the relevant sections of the textbook. I cannot stress enough how important it is to do this. Write a series of questions about the content you need to know from the relevant textbook sections, and then *answer your questions.* In answering your questions, use only as many key words (cues) and phrases as needed for you to recall a more detailed answer. This makes it easier for you to retrieve your answers later when you need them. All you need to do is recall the key words and phrases in your answers, and understanding will take you the rest of the way. You prepare for an examination by reading your questions and then determining if you can answer them without first looking at your abbreviated written responses. The questions you construct will most likely be similar to those you will be asked on an examination. This alternative to underlining is expanded in section 2, and includes examples of such questions.

Multiple Readings of Your Textbook Are Necessary

Understanding what you read in your textbook is an ongoing process, most likely requiring multiple readings. Your reading can be done in a cursory or a detailed manner, and which approach to use depends on whether it is the first reading or a subsequent reading. You may read a section of the textbook shortly before, or after, it has been lectured on in class; read it again one or more times as your progress through a unit; and read it again near the end of the unit, or as you do your final preparation for an examination. It is recommended that you read a section of a textbook three times, not necessarily the same way each time. More is said on multiple readings of your textbook in section 2.

The Path You Take in Reading Your Textbook Is Not Always Straight

Textbooks are written in a logical way. Authors have structured their writing so that one idea flows into another in a systematic and linear way. Unfortunately, our minds are not always structured that way. That is, in this rational flow of ideas, you may get confused at a certain point and think that you cannot proceed with your reading until you consult someone about your difficulty. You may be confused about a line in the proof of a theorem, the meaning of a definition, or a step in the solution of an example. Skipping over this and continuing with your reading may not be a problem. Mark the confusing text so you can return to it later. You may have to do this several times as you read a section of the textbook. In continuing your reading, you might learn something that relates to the confusing text that will help you understand it. At that point you can go back and reread what you bypassed. You can also go back to this later when you have gained more knowledge from attending class, working exercises, or through discussions with your classmates or instructor. This is a reason why multiple readings of a textbook section are a good idea, to say nothing about the reinforcement that occurs from repeated exposure.

All That You Read Is Not of Equal Value

At a certain time in a unit of study, studying specific content in the unit has greater value to you than studying other content in the unit. If time were not a consideration, it would be ideal to learn all of it equally well. But often time is of the essence; especially when an examination is impending. Therefore, focusing more of your study in the textbook on content of greater value to you is a wise thing to do. You may ask, "What content should receive less attention at this time?" Eventually, as your course progresses, you learn this based primarily on the expectations of your instructor. Nonetheless, here are some questions you can ask yourself to judge if specific content requires more or less of your focus: (1) Is it an extremely difficult proof? (If so, give it less focus.) (2) Is it a concept, principle, or application that seems to be an optional topic, based on where and how it appears in the section, whether it has been discussed in class, and whether exercises have been assigned that relate to it? (If so, give it less focus.) Please don't get the wrong idea here. I am not suggesting that you should be selective in what you read in your early readings of a section of your textbook. That would be a mistake since it will interfere with your learning. What I am saying is that the time will come in the study of a unit when your study will have to be more focused, and at that time it is prudent to be selective choosing the content you will spend more time studying.

Read Sentences by Phrases

You read for ideas and most ideas are expressed using multiple words. To focus on the ideas you need to *think in phrases*, which means that you will have to *read in phrases*. When you read, fix your eyes on phrases rather than on individual words. This takes practice, but your efforts will be rewarded. Here are three examples:

Example 1. *The text you read*: The axis of symmetry of the parabola $y = ax^2 + bx + c$ is the vertical line passing through the vertex.

There are two key phrases expressed here, each referring to the same idea (denoted by the word "is"), but using different words:

the axis of symmetry of the parabola $y = ax^2 + bx + c$
the vertical line passing through the vertex

Reread the complete statement, but this time fix your eyes on each phrase as you proceed.

Example 2. *The text you read*: The graph of the cotangent function can be obtained from the graph of the tangent function by means of the identity $\cot x = \tan(x - \frac{\pi}{2})$.

There are four key phrases expressed here:

the graph of the cotangent function
can be obtained
from the graph of the tangent function
by means of the identity $\cot x = \tan(x - \frac{\pi}{2})$

Reread the complete statement, but this time focus your eyes on each phrase as you proceed.

Example 3. *The text you read*: The product of two differentiable functions u and v is differentiable and $\frac{d}{dx}(uv) = u\frac{dv}{dx} + v\frac{du}{dx}$.

There are three key phrases here:

the product of two differentiable functions u and v
is differentiable

and $\dfrac{d}{dx}(uv) = u\dfrac{dv}{dx} + v\dfrac{du}{dx}$.

Reread the complete statement, but this time focus your eyes on each phrase as you proceed.

Go to your textbook and spend some time reading a page by training your eyes to fixate on groups of words that convey an idea. Always make an effort to do this when you read. This method of reading will become second nature to you. As you practice more you will be able to fixate on larger groups of words, thus increasing your reading pace and understanding.

Read Paragraphs for Units of Thoughts

Generally, a well-written paragraph has one main thought that comes from uniting other thoughts in the paragraph. Your goal in reading a paragraph is to locate and understand that thought. However, mathematics textbook authors often have more than one main thought in a paragraph, which makes it more difficult to locate all of them. To locate and understand a main thought, you will read phrases contained in the sentences comprising the paragraph, and unite them as you proceed. At the end of the paragraph you may find a summary statement or a statement that leads into the next paragraph. Reading a paragraph to locate and understand the unity of thought requires intentional practice. Two examples are given on reading a paragraph for unity of thought. These paragraphs are taken from David Cohen's precalculus textbook.[1]

> **Paragraph 1.** Why are the graphs of f and f^{-1} always mirror images of each other about the line $y = x$? First, recall that the function f^{-1} switches the inputs and outputs of f. Thus, (a,b) is on the graph of f if and only if (b,a) is on the graph of f^{-1}. But (as we know from Section 2.4) the points (a,b) and (b,a) are mirror images in the line $y = x$. It follows, therefore, that the graphs of f are f^{-1} are reflections of each other about the line $y = x$.

The first line of the paragraph gives the main or unifying thought of the paragraph, which is a question asking you to justify why the graphs of f and f^{-1} are mirror images of each other about the line $y = x$ *(the first piece of information)*. The next three sentences give the justification, and each one contains phrases conveying ideas *(the second piece of information)*. The last line lets you know that the justification is completed *(the third piece of information)*. These are the three pieces of the paragraph that are united to give unity of thought to the paragraph.

> **Paragraph 2.** The example that we have just concluded serves to remind us of the difference between a conditional equation and an identity. An identity is true for all values of the variable in its domain. For example, the equation $\sin^2 t + \cos^2 t = 1$ is an identity: It is true for every real number t. In contrast to this, a conditional equation is true only for some (or perhaps even none) of the values of the variable. The equation in Example 1 [i.e., $\sin x + \cos x = 1$] is a conditional equation; we saw that it is false when $x = \dfrac{\pi}{4}$ and true when $x = \dfrac{\pi}{2}$. The equation $\sin t = 2$ is an example of a conditional equation that has no solution. (Why?) The equations that we are going to solve in this section are conditional equations that involve the trigonometric functions. In general, there is no single technique that can be used to solve every trigonometric equation. In the examples that follow, we illustrate some of the more common approaches to solving trigonometric equations.

The unifying thought of this second paragraph is to make clear the difference between a conditional equation and an identity, which is done by defining these two entities *(the first piece of information)*. Examples are then given of a conditional equation and an identity *(the second piece of information)*. This is done so the reader will know what needs to be done to conditional trigonometric equations that appear further on in the section *(the third piece of information)*. These are the three pieces of the paragraph that are united to give unity of thought to the paragraph.

Take time now to read some paragraphs in your mathematics textbooks, identifying the unifying thought for each paragraph, and the pieces of the paragraph that are united to give this unit of thought. If there is more than one unit of thought, give those also.

2. A Method for Reading a Mathematics Textbook

Many students who read content from other disciplines have used the method presented in this section to read a mathematics textbook. There are six components to the method, and exactly how you apply each component is your decision. The components comprising the method follow:

1. **Survey (S)**—gain an overview of the section.
2. **Question (Q)**—write questions, based on your survey, other content in the section, or your class work.
3. **Read (R)**—carefully read the section.
4. **Recite (R)**—recite your answer to each question aloud, silently to yourself, or write it.
5. **Record (R)**—record your answer to each question using cue words or phrases that will help you recall your answer later.
6. **Review (R)**—review your answers to all your questions on the section.

An acronym for this reading technique is SQ4R. SQ4R may be a new method of reading for you. If so, it may take you awhile to learn it, but stay with it for these two reasons: (1) it is an active process, using more senses, resulting in increased learning and retention, and (2) it can be used to read any content, regardless of the discipline. As you get better at it, you might want to make minor changes based on your learning style and the nature of the content. More details are given on each of these six components.

S—SURVEY (Gaining an Overview)

Surveying, or gaining an overview of a chapter or a section of your mathematics textbook, is done prior to reading it carefully. We direct our discussion of this component to a section of a chapter, rather than to an entire chapter since the focus of mathematics study is basically one section at a time. (Surveying an entire chapter is not discouraged, but it should be completed in a more cursory way than surveying a section of the chapter.) A survey is done somewhat quickly, yet is valuable to do since it gives you a general idea of the structure and content of a section.

Without the guidance of a survey, your reading of a textbook can be misdirected, resulting in a considerable waste of time. More can be gained from a survey of a section if the section is well written. Unfortunately, not all mathematics textbooks are well written. A major shortcoming of some textbooks is that the presentation of content is not always placed into a meaningful structure. A quality development of content lets the reader know, at appropriate times throughout a section, what will be presented, how it relates to previously presented content, and how it will be used. Since not all sections are structured the same way, what you include in your survey of one section may not be possible to include in another. A survey or overview generally consists of—

- Reading the textbook authors' content objectives for the section
- Reading major and minor titles and headings
- Reading introductory and summary paragraphs of the section and subsections
- Reading statements of definitions and theorems
- Reading boldface headings of charts, tables, diagrams, and graphs
- Noting what is addressed by the examples

In a survey *you do not attend to the details* in the section; those are reserved for a thorough reading of the section.

What Surveying Accomplishes

You will be motivated to use the survey technique if you know what it can accomplish. You may spend from 10 to 30 minutes to conduct a survey, depending on how much there is to survey, yet you can accomplish these objectives in that short time interval:

- Knowledge of what the author is trying to accomplish in the section, including the organization of the section
- A "big picture" context in which to place the details that you will examine in a more careful reading
- Inspiration to know more about the content you surveyed
- Information that will help you construct questions that will direct a more careful reading
- Knowledge of how the section relates to previous or future sections

What it means to understand mathematics is discussed in section 2 of chapter 9. Appearing in that section is this statement: *Understanding something comes from connecting or relating that thing to other things we know.* With this description of understanding, it is readily apparent from reading the above list that doing a survey will increase understanding. You don't have to survey a section, but considering the benefits, why wouldn't you?

Part of a good survey is to devise questions based on your survey that you will want to answer when you do a more careful reading of the section. The question component of SQ4R is now addressed, followed by an example of a survey, including questions that were posed based on it.

Q—QUESTION (Write Questions Based on Your Survey)

It is important to write questions as you do a survey, and then use them to direct your careful reading of the textbook. A primary goal of a *careful* reading is to answer these questions. The questions you asked yourself based on your survey will be the more important questions, since a properly completed survey reveals the major ideas of the section and how they are ordered relative to each other. This is not to say that you won't be asking other questions that arise as you follow your survey with a careful reading. Be on the lookout for additional questions during a more careful reading.

There are excellent reasons for reading an assignment to answer questions. Included among them are:

Better understanding and retention—you read with the purpose of seeking out answers, which makes your reading a more focused and active process.

Increased motivation to read—your curiosity is piqued since you want to find the answers you seek.

Reading is more directed—the details that help answer important questions are the more important ones in the section.

Characterize the role of content—you have identified the content that answers specific questions.

More efficient learning—you are answering questions as you read, as opposed to reading and then perhaps answering questions that arise later.

Better preparation for examination questions—the questions you most likely will be asked on an examination are among the important ones on the unit, which you have already posed and answered based on your survey. You will only need to briefly review them when examination time arrives.

Example of a Survey, Including Questions That Arise

This is my survey of the first section of a chapter of a precalculus textbook, titled, "Trigonometric Functions of Real Numbers." The chapter appears in David Cohen's precalculus textbook.[2] I will tell you what I read, and then give you my thoughts as though I experienced the ideas for the first time.

I noticed that the first section in the chapter is headed, "Radian Measure." Prior to reading this section, I read the "Introduction" to the chapter, which gave an overview of the chapter. The part of the "Introduction" that interested me the most were these remarks of the author that applied to the first section of the chapter:

> the trigonometric functions play an important role in many of the modern applications of mathematics. In these applications, the inputs for these functions are real numbers rather than angles, as in the previous chapter. In order to make the transition from angle inputs to real-number inputs, we introduce in Section 7.1 radian measure for angles.

These remarks tell me that certain applications of trigonometric functions require a new form of input for trigonometric functions, and that a new form of angle measure is introduced in the first section. I posed these questions that I will want to answer as I do a more careful reading: (1) Why do specific applications of trigonometric functions require inputs for trigonometric functions that are different from angle inputs? (2) What is the radian measure of an angle?

Here is the introductory paragraph of the section:

> For the portion of trigonometry dealing with angles and geometric figures, the units of degrees are quite suitable for measuring angles. However, for the more analytical portions of trigonometry and for calculus, radian measure is used.

From this paragraph, I posed these questions: (3) What is meant by the "analytical portions" of trigonometry? (4) Why does calculus require the use of radian measure for trigonometry functions?

I then read the definition of the radian measure of an angle. The author proceeds to state the definition, and includes an accompanying diagram to aid understanding. I didn't spend much time with this definition, leaving that for a later detailed reading. Two solved examples were given, displaying a process for finding the radian measures of two angles. After noting this purpose of the examples, I dispensed with reading the solutions until a later detailed reading.

Next in my survey, I noticed a boxed-off statement that related degree measure to radian measure, namely, $180° = \pi$ radians. I posed these questions: (5) How was the relationship, $180° = \pi$ radians, obtained? (6) How will the relationship, $180° = \pi$ radians, be used? Examples were then given for converting radian measure to degree measure and vice versa. This gave rise to this question: (7) What need is there to convert radian measure to degree measure, and vice versa?

An example followed where trigonometric functions of radian measure were calculated. This was followed by some remarks on how to use a calculator to find trigonometric functions of radian measure. Again, I dispensed with reading the details. I wrote this question: (8) How will I use a calculator to find trigonometric functions of radian measure?

As I progressed with my survey, I noticed a formula for finding, in a circle of given radius, the length of an arc of a circle determined by a central angle of radian measure. It was intriguing to me that one could calculate the length of an arc of a circle knowing these two associated quantities. This question came to mind: (9) How is the formula developed for determining the length of an arc of a circle, given the radius of the circle and the radian measure of the central angle that determined the arc? Two examples followed that applied the formula for finding an arc length of a circle.

Next, a formula is given for finding, in a circle of given radius, the area of a sector of a circle with a given central angle. I began to see that having radian measures allowed me to determine some significant things, and that interested me. I was curious as to how this formula could be proven, but I left the author's proof for a more careful reading. However, my curiosity prompted me to write this question: (10) How is the formula developed for finding, in a circle of given radius, the area of a sector of a circle with a given central angle? An example was then given to show how the formula is applied to find an area of a given sector.

Finally, in the section I was surveying, definitions were given of the concepts of angular and linear speeds of a point on a circle that is revolving at a constant speed about its axis. Both concepts were defined by formulas that are used to determine them, followed by applications of these formulas to find angular and linear speeds. These questions arose: (11) How can I make plausible the formulas that defined the concepts of angular and linear speeds of a point on a revolving circle? (12) What practical use is there for the formulas that give angular and linear speeds of a point on a revolving circle? There were some geometric figures associated with the examples, but I decided to study them later during a more careful reading. There were no tables or graphs with boldface titles, so I could not read these in my survey. Likewise, there was no summary of the section to read.

While doing this survey I avoided attention to the details of the development. My main purpose was to get a good sense of the main content of the section and how it was organized. I constructed 12 important questions that I would use to direct a more careful reading. You might have constructed other questions, fewer questions, more questions, or worded your questions differently. I will do a more detailed reading to find answers to my questions, and to learn about other aspects of the section. I am well positioned to focus on understanding the details since I now have a structure in which to place them. Most likely I will have more questions as I complete a thorough reading of the section.

Here are the 12 questions that I posed based on my survey:

1. Why do specific applications of trigonometric functions require inputs for trigonometric functions that are different from angle inputs?
2. What is the radian measure of an angle?
3. What is meant by the "analytical portions" of trigonometry?
4. Why does calculus require the use of radian measure for trigonometry functions?
5. How was the relationship $180° = \pi$ radians obtained?
6. How will the relationship $180° = \pi$ radians be used?
7. What need is there to convert radian measure to degree measure, and vice versa?
8. How will I use the calculator to find trigonometric functions of radian measure?

9. How is the formula developed for determining the length of an arc of a circle, given the radius of the circle and the radian measure of the central angle that determined the arc?

10. How is the formula developed for finding, in a circle of given radius, the area of a sector of a circle with a given central angle?

11. How can I make plausible the formulas that defined the concepts of angular and linear speeds of a point on a revolving circle?

12. What practical use is there for the formulas that give angular and linear speeds of a point on a revolving circle?

You can see how helpful the survey technique will be to a more thorough reading of the section. The questions you compose during your survey will be meaningful to you because they are *your* questions. A survey won't make much difference in understanding if the section is poorly written, or if the survey is poorly done. *A major goal of a more careful reading is to answer your questions resulting from your survey.* The questions, and your answers to them, can be used later to review the section in preparation for an examination.

I could give examples of surveys for different mathematics subjects, which would vary somewhat from the example I just gave. The example given here should help you conduct your own surveys, regardless of the mathematics course you are in. You survey to gain a sense of the big picture, avoiding any focus on details, but looking at virtually everything else. This takes time, but it is time well spent. With practice, you will get better at it. The key to surveying is to create questions that you will answer in your detailed reading.

When you are ready to do a more detailed reading, you may have additional questions from other sources. These sources include class notes, instructor study guides, review questions from the textbook exercises, and conversations with classmates. For example, if you were confused about something your instructor said in class, or had difficulty with a type of exercise you were working on, then express this difficulty in the form of a question. You may have already read the section before these questions arose, but they are questions you can use in subsequent readings of the section. Remember, a key to learning is questioning, and then getting answers to your questions.

R—READ (Read the Section Carefully)

R—RECITE (Say Aloud the Answer to a Question)

R—RECORD (Write an Outline of Your Answer to Each Question)

R—REVIEW (Review Your Answers to All Your Questions on the Section)

Read the Section Carefully

Once your survey of the section is completed you are ready to do a thorough reading of the section. Since you are now including details in your reading, you will read slowly and thoughtfully, considering all the content including historical notes, examples, graphs, tables, symbols, definitions, theorems, and logical arguments or proofs. Keep paper and pencil handy so you can work on details that are confusing. You read to answer the questions you have created from your survey and from other sources. If in your reading you see a key question answered that you do not have, write the question down.

The questions you created from your survey most likely arose in the order that the content was developed in your textbook. Begin your reading and read until one of your questions has been answered. It is not far-fetched to say that you should understand every word of every sentence in every paragraph. To know that a question of yours has been answered, you have to understand the relevant examples, graphs, illustrations, tables, symbols, definitions, theorems, and logical arguments or proofs. *How to study each of these for understanding is covered in the next chapter.* If you need help to understand content that answers a question, and that help is not immediately available, continue on with this process, and get help later with those questions you were not able to answer. It is possible that a question you asked is not answered in the section. If you still believe the question is important enough to answer, then get help with it.

Recite by Saying, Thinking, or Writing the Answer

Once you discover the answer to a question, it is important to immediately spend some time reinforcing the answer. To do this, you need to *recite* the answer; that is, say the answer aloud. The advantage of saying it aloud is that you can hear it, and that will help you decide whether it sounds right to you. We usually know when our speech is cumbersome and does not convey what it should. If you are in a location where you cannot say it aloud, mentally recite the question and its answer. The process of reciting aids in understanding and retention. However, this will happen to a far greater degree if you use your own words. It is tempting to use the words of your authors, but this will minimize your thinking and your answer will be less a part of you. It will be more difficult to recall the answer later on.

Record the Answer Using Cue Words or Phrases

You have to approach your reading with the expectation that you will remember what you read. Tell yourself frequently that you are reading to remember for the long term as well as the short term. Reciting your answers to questions is a key way to support this. Additionally, you understand and remember better if you *record* or write your answer to a question. We often think we understand something by quickly going over it in our mind; but when we try to express our thoughts in writing we realize they are not as well formed as we thought. Eventually we are able to formulate a suitable written response. However, *the answer you record next to the question should not be a detailed response.* Writing too much interferes with the flow of your thoughts as you are reading the section. You want to hold disruptions to a minimum. It is better to write down some key words or phrases, *called cue words or phrases*, that you will use as prompters in deriving the answer. Remember this set of cues and you should be able to derive the answer to the question if the question appears on an examination.

Many students read their mathematics textbook improperly. They don't survey and devise questions based on it; hence, they can't look for answers as they read, and consequently they cannot recite and record answers. Instead they do many futile things such as (1) starting with the exercises and only reading what they think they need to work them; (2) underlining too much, which prevents separation of the major ideas from the minor ones; (3) taking too many notes, which makes recall difficult, and (4) skipping the more difficult developments. These actions have little to do with learning. The preparation of these students for an examination involves similar processes. They incorrectly think that being able to recognize an answer, not even sure of the question that it answers, is the same as knowing the question and being able to recall or reconstruct an answer based on cue words or phrases identified with the answer.

The following is an example of an answer to a question and how it is recorded using cue words or phrases. I created this earlier when surveying a section of a textbook on radian measure:

Example. How is the formula developed for finding the area of a sector of a circle of given radius and central angle?

My answer. Consider a sector of a circle of radius r and central angle of radian measure θ. The ratio of the area A of the sector to the area of the circle, is equal to the ratio of the radian measure of the central angle of the sector to the radian measure of the central angle of the circle, or $\dfrac{A}{\pi r^2} = \dfrac{\theta}{2\pi}$. Solving for A gives $A = \dfrac{1}{2} r^2 \theta$.

What I recorded.

Question. How is the formula developed for finding the area of a sector of a circle of given radius and central angle?

Cues for giving the answer.

Start with a circle of radius r and a sector with radian measure θ.

Let A be the area of the sector of the circle.

The ratio of areas of sectors is equal to the ratio of their respective central angles.

The radian measure of the central angle of the circle is 2π.

I should be able to write out a complete answer for the question based on being able to recall the cues given. If I can't write the answer when asked to do so on an examination, it is most likely that I didn't study to remember the cues, or during my study I could not give the answer when having the cues.

Review Your Answers to All Your Questions in the Section

When you have finished a careful reading of a section, it is wise to review the entire section by reading each question and reflecting on the answer (preferably without first looking at your cues). It might seem strange to do this so soon after you did it as you were moving through the section. You might feel otherwise if you realize that approximately half of what you read is forgotten soon after you read it. Short reviews conducted early on improve retention. Finally, writing a *summary* of the entire section, incorporating your questions and their answers, is a wonderful way to conclude your reading of the section.

You may believe that using the method of SQ4R in reading your textbook is too time-consuming. Granted, it may take you several times of employing it to become efficient in its use. However, once that happens, *and you see how well you are learning*, you will find it difficult not to use it. Furthermore, the extra time you spend employing the method will reduce the time you have to spend in remediation, learning subsequent content, and preparing for examinations. This is because you are learning well as you progress through a unit.

Reading Specific Types of Mathematics Content

The *basics* of reading a mathematics textbook are covered in chapter 15. It does not cover how to read specific types of mathematics content such as definitions, theorems, and examples. There is a methodology to reading particular mathematics content and the purpose of this chapter is to call it to your attention. The sections of the chapter are these:

1. Reading Definitions of Concepts
2. Reading Mathematics Terminology and Symbols
3. Reading Statements of Theorems
4. Reading Proofs of Theorems
5. Reading Graphs and Tables
6. Reading Examples

1. Reading Definitions of Concepts

Some textbook authors do not make it clear when a concept is being defined, either by writing the word, "definition," or highlighting it another way. The definition of the concept may inconspicuously appear as another sentence in a paragraph; that is, the word or phrase naming the concept is given, followed by its definition. This makes it difficult to locate a definition or to know if a statement you are reading is a definition. The statement may be a *description* of the concept, as opposed to a *characterization* of it, which a definition provides.

For example one way to describe a prime number is to say it is a whole number. We can also describe it by saying it is not zero, and that it has at least one divisor. Do you now know what a prime number is?

Of course not—many non-prime numbers fit the descriptions I gave. I need to describe a prime number in a way that makes clear what it is, or is not. When that is done, the description is called a definition of a prime number—it leaves no doubt about what constitutes a prime number. To let you know that a definition of a concept is about to be given, many authors write the name of the concept in bold type or italics. This is especially true for less major concepts, reserving the word "definition" for more major concepts.

Concepts are the most basic mathematics subject matter, and success in a mathematics course requires you to know and understand them. Your everyday life would be more difficult if you didn't understand the myriad of concepts you face, such as chair, light, automobile, toothpaste, anger, compassion, diet, water, and grandchild. It is no different in mathematics. Mathematics concepts abound in quantity and variety, and these are examples: prime number, diameter of a circle, parallel lines, obtuse triangle, least common multiple of two numbers, function, inverse of a function, graph of a function, derivative of a function, definite integral of a function, surface area of a solid of revolution, and partial derivative of a function of two variables.

The importance of learning mathematics concepts is easily understood when you realize that the content of a mathematics course consists of concepts, theorems that relate them, and problems that apply them. In a real sense, *it's all about concepts*. Concepts are the stars of mathematics. Previously introduced concepts continually appear throughout a course, and new concepts are continually being introduced. A concept is defined by a precise statement, which gives it meaning. Words are carefully chosen to define a concept since it has to be perfectly clear what it means. Mathematics cannot afford a lack of precision and clarity about its basic subject matter. Unfortunately, many college mathematics students slight their understanding of definitions of concepts. Either they are unwilling to do what is necessary to understand them, or they are naïve about the importance of understanding them or what is required to understand them.

In college mathematics, many definitions are difficult to understand. Immediate understanding is not expected; hence, do not get anxious if you are confused when you first read a definition. Understanding evolves if you work at it, and for the more complex concepts, understanding may take days or weeks. An understanding of a more complex concept such as the definite integral, which appears in introductory calculus, is a gradual process. The many ideas that comprise it have to germinate in your mind. It may be in a subsequent course that a thorough understanding of it is first realized. Fortunately, future work with a concept keeps it before you and helps you understand it, as well as the use of means other than its definition to learn it. But you have to be persistent in your desire to learn it. Instructors and authors of mathematics textbooks know the concepts that are difficult to understand and adjust student expectations accordingly.

A Closer Look at Definitions

An aspect of understanding the definition of a concept is to understand the words that comprise the definition. These words are (1) labels for previously defined concepts or for concepts that are accepted as being undefined, or (2) typical words commonly understood (but perhaps in an imprecise way). There is a logic to definitions that is conveyed by use of words such as "for every," "for all," "where," "whenever," "if . . . then," "and," and "or." If you don't understand their logical use, get help. The following are three examples of definitions of concepts:

    Definition 1. The *domain of a real-valued function* is the set of all real numbers for which the function is defined. (To understand this definition you need to understand the concepts of a real number, function, and what it means for a function to be defined.)

    Definition 2. A *diameter of a circle* is a chord of a circle that bisects the circle. (For this definition to make sense, you need to know the definitions of a circle, a chord of a circle, and what it means for a circle to be bisected.)

    Definition 3. (*Definite integral of a function f from a to b*). Suppose f is a function defined on a closed interval $[a,b]$ and P is a partition of $[a,b]$ with partition points $x_0, x_1, x_2, \ldots, x_n$ where $a = x_0 < x_1 < x_2 < \ldots < x_n = b$. Choose points x_i^* in $[x_{i-1}, x_i]$ and let $\Delta x_i = x_i - x_{i-1}$ and $\|P\| = \max\{x_i\}$. The definite integral of f from a to b is $\lim\limits_{\|P\|} \sum_{i=1}^{n} f(x_i^*)\Delta x_i$, if this limit exists. It is denoted by $\int_a^b f(x)dx$.

The definite integral is a major concept in a first calculus course and is used in many significant applications of calculus. Its definition is given here to illustrate how complex a definition can be. Note the many concepts and symbols that appear in the definition, which are necessary to understand before you can understand the definition. However, they are not sufficient to understand this concept; that is, you also need to understand how they all fit together.

The sequence of words and symbols defining a concept is a first step toward understanding it. Other methods are now given to help you understand definitions.

Interpreting Definitions Geometrically

Some definitions have a geometric interpretation. Restating the definition in the language of geometry can make it more understandable, especially if you view graphs that show relationships among the concepts expressed in the definition. A geometric interpretation of a concept is often useful in solving problems that involve the concept. When a new concept is defined, look for its geometric interpretation, if there is one. What follows are two definitions of concepts from precalculus, each followed by a geometric interpretation of the concept, and an exercise that helps relate the definition to the geometric interpretation.

Definition 1. *(Increasing Function).* A function f is increasing on an interval I if for any two points x_1 and x_2 in I, $x_1 < x_2$ implies that $f(x_1) < f(x_2)$.

Geometric Interpretation of an Increasing Function. The graph of f is rising as we move from left to right or in the positive x direction.

Exercise. Draw several different graphs of functions that satisfy this geometric interpretation and verify for each one that the definition applies. Then draw several graphs of functions that do not satisfy this geometric interpretation and show that the definition does not apply.

Definition 2. *(One-to-One Function).* A function f is one-to-one if for x_1 and x_2 in the domain of f, $x_1 \neq x_2$, implies that $f(x_1) \neq f(x_2)$.

Geometric Interpretation. A function is one-to-one if and only if each horizontal line intersects the graph of the function in at most one point.

Exercise. Draw several graphs of functions that satisfy this geometric interpretation and verify for each one that the definition applies. Then draw several graphs of functions that do not satisfy this geometric interpretation and show that the definition does not apply.

In the two exercises that were just stated, you are asked to give geometric *examples* and *non-examples* of the definition. This sets the stage for another method to use in understanding a concept.

Finding Examples and Non-Examples of a Concept

In your attempt to understand a concept you need to do more than read the definition of the concept. How will you know that you understand it? One way to help you understand a concept is to give examples and non-examples of the concept as demonstrated for the following two concepts:

Definition 1. A prime number is a whole number with exactly two divisors.

A test to determine if you have a good grasp of a prime number is to apply the definition to come up with several numbers that are prime (called *examples*) and several numbers that are not prime (called *non-examples*). For instance, 2, 5, and 17 are prime since they each have exactly two divisors (the divisors of 2 are 1 and 2, the divisors of 5 are 1 and 5, and the divisors of 17 are 1 and 17). The numbers 0, 1, 14, and 12 are not prime (these are non-examples), since they each have more than two divisors (0 has an infinite number of divisors, 1 has one divisor, 14 has 4 divisors, and 8 has 4 divisors.) Notice that I did more than list examples and non-examples. I also *explained* why the examples satisfy the definition and why the non-examples do not. You will also want to justify your examples and non-examples because that will help you know if you understand the definition.

Definition 2. A quadratic function $f(x)$ is defined to be a function of the form $f(x) = ax^2 + bx + c$, where a, b, and c are real numbers and $a \neq 0$.

What follows are varied examples of quadratic and non-quadratic functions, with accompanying reasons:

Examples of quadratic functions (with accompanying reasons).

(a) $f(x) = \dfrac{1}{2}x^2$.

(b) $f(x) = -4x^2 + 200x$.

(c) $f(x) = \dfrac{4}{\sqrt{2}}x + 6 - \dfrac{x^2}{2}$.

(d) $f(x) = -7 - \sqrt{5}x^2$.

These examples are all of the form $f(x) = ax^2 + bx + c$, $a \neq 0$ as they appear, or they are equivalent to functions of this form. Note, for instance, that $f(x) = -7 - \sqrt{5}x^4$, is equivalent to $f(x) = -\sqrt{5}x^2 + (-7)$, which is of the form $f(x) = ax^2 + bx + c$, $a \neq 0$. (Note that $b = 0$, which does not violate the definition.)

Non-examples of quadratic functions (with accompanying reasons).

(a) $f(x) = -4x - 4$ (since $a = 0$).

(b) $f(x) = \sqrt{4x^2} - \dfrac{x}{3}$ with $x > 0$ (since $f(x) = \sqrt{4x^2} - \dfrac{x}{3}$ is equivalent to $f(x) = \dfrac{5x}{3}$; hence, $a = 0$).

(c) $f(x) = 9x^3 + 4x^2 + x - 8$ (since the highest exponent on x is greater than 2).

The diversity of examples and non-examples were deliberately chosen and you should do the same for your definitions. The greater the diversity of your examples, the more you are assured that you understand the concept being defined.

Memorizing Definitions of Mathematics

It is recommended that you memorize the precise wording of a definition as given by your instructor or presented in your textbook. That you should also understand it goes without saying, but I felt compelled to say this one more time. Considerable thought is given by mathematicians when constructing definitions. They are careful to include appropriate words, grammar, and logic. Words in a definition are carefully chosen to ensure that they precisely and concisely characterize the concept being defined. It is possible, but often not easy, to give an equivalent definition using your own words. The slightest alteration can change the meaning of the concept.

Whenever I am working with a concept and have some doubt as to its exact meaning, I bring its definition to the forefront of my mind and repeat it to myself. I know the importance of understanding definitions so I make them part of my "memory repertoire" as I study a unit. You need to approach your study of mathematics knowing the importance of remembering important content, which includes definitions of concepts.

The words of a definition are the basic means of communicating the meaning of the concept to yourself and to others. I continually stressed this to my students, yet far too many did not heed my advice. This was at their peril since it negatively affected their learning of mathematics in numerous ways, many of which they were unaware of. The definitions they gave me on examinations would make a mathematician cry. Some of them would give the statement of a theorem, thinking they had given the definition; others would give the first part of the definition and finish with the conclusion of a theorem, or write something that was absolute nonsense. There are some educators who say it is meaningless to expect students to memorize definitions. They are correct if they mean memorization without understanding. However, *memorizing a definition with understanding is necessary to succeed in mathematics.* If you don't know the definition, how can you know the concept, communicate it to others, know when you have applied it properly, or understand the statements or proofs of theorems that use it?

Your Instructor's Definition and the Definition in the Textbook May Not Be Worded the Same

If your instructor's definition of a concept is worded differently from the wording that appears in your textbook, which one should you use? First of all, check to see that they are equivalent definitions. That is, determine that they are saying the same thing, even though the wording is different. If they are equivalent, then either one can be used. For example, here are two equivalent ways to define a prime number: (1) It is a whole number greater than one whose only divisors are 1 and itself. (2) It is a whole number with exactly two divisors. You can see that each of these definitions gives the same set of numbers that we call prime, namely $\{2,3,5,7,11,13,17, \ldots\}$. Memorize the one that is easiest for you to understand, apply, and remember. You benefit from knowing both definitions. If your instructor says that you must use his or her definition, then you don't have a choice.

2. Reading Mathematics Terminology and Symbols

The terminology you come across in reading mathematics is of three types: (1) special mathematics words being introduced for the first time, (2) special mathematics words that are prerequisite knowledge, and (3) commonly used words. In defining a mathematics concept all three types are used. Also, the specialized mathematics *symbols* that are used to define concepts are part of the vocabulary. Hence, to understand a concept you need to understand the mathematics vocabulary and the symbols that are used in its definition. For example, it is important to know that symbolic names or labels for the composition of functions f and g are $f \circ g$ and $f(g)$. Similarly, the definite integral of f from a or b, is labeled $\int_a^b f(x)dx$, and $x^{\frac{a}{b}}$ has labels $(x^{1/b})^a$, $(\sqrt[b]{x})^a$, or $\sqrt[b]{x^a}$. Symbolic form is the most highly condensed and concise form of expression in mathematics. It is *absolutely mandatory* that you can read it, write it, interpret it, give equivalent forms for it, and give its uses. Chapter 19 is devoted to learning symbolic form.

Reasons to Learn Mathematics Terminology

When diagnosing mathematics students' difficulties in understanding mathematics, language problems seem to go unnoticed. You won't succeed in a mathematics course if you are not, to a suitable extent, fluent in the use of mathematics terminology because—

- You cannot separate mathematics understanding from the language used to learn it.
- It is the major way mathematics is communicated to you.
- It is the major way you will communicate mathematics to others.
- Your knowledge of mathematics is assessed by your oral and written use of mathematics terminology.

It's interesting to note the *dual role* played by terminology in a mathematics course: (1) It is part of the content of the course, as well as (2) the means to convey this content. This dual role should be motivation for you to double your efforts to learn it.

Examples of Language Used in Mathematics

There are two ways you can have difficulty with special mathematics words: (1) you know the word or phrase that names the concept, but don't understand the concept, or (2) you understand the concept, but don't know the word or phrase that names it. The same holds true for the symbolic names for the concept. Examples of *special mathematics words or phrases* are:

hypotenuse, parabola, quadratic formula, closed interval, domain of a function, transcendental function, derivative of a function, definite integral of a function, geometric series, vector, gradient, factorial, complex number, conic section, Arctangent, cofactor, Lagrange multiplier, polar axis, solid of revolution, concavity, centroid, implicit function.

You can have difficulties with *words commonly used outside of mathematics that have precise meaning in mathematics*. Examples of these words are:

equivalent, continuous, limit, slope, reflection, period, translation, average, work, divergence, moment, and partition.

Additionally, there are *commonly used words or phrases in mathe-*
matics exercises that convey what you are being asked to do. Your success is compromised if you don't understand their use in mathematics.
Included among these are:

> verify, simplify, consider, equate, show that, combine, distin-
> guish, judge the meaning of, test, what persuasive technique,
> derive, refer, conclude, specify, evaluate, illustrate, identify,
> complete, denote, summarize, substitute, let, transform,
> demonstrate, give, verify, compare, deduce, assume, set up,
> eliminate, convert, develop, argue, describe, suppose, make a
> distinction, develop, formulate, judge, criticize, defend, carry
> out.

You need to understand phrases such as "at most," "at least,"
"if . . . then," "if and only if," and the precise use in mathematics of the
connectors "and" and "or."

Abbreviations are often used in mathematics, including:

> inverse (inverse function), cos (cosine), log (logarithm base
> ten), integral (definite or indefinite integral), sinh (hyper-
> bolic sine), lim (limit), csc (cosecant), rad (radian), ln (log-
> arithm base e).

You have to understand all types of mathematics terminology used
in your mathematics course, including abbreviations and symbolic
form. Vague or incomplete understanding will cause you problems.
Take the initiative to ask questions on the language you don't understand.

How to Learn Mathematics Terminology

You will want to involve as many senses as you can to learn a new
mathematics word or combination of words that label a concept. You
need to see it, say it, hear it, and write it. This has to be done numerous
times throughout a unit of study and in varied contexts. Here are ways
to be more fluent in your use of mathematics terminology:

Work in groups to discuss your assignments, your textbook
reading, and your class notes. Mathematics can be viewed as a foreign language and you know the challenges to understand and speak a

foreign language. You have to be engaged in oral work in mathematics. We learn to speak and understand everyday language because we are continually exposed to it, whether we seek it out or not. How often is mathematics language spoken on television, at a sporting event, in a church service, or around the dinner table? Mathematics conversation is generally restricted to an academic setting, and at best, mathematics students minimally engage in it. You have to avail yourself of every opportunity to hear and speak it—in and out of the mathematics classroom. Make a point to use new mathematics terminology in your conversation as soon as you can and whenever you can. Initially this may be difficult to do, but eventually it becomes habit. Perhaps you and a classmate can agree to meet frequently to practice using mathematics terminology. Many college students speak everyday language at a college level, but speak mathematics language at an elementary or middle school level. It doesn't have to be this way. To be fluent in the language of mathematics requires frequent use of mathematics language.

Write mathematics and write about mathematics. Understanding and using mathematics terminology happens in an environment where mathematics terminology is present. *You* create this environment when you discuss mathematics, when you write mathematics, and when you write about mathematics. The technical aspects of writing mathematics are the content of chapter 20, and writing about mathematics is the content of section 2 of chapter 14.

Read mathematics. Reading mathematics, occasionally with a partner, can help you learn mathematics terminology. Pause periodically to talk to someone about what you have read. *Use the language you are reading* to talk about what you find interesting, confusing, and how it might apply. Your reading will help your oral discussion, and vice versa. If you are inexperienced in reading mathematics textbooks, then you will have more difficulty learning mathematics terminology. The solution is to gain more experience reading mathematics, not to avoid it. Students who have difficulty reading mathematics tend to avoid reading it, which is the exact opposite of what they need to do. If you need help with your reading, get it. There is considerable advice throughout this book that will help you become a better reader of mathematics. For example, section 3 of chapter 9 discusses how you can improve your thinking in mathematics by working with a classmate as you read mathematics. The techniques described there will also help you learn mathematics terminology.

Write the verbal and symbolic name of a concept on one side of a 3×5 card and its definition on the other side. Make a card for each new concept. Here's how to use this stack of cards in your study: Read the verbal and symbolic name on one side of the card, and see if you can give the definition, which is on the other side of the card. At times, reverse the procedure (i.e., read the definition on one side of the card, and see if you can give the verbal and symbolic name on the other side). Do this frequently as you progress through a unit, beginning when you are first exposed to the definition of a concept. Eventually, you will be able to dispense with the cards. These cards will be useful again when you review content that is required for subsequent units or for a final examination.

3. Reading Statements of Theorems

As you read your mathematics textbook you may be confused as to why the term, Theorem, is used at times to label a statement that can be proven true, and at other times a statement that can be proven true is not labeled with "Theorem." You may also wonder why some statements that can be proven true are referred to as "Corollaries." Methods for understanding these statements—regardless of how they are labeled, and the importance of knowing their uses or applications, are now discussed.

Describing Theorems, Corollaries, and Propositions

Many statements appear in a mathematics textbook that can be proven to be true. The more important statements, those that apply widely to solving problems or to gaining a better understanding of the structure of the subject, are typically labeled "Theorems." However, some textbooks authors may use the words "Law," "Principle," "Rule," or "Test," as labels for a statement that can be proven true. The less important statements that can be proven to be true are often not given the lofty name of "Theorem," although it would be proper to call them such. They may be referred to as propositions or properties and their statements and proofs often appear without fanfare in the textbook, although some may be highlighted in some manner. They may be less important to the development of the content, but are still important. Their proofs are often more simplistic than proofs of the so-called theorems. Your success in understanding mathematics depends on learning the statements and proofs of many theorems and of these so-called propositions. Granted, there may be a proof of a theorem that is very difficult to understand. Your instructor knows this and adjusts his or her expectations accordingly. *From now on I will refer to any statement that can be proven true as a theorem, whether or not it is one of the more important statements.* Labeling of statements is inconsistent across textbook authors. You don't need to be too concerned about how a statement is labeled, since its label is far less important than its content.

The following are statements of theorems (the first two from precalculus and the last two from calculus):

1. A function h has an inverse function if and only if it is a one-to-one function.
2. The graphs of g and g^{-1} are symmetric about the line $y = x$.
3. Let f be continuous on $[a,b]$ and differentiable on (a,b). Then there is a number c in (a,b) such that $f'(c) = \dfrac{f(b) - f(a)}{b - a}$.
4. If $g(x) = c$, c a constant, then $g'(x) = 0$.

Theorems play a central role in a mathematics course since they *relate* the basic subject matter of the course; namely, the concepts of the course. The importance of the concepts of a course to the theory of a course, and to solving problems, comes from *relating* them. A necessary aspect of understanding the statement of a theorem is to understand the definitions of the concepts appearing in the theorem. In the following theorems, the concepts being related are identified:

1. A function h has an inverse function if and only if it is a one-to-one function. (The concepts being related are *inverse function* and *one-to-one function*.)
2. $\log_b PQ = \log_b P + \log_b Q$, if P and Q are positive real numbers. (The concepts being related are the *logarithm of the product of two numbers,* and the *logarithms of its factors*.)
3. In a circle of radius r, the area of a sector with central angle of radian measure θ is given by $A = \dfrac{1}{2}r^2\theta$. (The *area of a sector of a circle* is being related to the *radius of the circle* and the *central angle of the sector*.)

A *corollary* to a theorem is a theorem whose proof is easily or quickly deduced from the theorem. Stating it another way, the major step in the proof of the corollary is the use of the theorem that it is a corollary to. A corollary is often a special case of the theorem. The following calculus theorem is a corollary to The Mean Value Theorem whose statement was given a few paragraphs ago:

If $g'(x) = 0$ for all x in an interval (a,b), then g is constant on (a,b).

The key step in the proof of this corollary is justified by The Mean Value Theorem. This corollary is used in the proof of this theorem, which provides a method for finding all the antiderivatives of a function:

If $f'(x) = g'(x)$ for all x in an interval (a,b), then $f(x) = g(x) + C$, where C is a constant.

Methods for Understanding Statements of Theorems or Propositions

Reading or memorizing the statement of a theorem without thoroughly understanding the statement is a fruitless activity. Here are some strategies you can use to help understanding:

Restate the theorem in your own words. It is difficult to claim something as your own if you have not invested yourself in it. Restating a theorem in your own words is investing yourself in the theorem. This does not mean that you won't use some of the words of your instructor or those that appear in your textbook. What it does mean is that you will work to formulate a wording that helps you understand the theorem, or makes it easier for you to remember and apply it. You may change or add a few commonly used words, eliminate some symbols, or change the order in which things are stated. You have to be careful not to alter the statement so that you change its meaning. Here are examples of statements and their restatements, which use fewer symbols:

Statement 1. If $G(x) = C$, where C is a constant, then $G'(x) = 0$.

Restatement. The derivative of a constant function is 0.

Statement 2. $\sin(s + t) = \sin s \cos t + \cos s \sin t$.

Restatement. The sine of the sum of two real numbers is the sine of the first number times the cosine of the second number plus the cosine of the first number times the sine of the second number.

Statement 3. Suppose f is continuous on $[a,b]$. Then for any antiderivative F of f, $\int_a^b f(t)\,dt = F(b) - F(a)$.

Restatement. To find the definite integral from a to b of a continuous function f, find an antiderivative of the function, evaluate it at b and subtract from this its value at a.

For many students, especially those with a verbal learning style, these restatements are simple to remember and better convey the meaning of the statement, especially since fewer symbols are used that can obscure meaning. The point to remember is that your restatement has to serve you. If your learning style adapts well to symbolic form, then the statements (1), (2), and (3) may serve you better than the restatements that are given.

Another way to restate a theorem is to restate all or part of it in the language of geometry; that is, restate it geometrically. This is not always possible to do, but if it is, it not only makes it more understandable, it provides a visual verification of the *plausibility* of the theorem. Here is a geometric interpretation of The Mean Value Theorem (if you have not had a first course in calculus, the mathematics used will be confusing):

The Mean Value Theorem. Let f be continuous on $[a,b]$ and differentiable on (a,b). Then there is a number c in (a,b) such that
$$f'(c) = \frac{f(b) - f(a)}{b - a}.$$

Geometric Interpretation. If the graph of f has no breaks and no sharp points on $[a,b]$, then there is a point $(c, f(c))$, for c belonging to (a,b), which is on the graph of f such that the tangent line to the graph of f at this point has the same slope as the line through the points $(a, f(a))$ and $(b, f(b))$.

Exercise (which shows its plausibility). Draw several different graphs, each of which are graphs of a function that satisfies the hypothesis, and verify that the conclusion holds by drawing a tangent line to the graph at a point $(c, f(c))$ whose slope appears to be the same as the slope of the line through the points $(a, f(a))$ and $(b, f(b))$.

Memorize the statement of the theorem. Restating a theorem in your own words helps you understand and remember it, and perhaps communicate it to others. This does not imply that you should not also memorize the statements of theorems given by your instructor or appearing in your textbook. The statements of theorems your instructor gives in class will often vary in language from equivalent ones given in your textbook (the same is true for definitions). Which ones should you be able to give? That's a decision you have to make based on which formulation serves you best.

If your restatement of a theorem has eliminated many of the symbols used in the other formulations you have, it is wise to also know a formulation that has more symbols in it. Most of your uses of a theorem to solve problems or to prove other results will involve using symbols. Hence, you need to be familiar with how these symbols are used in the statement of the theorem. Knowing the original statement of the theorem helps formulate a restatement, and vice versa. A good study aid is to make a 3×5 card with the statement of a theorem on one side, and your restatement of it on the other side. Study by looking at one side of the card, and then see if you can give the original statement of the theorem, or your restatement of it, on the other side.

Know the logical structure of the theorem's statement, including the hypothesis and the conclusion. When you memorize the statement of a theorem, pay attention to its logical structure. "Laws of Logic" are used in mathematics to give meaning to statements of theorems and their proofs. Key words in these Laws are: "or," "and," "if," "then," "if and only if," "for every," and "for all." Other non-mathematics words or phrases used include: "let," "suppose," "consider," "assume," "such that," "so that," "whenever," "where," "along with," and "satisfying the condition." It is important to know the specific meanings of these words when they appear in statements of theorems or definitions. The following are equivalent phrases (i.e., they say the same thing—they are interchangeable):

Suppose f is continuous on $[a,b]$
Let f be continuous on $[a,b]$
Assume f is continuous on $[a,b]$
Consider f to be continuous on $[a,b]$

These are also equivalent phrases:

such that f is an increasing function
so that f is an increasing function
where f is an increasing function
along with f being an increasing function
satisfying the condition that *f* is an increasing function.

Many theorems are stated in "if-then" form. That is, *if* such and such is true, *then* such and such is true. For example: "If a function *f* has an inverse, then *f* is a one-to-one function." What immediately follows the "if" in the statement, namely "a function *f* has an inverse," is called the *hypothesis* and what follows the "then" in the statement, namely, "*f* is a one-to-one function," is called the *conclusion*. The hypothesis and the conclusion may contain the connectors "and" or "or," one or more times, which makes them compound statements. This creates a more complex statement. When there is more than one statement comprising the hypothesis you will often hear this question, "What are the hypotheses?"

An important fact to know about a theorem is that if the hypotheses are true, then the conclusion is true. A proof of the theorem is a sound (logical) argument that shows how the conclusion can be *derived*, given the hypothesis. To make this argument, Laws of Logic are applied. Generally, a statement is not included in the hypothesis of a theorem if it is not needed to derive the conclusion. However, there are times when more information is included in the hypothesis than what is necessary so the conclusion is more easily derived from the hypothesis. This is the case for theorems whose proofs, without this added information, may be too difficult to understand for the intended readers.

Change parts of the hypothesis of the theorem. If the theorem relates to functions, you can be assured that the conclusion is true for a *specific* function if the hypothesis is true for that function. That is, it is impossible to find a function for which the hypothesis of the theorem is true and the conclusion is false. This is real security! However, if the hypothesis is false for a specific function, then the conclusion may or may not be true for that function. Nothing is guaranteed if you have a false hypothesis. For the following theorem, a function is defined in *example 1* where the hypothesis is false and the conclusion is false, and another function is defined in *example 2* where the hypothesis is false and the conclusion is true. (You may not understand the theorem and the examples if you haven't had the appropriate mathematics courses.)

The Intermediate Value Theorem. If a function *f* is continuous on a closed interval $[a,b]$ and D is a number strictly between $f(a)$ and $f(b)$, then there exists a number c belonging to the open interval (a,b) such that $f(c) = D$. (That is, if *f* is continuous on $[a,b]$, then *f* takes on every value strictly between $f(a)$ and $f(b)$.)

Example 1. (false hypothesis and false conclusion).

$$f(x) = \begin{cases} x, & \text{if } 0 \le x < 1, \ 1 < x \le 2. \\ 2, & x = 1. \end{cases}$$

Note that f is defined on [0,2] but is not continuous on [0,2] since it is not continuous at $x = 1$ (there is a break in the graph of f). Thus, the hypothesis of the theorem is false. The conclusion is also false since f does not take on the value 1, which is between $f(0) = 0$ and $f(2) = 2$. This is easy to see from the graph of this piecewise-defined function.

Example 2. (false hypothesis and true conclusion).

$$g(x) = \begin{cases} x, & \text{if } 0 \le x < 2. \\ \dfrac{3}{2}, & \text{if } x = 2. \end{cases}$$

The function g is defined but not continuous on [0,2]. However, it takes on all values between $g(0)$ and $g(2)$. This is easy to see from the graph of this piecewise-defined function.

It is a good exercise in calculus-track courses to create examples of functions where a statement in the hypothesis is not true and the conclusion is false. This tells you that you cannot say the conclusion is true for *all* situations unless you know the hypothesis is true for those situations. For some theorems, textbook authors will have exercises that ask you to do this, but it is something you should do regardless. Another thing you can do to help you understand if each statement in the hypothesis is necessary, is to determine where each is used in the proof of the theorem. There are theorems where this is not easy to do, but it is worth the effort. It is more important to do this in higher-level mathematics courses.

Know the uses or applications of theorems. You need to know the theoretical and practical uses or applications of the theorems you study. This has to be part of your ongoing study for two reasons: they are a critical part of understanding mathematics, and to prove theorems and work problems you need to use them. Hence, the uses or applications of theorems have to be at your fingertips. When you are exposed to applications in class or in your textbooks, be cognizant of the theorems that were applied, and keep a written record of these relationships. One suggestion is to take a 3×5 card and write the name and statement of the theorem on one side of the card, and its uses on the backside. Add additional uses as they arise in the course. If a theorem gives a technique for working certain types of problems, you can also include some of these problems on the backside of the card. You have now created a great study device—look at one side of a card and see if you can give the information on the other side. Reviewing these cards as you progress through a unit will help you with your reading, conversations, and assignments. It also makes it easier to study for unit examinations and the final examination. You can take your deck of cards for a unit or for the course, shuffle them, and then work with them as suggested.

Example. On one side of a card write:

A formula for changing the logarithm of a function in *base b* to the logarithm of a function in *base a.*
On the other side of the card write the formula, followed by an example of its use:

$$\log_a x = \frac{\log_b x}{\log_b a}.$$

$$\log_2 100 = \frac{\log_{10} 100}{\log_{10} 2} = \frac{2}{\log_{10} 2}.$$

4. Reading Proofs of Theorems

One of the great contributions of mathematics to knowledge involves its methods of proof. Proof is a powerful means of justification. It makes mathematics what it is, and if you are not engaged in the proofs of theorems, to an appropriate extent, you are compromising your understanding. Mathematics is unique in its methods of proof. Its means of validation is, for example, unlike chemistry, where the basic means of validation is experimentation. If your goal is to understand mathematics well, then understanding proofs of theorems becomes indispensable. To understand something, you have to relate or connect it to other things and that is what a proof accomplishes. In the proof of a theorem you begin with the hypothesis and make a series of relational statements that lead you to the conclusion. Regrettably, understanding proofs of theorems is an area that is severely slighted by many students, most notably those who don't do well in their mathematics courses. The benefits of understanding proofs are many and include these:

To help you determine the theorems that can be used, or how they can be used, to solve problems in the course. It is often the case, after you read a problem, that you are at a standstill as to what to use to solve it. You have a greater chance of coming up with solution strategies if you understand proofs of theorems since specific statements in a proof may very well relate to what you are being asked to do in the problem.

To help you solve problems in disciplines that use mathematics. In my early years of teaching I frequently taught calculus courses and knew well how to apply calculus to solve certain types of science problems, whether in chemistry, physics, or engineering. Because I was familiar with specific science concepts and with the proofs of calculus theorems, I had little difficulty seeing how the more generically stated concepts or theorems applied in these specific situations. Then I started getting questions from undergraduate students in other disciplines such as biology or business, or even from college graduates on how they could use calculus in their jobs to solve problems that arose in their work. I initially struggled with these problems since I didn't understand specific concepts in disciplines from which the problems came. I couldn't see how they related to the more generically stated mathematics concepts and theorems. It was only after I learned more about the meaning of certain concepts in specific disciplines, and reflected more on the proofs of theorems, that I saw how they could be used. This depth of understanding will stand you in good stead when you have to apply mathematics.

To increase your interest and enjoyment. Working to understand a proof is like playing a card game or working a puzzle. It is a challenging activity. For many students, it makes learning mathematics enjoyable. They accept the challenge of deciphering a proof, and marvel at the ingenuity of the "proof-maker" who logically derives the conclusion from the hypothesis. You may not be able to construct a proof of a specific theorem, but lots of satisfaction and knowledge comes from understanding someone else's proof of that theorem.

To help you review knowledge. Since a proof involves the use of previous concepts and theorems, you will have to recall and understand them to understand the proof.

To increase your ability to think logically or analytically. When employers came back to my campus, it was exciting to hear how they liked to employ graduates with a mathematics background (particularly, mathematics concentrators), almost regardless of the types of jobs they wanted to fill. They stated that employees with this background can think logically and use this method of thinking to approach problems, regardless of the problem. An employer would not be able to make this comment if the employee did not work with proofs in their college mathematics courses, and eventually understand them. The more you do this in your mathematics courses, the more you will perfect the skill of thinking logically. (See section 3 of chapter 9 for more comments on growing in this method of thinking.)

Your Understanding of Proofs of Theorems May Not Be Required by Your Instructor

You may have the misfortune to have an instructor who does not expect you to know proofs of theorems. It doesn't mean that this instructor does not prove theorems in class; but it does mean that you will not be evaluated on your ability to understand or construct proofs. You may then think that this is an optional objective for you and dispense with learning proofs of theorems or working exercises that require you to prove theorems. From all that has been said you should know that this would be a huge mistake. If your goal is to understand mathematics, it cannot be an optional objective for you. This is not to say that you should be held accountable for the proofs of all the theorems in the course. Far from it! There needs to be an appropriate balance between the proofs you are, and are not, expected to know, and this has to take into consideration the time you have and what can be gained by knowing a specific proof. In the final analysis, you have to make the decision—*you* are responsible for your learning, not your instructor or anyone else.

Understanding the Proof of a Theorem Takes Effort

If you are not used to understanding proofs of theorems, it may take time for you to develop this skill. Perseverance, questioning, and getting help when you need it are the keys to success. Knowing what this type of learning can do for you should motivate you. When you work through the proof of a theorem, have pencil and paper handy. Most proofs you read are incomplete for one or more of the following reasons:

Steps are left out that justify going from one step to the next. For example, a statement in the proof is recast in another form, but steps are not given that show that this form is equivalent to what you had. You need to fill in these steps.

The theorems that justify key steps of the proof are not named. You need to supply the missing names.

It is not shown that all the hypotheses of a previous theorem are satisfied, when that theorem is being used to justify specific steps in the theorem whose proof is being studied. You need to show that these hypotheses are satisfied.

Be thankful that all proofs are incomplete because it provides more opportunity to learn. By filling in the missing steps and justifications you are, in some sense, a co-author of the proof. To fill in the missing details you will need to look back in the textbook to re-familiarize yourself with theorems and definitions, consult your class and textbook notes, or consult others. Over time, you will get better at filling in the details. The gaps you need to fill in between two statements in the proof may be big or small. Make a note when you cannot fill in a gap between two steps of the proof, and then get help when you can. Meanwhile, continue on with the proof. Eventually, you need to understand the justification *of each step* of the proof. That is what it means to understand the proof.

When you finish reading through the proof, make an outline of the key steps of the proof. Then close the book and see if you can reproduce the theorem, using your outline if needed. Go back to the book for a particular step if you need to, but eventually you should come up with the proof. At times I would ask students on an examination to prove a theorem that was proven in their textbook. (In a study guide for the examination I would mention the theorems I wanted them to be able to prove.) It was always clear to me when a student did not understand a proof. He or she would attempt to write verbatim the proof given in the textbook, and would almost always leave out key statements or misstate a statement. Memorizing the proof of a theorem, without understanding it, is problematic at best. You need to reconstruct a proof by remembering key steps in the proof and then fill in missing steps based on your understanding of the proof. Your proof will probably not be stated verbatim to the proof in the textbook (be consoled by that), but it will be a valid proof, and most importantly, it will be yours.

An example is now given of (1) the statement of a theorem, (2) a proof of the theorem as it may appear in a textbook, (3) a proof of the theorem with the missing steps filled in, and (4) an outline of the proof of the theorem (which you will use to recall the proof of the theorem).

Theorem. If the slopes of two lines are the same, then the lines are parallel.

Proof (that might appear in a textbook). Let the two lines be L_1 and L_2, each with slope m. Let the equations of L_1 and L_2 be $y = mx + b_1$ and $y = mx + b_2$, respectively. If they are not parallel lines, then they have a point in common, say, (x_0, y_0). Hence, $y_0 = mx_0 + b_1$ and $y_0 = mx_0 + b_2$. Therefore, $b_1 = b_2$. This means L_1 and L_2 are the same line since they have the same slope and y-intercept. Hence L_1 and L_2 are parallel.

Proof (with the missing steps filled in). Let the two lines be L_1 and L_2, each with slope m. Using the slope-intercept form of each line, let the equations of L_1 and L_2 be $y = mx + b_1$ and $y = mx + b_2$. $b_1 \neq b_2$ since the lines are not the same. They are parallel or they are not. If they are not parallel, then they must intersect in one point, say, (x_0, y_0). The coordinates of this point satisfy each equation. Hence, $y_0 = mx_0 + b_1$ and $y_0 = mx_0 + b_2$. Now, $y_0 = mx_0 + b_1 = mx_0 + b_2$ implies that $b_1 = b_2$. But this is a contradiction that $b_1 \neq b_2$. Hence, by the Method of Indirect Proof, L_1 and L_2 are parallel. (Note: the first proof leaves out a lot of justification. The second proof justified all of the steps of the textbook proof.)

Outline of the proof.
1. Introduce two distinct lines with the same slope and different y-intercepts using their slope-intercept forms.
2. They are or are not parallel.
3. Assume they are not parallel.
4. Derive the contradiction that they are the same line.
5. Accept the only other possibility, which is that they are parallel.

If you have studied the proof of the theorem, reviewed your proof outline periodically, and practiced deriving a proof from the outline, you will be in good shape when examination time comes.

To learn the proof of a theorem, a good study technique is to place the statement of the theorem on one side of a 3×5 card, and an outline of the proof on the other side. The outline will consist of the major steps of the proof. Then, periodically throughout a unit, read the statement of the theorem and see if you can give the outline appearing on the other side. If so, write out a complete proof, not using any notes. If you cannot recall the outline of the proof, read it on your card, and study it to remember it. Now work on giving the complete proof, filling in the necessary details, relying on your *understanding* of the proof.

Giving Your Proof of a Statement

There are times in an assignment when you will be asked to prove a statement that you have not seen proven. Sometimes you will be asked to prove a second part of a theorem that is analogous to the first part that has been proven in the textbook. In that situation, all you need to do is come up with statements in your proof that are analogous to those in the first part of the proof. These are excellent learning assignments. The following theorem from calculus, called the Second Derivative Test, is an example:

> **First part proven in the textbook.** If $f'(c) = 0$ and $f''(c) < 0$, then f has a relative maximum value at c.
> **Analogous second part that you are to prove.** If $f'(c) = 0$ and $f''(c) > 0$, then f has a relative minimum value at c.

The symmetry of these two statements implies that their proofs will be similar. When making the corresponding changes, don't do it mindlessly.

Understanding the proofs of theorems given in your textbook, or those given by your instructor in class, is a necessary step to be able to supply your own proof to a theorem whose proof you have not seen. However, it is one thing to understand someone else's proof, and another thing for you to construct your proof. The latter is a more difficult task but becomes easier with practice. Most likely the statements you may be asked to prove have proofs that are within your grasp. Nonetheless, they still may be a challenge. Constructing a proof is a problem solving process. Many of the skills used to construct proofs are skills that are used in problem solving.

Words of Caution on Reading and Understanding Proofs

Now that I have said what I wanted to say about reading and understanding proofs, you have to use some common sense in the amount of time you spend on learning proofs. You only have so much time to

study in your mathematics course; hence, you have to use that time ju-
diciously, choosing the study activities that will provide the most help.
When time is of the essence, study proofs of theorems that shed the
most light on the concepts the theorems deal with, and less time on
those that have subtle twists that are not useful for you to understand.
The more you progress in mathematical maturity, the more you will be
able to ascertain this as you read proofs of theorems. Hopefully, your
instructor will help you make these determinations—ask him or her for
this help. There comes a time when the amount of additional understanding obtained by spending more time isn't sufficient to justify the
additional time spent.

5. Reading Graphs and Tables

Mathematics textbooks display many graphs and tables. Over the years
there has been an increased use of graphs and tables in mathematics
textbooks. The more innovative textbooks of today are replete with
graphs and tables. It is not coincidental that this change coincided with
the invention and affordability of graphics calculators, which also aid
in the construction of tables. There is a simple explanation for this
change, and that is to help students attain a better understanding of
mathematics. Many years ago most mathematics textbooks were more
difficult to understand, relying in a major way on words and algebraic
statements. The presentation was heavily laden with abstraction,
mainly through serious use of symbolic form. More diverse ways were
needed to communicate mathematical ideas, especially those associated with functions, which is what mathematics is primarily about. The
functions of precalculus and calculus are primarily functions of numbers, but functions of vectors are introduced later on in the calculus sequence. Functions can be defined in a variety of ways—verbally,
graphically, and tabularly. Hence, the Rule of Three came into play,
and that is to analyze, when feasible, concepts and theorems involving
functions by these three means: (1) arithmetical, (2) algebraical, and
(c) geometrical. The change in textbooks was a greater increase in displaying and discussing graphs of functions, and tables of arithmetical
numbers.

How much do graphs aid understanding? This is easy to answer by
reflecting on the statement, "A picture is worth a thousand words."

What comes to mind when you think about some event that has happened? For example, suppose you went to a movie with a friend last evening and are thinking about that event. There is little doubt that visual images arise in your "mind's eye." You saw events as they unfolded. Maybe words also come to mind but they are coupled with the images. Why should it be any different when you think about mathematics? Life is a series of continuing stories, involving words and visual images. We live in a visual world. Similarly, mathematics is a series of continuing stories involving words and visual images. The mathematics world is visual and graphs are a big part of that.

The Rectangular or Cartesian Coordinate System is a wonderful invention that married algebra and geometry. This marriage allowed us to picture algebraic statements by plotting ordered pairs on a grid containing two perpendicular lines as axes. To learn mathematics well you have to see mathematics, which means that you have to carefully study the variety of graphs that appear in your textbook. More is said on this later.

The main reason why tables are so in vogue these days in mathematics textbooks is that their displays of specific numbers are easily grasped by the reader. A number, whether it is a whole number, an integer, a rational number, or real number, is an abstract entity, yet your extensive use of numbers over the years have made them *concrete* to you—they have become old friends. Tables containing numbers are used in a variety of ways to accomplish specific learning objectives. Studying tables of numbers, as well as constructing your own tables of numbers, enhances your understanding. More is said on this later.

What Can Be Accomplished by Studying Graphs?

Here are ways in which graphs enhance your learning:

They aid in understanding mathematics concepts and statements of theorems. The words used to define concepts or state theorems, together with seeing examples or applications of them, are often not enough for you to gain a good understanding of them. A bright light is turned on for you when a concept or theorem is pictured graphically.

These precalculus and calculus concepts, among others, become more understandable through the use of graphs (and, for many of them, through the use of tables of numbers):

$g(x)$ is increasing on $[a,b]$
Trigonometric functions
Limit of $f(x)$ as x approaches a
$h(x)$ is continuous at a
Relative minimum of $f(x)$ at a
Definite integral of f from a to b
Δx and dy
Antiderivatives of a function h
Errors in using the Trapezoidal Rule
Partial derivative of $f(x, y)$ at (a,b)

These precalculus and calculus theorems, among others, become more understandable through the use of graphs:

Sine function is periodic with period 2π
Graphs of g and g^{-1} are symmetric about the line $y = x$
Tangent Line Approximation Theorem
Intermediate Value Theorem
First Derivative Test for Local Maxima and Minima

They are useful in recalling knowledge. You can easily recall a graph and, due to its compactness, retrieve a lot of information with a minimum amount of work. Are you learning graphs of functions so you can easily recall them to answer questions? For example, what comes to mind when you have to think about the cosine function, assuming you had trigonometry? If you are not visualizing its graph in your mind's eye, then I wonder about the role that graphs play in your study of mathematics. See how quickly you can answer these questions by visualizing, in your mind's eye, the graph of the cosine function: What is the domain and range of this function? Is it a continuous function? What is its period? Is its graph symmetric to one of the axes or to the origin? What is its maximum value? What is its minimum value? Where is it increasing? These and other questions can be answered almost immediately by visualizing the graph of the cosine function, including key values on its axes.

Through the use of graphs a lot of information can be shown in a small space. Are you amazed at a classmate's ability to quickly answer questions that he or she is asked? You would be less amazed if you knew that all he or she probably did to answer these questions was to visualize a graph. Graphs are useful to quickly reconstruct knowledge you have forgotten. For example, suppose you are asked where the cosecant function is undefined, and you can't recall the graph of the cosecant function. This can be quickly answered by knowing these three things: (1) csc x is 1/sin x, (2) 1/sin x is undefined when sin $x = 0$, and (3) the graph of sin x. In the blink of an eye you can quickly give the answer of $n\pi$, n an integer. Wow, what a memory! No, it was a quick reconstruction of knowledge that appeared to come completely from memory.

They show the plausibility of theorems. Basically, proofs are done algebraically; however, analyzing their graphs can show the plausibility of many of them. Formal proof is downplayed in the more innovative textbooks in lower division mathematics. Instead, more emphasis is placed on using graphs to show their plausibility. There is good reason for this since a graphical explanation of a theorem may assist your understanding more than a non-graphical one. The best of both worlds is to show the plausibility of a theorem using geometry, and prove it using algebra.

Examples of precalculus and calculus theorems whose plausibility is enhanced by graphs include these:

Maximum value of $y = a(x - h)^2 + k$, when $a < 0$, is h
Two non-vertical lines are parallel if they have the same slope
The Mean-Value Theorem
$\Delta y \to dy$ as $\Delta x \to 0$
Newton's Method of Approximating Roots of a Function
$f(x)$ is increasing on (a,b) if $f'(x) > 0$ on (a,b)

They show similarities among a class of functions. Graphs of several or more functions that belong to a specific class of functions illustrate very vividly the similarities and differences of these functions. Knowing the basic look of a function belonging to a class of functions will serve you well; hence, you need to approach your study of them with the goal of remembering them.

Examples of classes of functions whose graphs you need to know include these:

$y = Ax$ $(A \neq 0)$.

$y = Ax^2$ $(A \neq 0)$.

$y = Ax^3$ $(A \neq 0)$.

$y = (x - h)^2 + k.$

$y = |x|.$

$y = \log_a x$ (for $0 < a < 1$, or a>1).

$y = A \cos x$ $(A > 0).$

$y = A \sin x$ $(A > 0).$

They can be used to construct graphs of related functions. For example, constructing the graph of the inverse of a function from the graph of the function.

They verify a need for all of the hypotheses of a theorem to ensure the truth of the conclusion. For example, give the graph of a function that does not satisfy the differentiability hypothesis of The Mean-Value Theorem, satisfies the continuity hypothesis, but the conclusion of the theorem is false.

You can learn about important features of related functions by being required to identify their graphs. For example, given three graphs, identify the graphs of $f(x), f'(x)$, and $f''(x)$.

They aid in problem solving. A key problem-solving tool is to draw a picture of the problem situation, which may mean drawing one or more graphs.

There is no doubt as to the value of graphs in learning mathematics. They are plentiful in your mathematics textbooks and are widely used in your instructors' lectures. Not only is it important to understand them and use them, but you also have to determine which ones are important for you to remember so you have them readily available when needed.

How to Study Graphs

There is not much more to say about how to study graphs. You need to realize their importance, which means that you need to study them carefully as you read your textbook and notes. *Graphs with titles* are especially important to learn and remember. We often equate the importance of people with their titles (e.g., Mary is CEO of X Corporation, or Marvin is a Bishop). The same is true in mathematics. Studying a graph includes studying its title, its labels, and any auxiliary lines or curves that have been added to it. The following are techniques to help you study graphs:

Relate, with understanding, the graph and the text that references it. Ask yourself the question, "What should I learn from this graph?" This may take some time to answer, but work to come up with an answer before you proceed with your textbook reading. Use your response as the title of your graph, if it doesn't already have one, or if its title is not very descriptive. For example, you might say (and record), "This graph shows the quantities Δx (change in x) and Δy (change in y) when a point on the graph of f moves from $(a, f(a))$ to $(b, f(b))$."

Attempt to reproduce the graph without looking at your textbook or class notes. Reproducing the graph includes sketching it with all its labels and auxiliary lines or curves. This is important to do immediately after you have been introduced to the graph, and before you continue with your textbook reading. Check to see if you have reproduced it correctly, and don't leave it until you can reproduce it correctly from memory.

Record your drawing of the graph and what you learned from the graph on a 3×5 card. Place the graph on one side of the 3×5 card with the question, "What should I learn from this graph?" On the other side of the 3×5 card write your answer to this question, which is now the title of the graph. You don't need to do this for graphs of minor significance. When reviewing, look at one side of the card and see if you can state what is on the other side.

When you have finished reading a section of your textbook, reproduce from memory, the important graphs of the section with all of their labels and titles. Repetition and review help cement understanding and reinforce memory. Forgetting happens quickly, and repetition and review will keep that from happening.

You need to identify the graphs that are important to remember so you can retrieve them when needed. In your attempts to understand better, you will also need to construct your own graphs that may be similar to, but different from those in your textbook or those presented by your instructor. A question to often ask yourself when you read or apply mathematics is this: "Is there a graph I can construct that will help me?"

Benefits of Studying or Constructing Tables of Numbers

Studying or constructing tables of numbers provides clarity. Abstraction abounds when explanations are given using words and symbols.

Many of us need something more concrete at the outset of an explanation to better understand the content presented. Besides graphs, appropriately constructed tables of numbers can provide this concreteness. I have often said this to myself when analyzing specific data in tables: "Aha, now I really understand this." This is not to say that I didn't have some understanding after reading a written explanation, but it was only after I looked at the situation in a more concrete way that my understanding deepened. I also realized some shortcomings of using table of numbers. Tables cannot display all values in the domain of a function; hence, you can incorrectly infer the truth about something based on a limited quantity of numbers comprising a table. Hence, there is also a need for algebraic arguments or explanations.

If you have difficulty understanding or relating concepts by reading statements with symbolic form, coupling this reading with an analysis of specific data in tables is especially important for you to do. Students with a sensory preference for learning especially benefit from working with tables of numbers. (See section 1 of chapter 13 on learning styles.) Beyond analyzing the tables that are in your reading, constructing your own tables when you need further clarification is a good habit to form. This is especially important if your textbook authors and your instructor do not make much use of tables. Don't hesitate to consult other textbooks to study the tables that appear there, especially if they are missing or poorly presented in your textbook.

Examples That Show the Value of Tables and How They Can Be Analyzed

The examples of tables given here are restricted to precalculus and calculus. Your background in the subject matter may be such that your understanding of the content presented is minimal at this time. But it may be enough to convince you that tables are important and can be used in a variety of situations. Reflect on the understanding you can gain from reading these tables (you may want to come back to these tables as your mathematics background knowledge increases).

Example 1. We know that if (a,b) satisfies a function f, then (b,a) satisfies the inverse function of f, namely, f^{-1}. This is verified in these two tables for $f(x) = \ln x$, and its inverse function $f^{-1}(x) = e^x$ (results are rounded-off):

Table A

x	0	0.5	1	1.5	2	12 . . .
e^x	1	1.65	2.71	4.48	7.39	162754 . . .

Table B

x	1	1.65	2.71	4.48	7.39	162754 . . .
$\ln x$	0	0.5	1	1.5	2	12 . . .

Use your calculator to verify the results in the table. Notice that e^x grows quickly and $\ln x$ grows slowly. Use your calculator to graph $\ln x$ and e^x. Note that their graphs are symmetric to the line $y = x$.

Example 2. Here we examine the behavior of $f(x) = \dfrac{x+3}{x-2}$, as x gets closer and closer to 2. We first examine it in Table A as x gets closer to 2 through values greater than 2, and then in Table B as x gets closer to 2 through values less than 2.

Table A

x	2.1	2.01	2.001	2.0001 . . .
$f(x)$	51	501	5001	50001 . . .

Table B

x	1.9	1.99	1.999	1.9999 . . .
$f(x)$	−49	−499	−4999	−49999 . . .

We ascertain from the tables that $f(x)$ is increasing without bound as x approaches 2 through values greater than 2, and $f(x)$ decreases without bound as x approaches 2 through values less than 2. Use your calculator to graph $f(x) = \dfrac{x+3}{x-2}$. Are the predictions supported by the graph? Realize, by dividing, that $f(x) = \dfrac{x+3}{x-2}$ can be written as $f(x) = 1 + \dfrac{5}{x-2}$. Can you ascertain what is happening to $\dfrac{5}{x-2}$ as x approaches 2 from the right? As x approaches 2 from the left? Note that the numerator is fixed and the denominator is approaching 0, through values less than 0 as x approaches 2 from the left, and through values greater than 0 as x approaches 2 from the right. We are analyzing the behavior of $f(x)$—arithmetically, geometrically, and algebraically. Note that the results from any one of these three means are compatible with the results from the other two means, which is expected.

Example 3. Table A is used to numerically estimate the right-hand derivative of $f(x) = 4x^3$ at $x = 2$. Recall that this derivative is the value approached by the values $\dfrac{f(x_i) - f(2)}{x_i - 2}$ as x_i approaches 2 from the right. Note that $f(2) = 32$.

Table A

x_i	$f(x_i)$	$x_i - 2$	$f(x_i) - f(2)$	$\dfrac{f(x_i) - f(2)}{x_i - 2}$
2.01	32.48240	0.01	0.48240	48.24
2.001	32.04802	0.001	0.04802	48.02
2.0001	32.00480	0.0001	0.00480	48
2.00001	32.00048	0.00001	0.00048	48

Using theorems of calculus, the derivative of $4x^3$ is $12x^2$, which, for $x = 2$, is 48. We see this in the table but realize that the results were rounded-off. If more decimal places were carried, the values in the last column would not be 48 but would get closer and closer to 48 if the chart were extended. Similarly, you will find out by use of a table that, by letting x_i approach 2 through values less than 2, the quotient values in the right-hand column of the chart are approaching 48. Thus, the left-hand derivative of $f(x) = 4x^3$ at 2 is also 48.

Example 4. The definite integral of a function can be estimated numerically. Table A is used to do this for $f(x) = x^2 + x$ over the interval $[0,1]$. This definite integral is denoted by $\int_0^1 (x^2 + x)dx$. To estimate its value, we divide the interval $[0,1]$ into n subintervals of the same width (Δx denotes the width of each interval). We do this three times, using smaller Δx each time. We also evaluate f at the right-endpoint of each subinterval (this is called a right-hand approximation of the definite integral). We do this for $n = 2$, 4, and 8. To get as close as we please to the actual value, we would have to continue making more subdivisions, letting Δx approach 0 in width (say, for $n = 16, 32, \ldots$). This example will only make sense to you if you have the necessary background. Notice that in the first section of Table A, $n = 2$, in the second section of Table A, $n = 4$, and in the third section of Table A, $n = 8$.

Table A (values are rounded off)

x	$f(x) = x^2 + x$	Δx	$(f(x_1) + f(x_2) + \ldots + f(x_n))\Delta x$
0.5	0.75	0.5	$2.75 \times 0.5 = 1.38$
1	2		
0.25	0.31	0.25	$4.37 \times 0.25 = 1.09$
0.5	0.75		
0.75	1.31		
1	2		
0.125	0.14	0.125	$6.05 \times 0.125 = 0.76$
0.25	0.31		
0.375	0.52		
0.5	0.75		
0.625	1.02		
0.75	1.31		
0.875	1.64		
1	2		

If we continue this table by next taking n to be 16, then 32, etc., we would see that the values 1.38, 1.09, 0.76, . . . would be approaching a particular value. Fortunately, calculus gives us a quick way to calculate that value: The definite integral we seek is $\int_{0}^{1} (x^2 + x)dx = \dfrac{5}{6} = 0.83$. Our last table value, 0.76, is not too far away from it (it is $0.83 - 0.76 = 0.07$ away, which is about an 8% error).

There are many other examples in mathematics where tables of numbers are useful to construct and analyze. Additional examples of how tables can be used to enhance understanding include the concepts of increasing and decreasing functions, continuity of a function, maximum and minimum values of a function, distance a moving particle travels (given an equation of its velocity at any given time), and the elusive but all important concept of the limit of a function, which was touched on in examples 2, 3 and 4. As you can see from the examples, you need an appropriate background to understand tables that appear in a course. Once you have the prerequisites for understanding specific tables, they are especially helpful in understanding the more difficult concepts that are often introduced in the course using more abstract language.

6. Reading Examples

Textbook authors know that most students need to see *examples* that relate to the content of the section being studied. These examples apply the concepts and theorems that are discussed in the section. There is a limit to the number of examples that appear in your textbooks, mainly dictated by an economic necessity to restrict the size of textbooks. Hence, it is likely that the examples appearing in textbooks are carefully chosen to give the most bang for the buck. Some of the exercises at the end of a section of the textbook will be closely aligned with examples; others will not. In your efforts to understand the concepts and theorems of the section and how they can be used, it is important that you understand the examples. Imagine someone describing a somewhat complicated card game to you with its rules and strategies. You will not know how well you understand the game or can play it until you play it. Most likely you will find out that you are more confused than what you thought. An intermediate step is to watch someone else play the game. This is the purpose of textbook examples—they are an intermediate step to help you work exercises. It has been said by many that, "Until you can do, you do not understand," and that surely applies to mathematics. *A primary purpose of examples and exercises in a section of the textbook is to help you understand the mathematics of the section.*

Do all students carefully work through the examples of a section in order to understand all the steps of each example? If you believe that then I suspect I can sell you some swampland in Arizona. The answer to my question is an emphatic, "No." How do I know this? When a student asked me a question in class or in my office on an assigned exercise he or she was struggling with, and I knew it related to an example in the textbook, I began asking the student this series of questions, in the given order, and stopped asking when I received a "No" to one of them:

1. Have you read the section in the textbook?
2. Are you aware that there is an example that relates to this exercise?
3. Have you read this example?
4. Did you understand every step of the example?

The number of times I heard a "Yes" to all of these questions was small. It became clear to the student what he or she needed to do after replying, "No," to a question. The student came to my office wanting help on an exercise, but the exercise was not the appropriate place to ask for help until he or she could say, "Yes," to questions 1-4. A primary role of mine as an instructor was to help students diagnose why they were experiencing difficulty (i.e., to help them ascertain deficiencies in their study process).

How to Read Examples

There are basic steps to take when reading examples, and they include these:

Read with anticipation. See if you can predict what the next step may be. This is active reading.

Read with a pencil in hand, working out details or intermediate steps that have been omitted. You have to be satisfied that you understand how each step of an example follows from previous steps. You may have to review the textbook or class notes to re-familiarize yourself with specific definitions, statements and proofs of theorems, graphs, tables, symbols, and terminology.

Obtain assistance if you need it. Note your areas of confusion and get clarification as soon as you can. Seek help from a classmate, your instructor (in or out of class), or a helper in a mathematics-tutoring center. *Resist this dangerous thought:* "Understanding this is probably unimportant."

When you are finished working through the example, close your book and see if you can write out the solution. Don't leave the example until you can do this with understanding.

As an alternative to reading through the example when you first come upon it, close your textbook and see if you can work it. If you get stymied at some point, open the textbook and read until you understand (but no further). Then close the book and continue working the example.

When you are satisfied with your understanding of an example, characterize the purpose of the example. For instance, if it is an example of finding a derivative of a function by using the definition of derivative, write this phrase in your notebook, as well as the statement of the example and page number on which it appears. You will then know where to look for it when you need it.

When you have finished reading the section, see if you can work all the examples that you deem important, without looking at the work of the authors. Remember, reading your textbook authors' examples with understanding, and then doing them yourself, are vastly different things. The latter is the most meaningful. If you work all the important examples again, as soon as you complete a section, your short-term memory will be reinforced and the risk of forgetting will be lessened.

Managing Problem Assignments in Mathematics

Success in a mathematics course depends on many things, and foremost among them are these items: (1) having an appropriate work ethic and excellent attitude; (2) having the prerequisites for the course; and (3) having a viable method for *managing problem assignments*. This chapter addresses the last item. (The first item is primarily addressed in chapters 7 and 10, although it appears in other chapters as well. The second item on prerequisites is the content of chapter 4.) Perhaps the one thing that contributes most to students' lack of success in a college mathematics course is the cavalier way they approach their problem assignments. A large number of college students mismanage problem assignments in their college mathematics courses; in many lower-division mathematics courses it can be from a third to one-half of those enrolled.

Do you ever think how wonderful it would be if you had no problem assignments? That all you had to do was attend class and read the textbook? That would make for a nice romp through the course, but you wouldn't learn much mathematics. A primary reason to work problem assignments *is to learn the theory of the course*. To learn the concepts, theorems, and techniques of mathematics you have to work with them, and that is accomplished by working problems. Hence, don't be surprised when you struggle solving problems. That is to be expected since you won't know the theory well enough, or how to put it together, to solve problems. It is in your struggles with problems that you learn the theory. *You may think that the focus is to solve problems, which requires you to make use of theory; but for college students it is the reverse situation. In college, the focus is to learn the theory, which requires you to solve problems.* Once you have completed your mathematics study, the focus shifts—there is a need to solve specific problems, and you look for the mathematics that will help you do this.

The sections of this chapter are as follows:

1. Cognitive Levels of Mathematics Problems
2. Handling Problem Assignments
3. Instructors' Assignments
4. Improving the Quality of Work on Assignments
5. Importance of Attending Class

An important question to ask and have answered, before reading advice on managing problem assignments in mathematics, is this: "What *types* of problems am I expected to solve?" This is the content of section 1.

1. Cognitive Levels of Mathematics Problems

The cognitive levels of mathematics problems that might comprise a problem assignment can be described as (1) knowledge, (2) comprehension, (3) application, (4) analysis, (5) synthesis, and (6) evaluation. Generally speaking, the thought process required to solve problems at these levels intensifies as the problem falls further to the right in this list. That is, the further to the right in the list the problem falls, the higher the cognitive level of the problem, or the more complex the mental behavior required to solve it. The synthesis and evaluation levels will not be discussed here because of their limited use in many lower-division mathematics courses, and to make the discussion more manageable. Many problems are at more than one cognitive level; for example, a problem could be at the comprehension level and also at the application level. The level at which it is categorized is the highest of these levels; hence, a problem at the comprehension and application levels would be categorized as an application level problem.

It is not easy to categorize the cognitive level of some problems since the thinking skills needed to solve them can be elusive. The thinking skills you use are dependent on your experiences, including the instruction you receive related to the problem. Fortunately, it is not that important for you to precisely determine the cognitive level of a problem. However, there are reasons why it is valuable for you to have *a sense* of the cognitive levels of problems, and they include these:

You become more aware that, in order to get a first-rate mathematics education, you need to work problems from a variety of cognitive levels. However, it is not necessary in your assignments to have an equal balance of cognitive levels (1) through (4); but, it is important to have worked an adequate number of problems at each cognitive level by the end of the course.

You are more willing to accept struggles with problems if you expect that to happen because of their cognitive levels.

You will be more receptive to being evaluated at each cognitive level.

You recognize the need to supplement your instructor's problem assignments if you are not expected to work problems at each cognitive level. The problems most likely not assigned will be at the application and analysis levels, and perhaps even at the comprehension level. Many instructors err by assigning too many problems at the computational level, and not enough at the other three levels. Following are brief descriptions of each cognitive level, including example problems at each level.

Knowledge Level

You are probably most familiar with this level since the problems that most likely dominated your assignments were at this level. These problems are primarily mechanical in nature, consisting of remembering or recalling previously learned information. They mainly required you to *remember or recall* specific facts, vocabulary, symbols, or patterns; or to *perform computational algorithms* that you had practiced over and over again. You watched your instructor work problems almost identical to these, you practiced them in similar contexts, and then you worked them again on examinations. Here are six examples of problems at the knowledge level:

1. Complete this definition: function f is continuous at a if . . .
2. If A is an acute angle and $\tan A = 2/3$, find $\cos A$ and $\csc A$.
3. State the rule for finding the derivative of the quotient of two differential functions.
4. Find the derivative of $g(x) = \dfrac{x^3}{\sqrt{x-1}}$, for $x = 4$.

5. Determine the horizontal asymptotes, if any, for $f(x) = \dfrac{x}{4x^2 - 1}$.

6. Evaluate $\int (4x^2 - \sqrt[3]{x} + 2)dx$.

Let us, for example, analyze what you need to know to work problem 5. It is presumed for problem 5 that you practiced similar problems, where you divided the numerator and denominator by the highest power of the variable (in this case x^2), and then found the limit of f as x increases or decreases without bound. It is also assumed that you know the limit of a constant divided by a whole number power of x, as x decreases or increases without bound, is 0. Because you practiced these things, you are able to work problems, similar to problem 5, in a very mechanical or rote way. Hence, they are at the knowledge level. The same can be said for the other five examples.

Comprehension Level

Problems at this level require you to show *understanding* of concepts and principles that appear in *unfamiliar* settings. Working these problems correctly implies that you grasp the meanings of definitions and theorems used to solve the problems. That is, you demonstrate an understanding of them as opposed to just giving a memorized response. Other problems at the comprehension level require you to summarize a particular body of knowledge, translate statements from one form into another (such as from symbolic form into verbal form, or from verbal form into pictorial form), or communicate mathematics in a meaningful way. Six examples follow:

1. The total surface area of a right circular cylinder is $14in^2$. Express the volume as a function of the radius.

2. Use differentiation to show that $\int \sec^2 5x\,dx = \dfrac{\tan 5x}{5} + C$. Why does your work show this?

3. Sketch a graph of $y = \sin x$, $x \in [0, \pi]$, and then use the graph to verify geometrically the conclusion of The Mean Value Theorem.

4. Explain why the following table does or does not define y as a one-to-one function of x:

    x: 2   3   7   12   15   17
    y: 5   2   8   15   8   12

5. Given three unlabeled graphs, identify them as f, f', and f'', and explain why you identified them as you did.

6. Explain why the composite function $g \circ f$ may not always exist for every value in the domain of f.
7. Explain the distinction between $h^{-1}(x)$ and $[h(x)]^{-1}$.

Application Level

For a problem at this level you have to *select*, from your repertoire of knowledge, *information you can use or apply* to solve it. The problem is uncomplicated and similar to, but different from, problems that you have studied. What you select and then apply to solve it may be a specific definition or theorem, perhaps yielding a method, technique, or algorithm that can be used. If it is difficult to determine what is needed to solve the problem; that is, if it requires special problem solving ability or insight, then the problem is at a higher cognitive level. In your efforts to solve a problem you may see that it is analogous to a problem you have already solved; hence, you apply knowledge that worked there to solve the new problem. The problem still qualifies as an application level problem. Six examples follow:

1. Find the area of the region enclosed by the curves $x = y^2$ and $x = -2y^2 + 3$.
2. Sawdust is falling off an elevator onto a conical pile at the rate of 3 meters per minute. The radius of the base of the pile is always equal to 1/3 the pile's height. How fast is the radius of the base growing when the radius is 3 meters?
3. Find the intervals on which $f(x) = 2x^2 - x^4$ is increasing or decreasing.
4. Find the point on the line $y = 2x - 3$ that is nearest the origin.
5. A cylindrical can without a top holds 40 *cm* of water. Find the dimensions of the can so the cost of metal to make the can is minimal.
6. Argue, analytically, that there is at least one value x such that $x + \cos x = 0$.

It is assumed that a student working problem 6 has been exposed to the Intermediate Value Theorem, and has applied it to solve similar problems.

Analysis Level

There are many adjectives or phrases to describe the types of problems that are at this cognitive level, which is the highest of the four levels.

They *require more complex behaviors* of the problem solver, including creativity, applying knowledge to unfamiliar situations, employing general methods for attacking non-routine problems, making discoveries, and constructing original proofs. Problems at this level are of a problem-solving nature. An important goal of instruction is to support students working at this more advanced level. Six examples follow (it is assumed that the students solving these problems have not had prior exposure to them):

1. Suppose a square and equilateral triangle have the same perimeter. Which has the larger area? Can you draw a similar conclusion for any rectangle and an equilateral triangle? Explain.

2. Use the rule for finding the derivative of the product of two functions to prove the integration by parts formula: $\int uv'dx = uv - \int vu'dx$.

3. Use The Mean Value Theorem to prove this theorem: If f is continuous on $[a,b]$ and differential on (a,b), then (i) for $f' > 0$ on (a,b), f increases on (a,b), and (ii) for $f' < 0$ on (a,b), f decreases on (a,b).

4. Recall the rule for finding the derivative of the product of two functions: Take the first function times the derivative of the second function, and add to this the second function times the derivative of the first function. Find a comparable formula for the derivative of the product of three functions.

5. Discuss the advantages and disadvantages of using the First and Second Derivative Tests for Local Maxima and Minima.

6. Show that a cubic polynomial can have at most three real zeros.

In conclusion, working problems at a variety of cognitive levels is important. This does not mean that the number of problems worked at each level has to be balanced—you most likely will be assigned more problems to work at the lower levels. Does your instructor assign problems at a variety of cognitive levels? How do you feel about this? Knowing the benefits of working problems at a variety of cognitive levels will increase your efforts to do so.

2. Handling Problem Assignments

Important as it is to attend mathematics class, the bulk of your time in a course needs to be spent on problem assignments. There are many issues to consider when it comes to handling problem assignments, and some of the basic ones are presented here.

The Need to Practice Solving Problems

I am sure that you often marvel at the clarity in which your instructor presents solutions to problems. You leave the classroom pleased with your understanding, possibly to the extent that you believe you hardly need to work on problems yourself. But then reality sets in when you attempt to work a problem assignment. You quickly become stymied. You feel helpless and wonder what is wrong with you. You question whether you understood your instructor's work in class. Nothing is wrong with you, and at the time, you probably did understand what your instructor was doing. However, all you had to do was understand your instructor's thoughts, not deduce them yourself, which is a far different matter.

Your instructor knows well the concepts, theorems, and techniques of mathematics, and has used them over many years to solve problems in class, many of which are similar to problems you have been assigned. Solving these problems is "child's play" for him or her, but it is a different story for you. You have to come up with a logical sequence of steps to solve a problem, and not knowing well enough the mathematics you have to use to do it makes it a formidable task. You cannot rely much on remembering your instructor's work in class because you forget quickly, and the problems you have to solve are not exactly the same. Your class notes will provide some help but they most likely don't contain your instructor's remarks that he or she did not write, which are part of his or her solutions.

It should be apparent that you need to *practice* working routine and non-routine problems. Suppose you wish to excel as a basketball player. You can read about how to play the game and watch people play it, but unless you are diligent in practicing playing the game, and in the right way, you will be mediocre at best. The game of mathematics cannot be played any differently. Mastery of the content cannot be accomplished without having a viable method to manage problem assignments. You will know that you understand the content of a unit when you have no difficulty working a variety of well-chosen problems that relate to the content. You have to be able to apply content before you can say that you understand it. *This comes with appropriate and sustained practice.*

You Have to Spend a Substantial Amount of Time on Problem Assignments

When it comes to completing problem assignments, *nothing works unless you do.* You must have, and take, the necessary time to work problems. The following students' remarks allude to this—the first two statements appeared on student evaluation forms, and the third statement was given when I asked students for advice to include in this book:

- ❖ *"Allocate ample time* daily *to work problems."* (Calculus I student)
- ❖ *"I have never had difficulty with the material but the time involved in studying is considerable."* (Calculus II student)
- ❖ *"Be prepared to spend an inordinate amount of time on homework."* (Calculus II student)

The amount of time that many lower-division college mathematics students expect to spend working mathematics problems is less than what is needed to be successful. They stop studying when they think they have spent sufficient time at it, whether or not they were successful in their study. They may say to themselves, "I should not have to spend any more time than this." Does that describe you?

Problem Solving Skills and Appropriate Problem Solving Attitudes Are Needed to Solve Problems

Success in solving problems at various cognitive levels requires various problem-solving skills, and appropriate problem-solving attitudes. You need to know what you can do when confronted with a problem (i.e., to know skills that you can apply in your efforts to solve the problem). Some problem solving skills will bear fruit; others not. It is easier to solve problems that are mechanical or algorithmic in nature and more difficult to solve those at higher cognitive levels. This fact is captured by a Calculus I student who said, "My main difficulty is with the story problems. The mechanics of integration and differentiation were really no problem after some experience with them." A necessary problem-solving attitude to have was expressed by another Calculus I student: "Don't get frustrated when the problems don't work out." The ability to solve problems has to be your ultimate goal when you study mathematics. Attributes of a problem solver are addressed in section 6 of chapter 21, and the problem-solving process and development of problem-solving skills comprises the content of chapter 22. Those two chapters will serve you well in becoming a better problem solver.

Consequences of Delaying Work on Your Assignments

There are several reasons why it is important to begin your assignments as soon after class as you can. A student gave this advice:

❖ *"I found it very helpful to do the homework no later than the night of the day it was assigned. This helped me because the ideas, skills, and purpose of the section were fresh in my mind from class."* (Calculus II student)

Get an early start after class on your assignment while the material presented and discussed in class is somewhat fresh in your mind. This will help you better understand what you are reading or rereading in your textbook, and reinforce what you learned in class (thus increasing long-term retention). An added bonus to getting an early start is that you become aware, early on, of what needs to be done, how long it will take, and what additional support you will need. You can then lay out a work schedule to accomplish what still needs to be done.

Not only is it important to get started early on your assignments, but it is also important to finish them in a timely manner. *Delay completing your assignments is a major cause of failure in mathematics.* This is such a major problem with students that I have thoroughly addressed it in section 3 of chapter 11. It is there that you will find out why you procrastinate (the reasons may surprise you), what mathematics students say about procrastination, and methods for decreasing procrastination. Perhaps you identify with the Calculus I student who said, "I was always playing catch-up on the homework."

Keep a Problem Notebook

It was not atypical for a student of mine to come to my office hours and waste time looking through many unbound pages of work, trying to locate a particular problem that he or she wanted to discuss. Keep a problem notebook for your assignments, preferably a three-ring binder. You can remove pages from it when an assignment is to be turned in, and later you can add pages of corrections to problems.

When working problems in your notebook, write down enough of your reasoning so that a helper can look at it to see what you did to determine if your work was done correctly. Don't hesitate to use scratch paper as you work. Write out your solution as you would write it for a test or for an assignment to be graded. If you can't complete the solution to a problem, write down as much as you can so a helper can see exactly where things started to come apart, and why. The primary goal of a helper is to help you help yourself. To do this, the helper has to know where you are on the problem before appropriate help can be given.

How Problems on Your Examinations Relate to Those in Your Textbook, and Those Presented in Class

Do not expect the problem examples in the textbook, the problems in the textbook exercises, the problems presented by your instructor in class, or the problems on your examinations, to necessarily be at the same level of difficulty or at the same cognitive level. The following comments from Calculus I students, who were evaluating their courses and instructors, indicate that they expected these problems to be similar in difficulty or cognitive level, but were not:

❖ *"The example problems were usually too simple in comparison with the exercises of the section."*

❖ *"There were problems in the book that were not covered in any of the examples—that's my main complaint."*

These two students are saying that some of the exercises were more difficult, or perhaps not closely related to the example problems. That is to be expected and should be welcomed since a cookie cutter course does not serve you well.

❖ *"His tests were unfair (sometimes) because he put problems on them that he had not explained in class. However, it did make me think and study harder."*

Here is a student who may want the course to be a "repeat after me" course. That is, this student may be thinking, "I want problems on the examinations similar to those my instructor presented in class, which I practiced." This student's instructor is expecting his or her students to put some ideas together to solve unfamiliar problems. Fortunately, this student's last remark ("However, it did make me think and study harder"), indicates the value of the instructor's actions.

❖ *"Do as many problems as possible. The more you do, the better your chances that a similar problem will appear on the exam."*

It is difficult to argue with this student's remark. Not only are you more likely to see similar problems on the examination, but you will also learn more mathematics and become a better problem solver. That is why you are in the course.

❖ *"Tests were very hard in comparison to assignments."*

Either this student is not doing the more difficult assigned problems, or the instructor is not assigning difficult or challenging problems to prepare students for more challenging examinations. The student is at fault in the former situation, and the instructor in the latter situation. If your instructor is not assigning you the types of problems that you will face on examinations, then you should assign them to yourself (which is what you should do, regardless of whether they are on the examinations).

The following remark by a Calculus II student characterizes reasonably well what you can expect in college mathematics:

❖ *"Since I am coming back after a 15-year break, I feel I have a little maturity on my side. I would like to see more problems of a basic nature of every concept. Normally, we are given only one or two problems, and then they become too difficult to grasp sometimes. We spend years learning to add and subtract and then when we get to college we spend a few minutes on more advanced and difficult concepts."*

The college expectation that this student is facing is real and reasonable. Students coming to college need to come equipped to handle a faster pace, and one way to handle the fast pace is to put in the necessary time outside of class on their assignments, and obtain help when needed.

Read the Related Section of Your Textbook Before Working the Problem Assignment

Working your problem assignment *before* you have read the related sections of the textbook is like playing tennis without a racket. The primary mathematics you use to work problem assignments is the mathematics developed in the related sections of the textbook. Reading alternative textbooks can also benefit you since they most likely have different examples from those in your textbook. (To determine the benefits of reading alternative textbooks, and to find one that speaks to you, read section 2 of chapter 18.) It makes good sense to be equipped with knowledge as you begin your problem assignment.

3. Instructors' Assignments

Do not expect your instructors to be clones of each other. They will vary in almost all aspects of teaching, including the types of assignments they give. If it is your goal to have your instructor adapt his or her teaching style to you, you most likely will be disappointed. Your energies are better spent compensating for what your instructor is not giving you.

Nature of Instructors' Assignments

The typical assignment given by a college instructor can be characterized as "one-size-fits-all" assignment. That is, each student gets the same assignment. Hopefully, it is an expansive and challenging assignment that accommodates all students, but this is often not the case. Some assignments are so limiting that they have to be viewed as *less-than-minimal* assignments. Some instructors almost always assign problems of one type, mainly those that are computational in nature, forsaking those that require more understanding or problem-solving skills. Working only this type of problem may not have a negative influence on your course grade since that may be your instructor's expectation. However, it will have a negative influence on your understanding of mathematics, your achievement in subsequent mathematics or mathematics-related courses, and your career. There are mathematics instructors whose assigned problems cover a variety of cognitive levels, as they should, and you are at a disadvantage if you have not been prepared by previous instructors to meet these higher standards.

Some instructors' assignments are very loosely structured. They might say, "Look over the exercises in this section of the textbook and work a variety of them." This assignment assumes that you know the content of the section well enough to be able to discern the various cognitive levels of the problems, and that you realize the value of working a variety of them. Coupling this assumption with the staggering high number of exercises in lower-division mathematics textbooks, suggests that this type of problem assignment is problematic at best. Other instructors may assign no specific problems, but restrict the group of problems you are to look at. They may say, "Look at exercises 1–20 and work those you believe you need to." This is also problematic. Perhaps the majority of instructors will be quite specific, giving an assignment similar to this: "Work numbers 1, 2, 4, 7, 8, 11–14, 18, 23–26, and 30. It is highly likely that these instructors carefully selected problems based on specific criteria, but the quality of their assignments depends on the quality of their selection criteria.

There are reasons to believe that the most neglected student in our educational system, kindergarten through college, is the high-achieving student. Many of them should be challenged far more than they are. If you are a high-achieving student, then you need to go beyond a typical instructor's assignment. You need to stretch yourself in college so you are better equipped to stretch yourself in your professional career; thus becoming a more-valued employee. Some lower-division college mathematics textbooks are shy on problems at higher cognitive levels. However, most textbooks have many valuable and thought-provoking exercises that are not assigned, and you are encouraged to work some of them.

In summary, constructing appropriate problem assignments is one of the more important responsibilities of an instructor. After all, working on assignments will occupy the bulk of your time and is your major avenue to learning. You need to be aware of the characteristics of a good problem assignment, as presented in this chapter, and then *supplement* a problem assignment that is lacking, whether or not you are a high-achieving mathematics student.

Other Instructor Actions Regarding Assignments

In addition to the specific problems your instructor assigns, there are other instructor actions regarding assignments that may cause you consternation. Perhaps some comments on several of these actions most frequently mentioned by students, will give you a good perspective on them. Each statement is from a student who was evaluating his or her mathematics instructor:

❖ *"She waits too long to go over homework."* (Calculus III student)

One way college is different from high school is that there is considerably less class time available in college to learn a comparable amount of content. Thus, even if homework is discussed regularly in class, not much time can be devoted to it. Many college mathematics instructors choose to devote more class time to the development of the content than to giving feedback on homework. Also, instructors vary as to *when* they give this feedback—one or more class meetings may pass without addressing homework (although this is atypical). This is a problem if it is a common occurrence. If your instructor takes this to the extreme, talk to him or her about it. (The other extreme of consistently spending most of the class session on discussion of the previous assignment is equally bad.) You have to use a variety of means to get feedback on your assignments in a timely manner. Getting feedback during class time is just one of these means, but it cannot be the major one. More is said on this later in the section.

❖ *"Didn't always lecture on the material before assigning the homework."* (Calculus I student)

In college mathematics courses you should be expected to work on a problem assignment, even if the content it addresses has not been discussed beforehand in class. Many instructors, including me, believe our most important goal is to help you become an independent learner, and the expectation mentioned above supports this. Throughout this book I have addressed many ways to nurture independent learning, including the importance of reading a section of a textbook and working on some problems from that section, *before* your instructor lectures on the section in class.

❖ *"I also liked having the homework collected. It stimulated me to do it when I would have otherwise neglected it."* (Calculus II student)

Yes, we all seem to respond better when we are pressured to do something. There are two types of motivation: intrinsic motivation—coming from within you, and extrinsic motivation—coming from outside of you. Intrinsic motivation is necessary to becoming an independent learner. (Section 2 of chapter 11 is on motivation.) It is typical for college mathematics instructors who teach lower-division mathematics courses to grade very few assignments. There are good reasons why it is important to periodically have your assignments graded; but to motivate you to do well on them is not one of the better ones. Most of your learning in a mathematics course takes place in working assignments. You do them to learn—not because they will be graded.

4. Improving the Quality of Work on Assignments

There are a variety of issues for you to consider that will improve the quality of your work on assignments. Most of them are related to the feedback you seek on your work as you progress through an assignment.

Get Feedback on Your Solutions to Assigned Problems, and in a Timely Manner

Many students, in reviewing for an examination shortly before it is to be given, will try to get feedback on problems in their assignments that they could not work earlier in the unit. Students came to my office hours with questions on these problems that they should have had answered weeks before. This was far too late to be doing that. Reviewing for an examination should be exactly that—a review, which is not the same as learning that should have taken place earlier. Getting feedback for the first time on problems assigned earlier in the unit falls in the area of first-time learning, not reviewing. I would pass an instructor's office filled with students a half-day, a day, or two days before an examination, and hear the instructor lecturing to the group on how to solve some of the assigned problems. It was too late for that, and an instructor needs to help his or her students realize this. Get feedback in a timely manner.

Similarly, you cannot wait to do problems that have been assigned in a specific class session until after the next class session. Without conscientiously working on the problems prior to the next class session, you will be confused during that session, and quickly lose interest in it. Mathematics knowledge is hierarchical—understanding builds on previous knowledge learned. What you learn in a class session *is severely compromised* if you have not completed the necessary homework for that session.

Checking Answers and Solutions to Problems

If you are like most students you love to know whether you have solved a problem correctly by looking at answers or solutions found in other sources. These sources include your textbook (where answers appear at the back of the book), a solutions manual for the textbook, classmates, and your instructor. As an instructor I always made it a point in constructing an assignment to include problems from the textbook that did not have answers or solutions available to my students. Most textbooks and solutions manuals supply answers to either the even- or odd-numbered problems; hence, I would assign even- *and* odd-numbered problems. Early on in a course, after my students had worked on a problem assignment, they would ask me to give them the answers or solutions to *all* of the problems assigned. They could not believe it when I said, "No, I will not do that." Some of them became surly, even after I gave my reasons for not doing this. Their body language said, "I don't care what you say, you should not refuse such a reasonable request." A Calculus III student made this comment when evaluating her instructor: "He did not assign any odd-numbered problems so a student can know whether the answer is correct." Her comment was obviously meant to criticize her instructor. It appears that she did not want to work, or think of working, comparable odd-numbered problems that were not assigned (and then check the answers). Do you feel the same as this student?

There are good reasons why instructors, textbook authors, and authors of student solutions manuals do not provide answers or solutions to all of the problems that appear in a textbook. The reasons include these:

You need to know when you know, and when you don't know. Access to too many answers and solutions hinders this. It stifles development of *internal checks* to see if your thought process for working a problem is correct. If you have doubts about your solution, spend more time reflecting on the problem, perhaps rereading parts of your textbook or class notes, or have a classmate or your instructor look at your work and question you on some of your steps.

When reviewing your solution, ask and answer questions such as these:

- Am I comfortable with what I did?
- Am I apprehensive about my work?
- Can I justify each step?
- Is my logic correct?
- Is my computational work correct?
- Does the answer seem reasonable?

When taking an examination you will not have access to an answer key or solutions manual to see if you are correct, yet you need to have means to determine whether your work is correct. These means have to come from your knowledge base. Incidentally, getting an incorrect answer to a computational problem, based on making a small and mindless computational error, should be treated as basically insignificant in the learning process. It is another matter if such errors occur frequently.

Answer keys and solutions manuals will not exist in the workplace. In the workplace, as in college, you need other ways to determine if your work is correct. Typically, answers to the problems at your workplace do not appear elsewhere. You have been charged to come up with the answers.

Is it good to have answers and solutions to some of the problems that are part of your assignments? Yes! Can you have too many? Yes!

Accessing a Solutions Manual Too Quickly

It is a mistake to access a solutions manual too quickly when struggling with a problem. It is one thing to read and understand the solution to a problem, and another thing to construct your own solution. You don't want the former action to interfere with your ability to accomplish the latter one. Solutions manuals can be a wonderful resource if used properly, and a great detriment to your learning if used improperly. Guidelines for using the solutions manual that accompanies your textbook are given in section 1 of chapter 18.

Solving a Problem in a Piecemeal Way Does Not Mean That You Can Solve Similar Problems in a Non-Piecemeal Way

When you struggle with a problem, you will most likely go to your class notes, textbook, and individuals to eventually learn how to work it. If you do this, be careful not to believe that you have mastery of this type of problem. Eventually solving a problem over time, perhaps by piecing together information you gained from other sources, coupled with what you already know, is a good thing. But here is the quandary that can result: Piecing a solution together does not necessarily mean that you can work a comparable problem *with little hesitation*. In other words, you may have a good understanding of the various steps of the problem, as you learned them, but putting them together to solve a comparable problem without much delay is a different matter. You may think you know how to work this type of problem, but in reality you might not. You have an understanding of each of the pieces, but not necessarily a clear understanding of the whole solution. There is a big difference. Consequently, you need to solve in a non-piecemeal way (i.e., without the use of crutches), problems that are similar to ones you solved in a piece-meal way.

By examination time you need to be able to work problems without the solutions manual, notes, or other hints. You have to dispense with the use of aids. Until you can do this, you are not prepared to work this type of problem on an examination. You may say, "Everyone knows this." The actions of many students tell me otherwise.

Solve All Assigned Problems

You must have the attitude that you will never leave a problem for good until you have solved it, or at the very least, have understood a solution to it. My experiences showed me, time and time again, that a primary reason students get problems wrong on an examination is that they left related problems undone or incomplete from their problem assignments. It is enjoyable to be successful; therefore, students love to work more of the types of problems they know how to work. It is not fun to be unsuccessful; hence, the natural thing to do is to stop working on problems that are causing trouble. *If you want to learn, you have to continue working on the difficult problems, and cease working those you know how to do.*

It is more problematic than you may realize to leave problems uncompleted right up to the examination. You may need to know important concepts and generalizations to work these problems, and the fact that you cannot work these problems is a signal that you may not understand these concepts and principles. Consequently, you are more apt to miss several problems on an examination that require knowledge of these ideas. You are learning more than you think if you continue to work on a problem until you have solved it. Are you prone to say: "This is a difficult problem that most likely will not appear on the examination. I can leave it uncompleted." There is much to gain by completing the more difficult problems and there is no educationally sound reason to leave them uncompleted.

What Your Examinations Can Tell You about How You Handle Problem Assignments

When my students performed badly on an examination, some of them would come to my office and want to know why. They often asked, "Why can't I do better on examinations? I had no difficulty with the assignments." I asked them four questions:

1. Did you complete all of the assigned problems?
2. Did you complete them *correctly*?
3. How do you know you completed them correctly?
4. Would you show me what you did?

This is what I most often found, regardless of their answers to these questions:

- Some problems were correctly worked.
- Some problems were completely worked, but incorrectly.
- Some problems were worked up to a certain point, and then left unfinished.
- Some problems were not even attempted.

Almost all of these students seemed to ignore their spotty efforts as perhaps a reason for their poor performance on the examination. My question to these students, after I had helped them analyze their efforts on the assigned problems, was, "With these inconsistent results on the assigned exercises, why did you think you would do well on the examination?" Most of them had no answer.

An examination analysis was also completed for each student who wanted to discuss his or her poor performance on a test. The student and I jointly looked at the examination and identified those problems for which the student received specific amounts of credit. For sake of clarity, suppose each problem was assigned 5 points. We looked at those problems for which the student received (1) no credit or almost no credit (0 or 1 point), (2) about half credit (2 or 3 points), or (3) full credit or almost full credit (4 or 5 points). This analysis was instructive to the student for reasons that include these:

- The student was able to see the *types* of problems comprising each of categories (1), (2), and (3), and then reflect on why he or she performed that way.
- The student realized that his or her difficulties were the result of poor preparation, and why that was the case for each of categories (1), (2), and (3).
- The student realized that it was possible to do much better on examinations if he or she changed how problem assignments were completed.

Do you analyze your examinations this way after you get them back? Don't hesitate to ask your instructor for help to do this.

5. Importance of Attending Class

The benefits of being in class are many, and it is your responsibility to attend. (For a discussion on attending class see item 6 of section 1 in chapter 12.) Attending class can help you with your problem assignments in at least three ways:

In class you receive the assignment with accompanying remarks. If you are not in class when an assignment is given, you will most likely receive it later than you should. Many students who miss class do not get the assignment until the next class period. The situation then becomes problematic, for obvious reasons. Furthermore, instructors frequently make remarks about assignments when they are assigned. Students who miss class don't hear these remarks, may not find out about them from classmates, or may have them misinterpreted by classmates.

Your instructor works examples in class. Your instructor's examples worked in class should not, typically, be the examples in the textbook. Hence, attending class means you have additional problems to study. Since instructors are talking as they solve problems in class, they are giving considerably more information than what may appear with the solutions to the examples in the textbook. In the main, your instructor's spoken remarks refer to how the mathematics, developed earlier in the class period, is applied to solve these problems. Furthermore, your instructor will respond to questions about the example problems. Why would you not want to be in class for all of this? If you answer by saying, "I'll get class notes from one of my classmates," realize that class notes are a small part of what takes place in class. Most students who take notes do not record the spoken, unwritten remarks of their instructor and classmates. Most students only record what is written on a board or screen, and that is frequently written incompletely or incorrectly.

You have your class notes to refer to later when working your problem assignment. It is important to take good notes on the examples worked in class that relate to the new content, and on the problems worked from the previous class period assignment. Your notes then become a valuable resource for working problem assignments. I was shocked when students in my classes did not take notes on the solutions to the examples worked in class (some took no notes on anything that was done in class). The reasons for this behavior can include these: they have a photographic memory, they believe it will not be of any use to them, or they just don't want to put forth the effort. Hardly ever is it the first reason.

I asked my teaching colleagues this question, "Did you miss mathematics class often when you were an undergraduate student?" Their response was an overwhelming, "No." Don't you think that if instructors (many with doctorates in mathematics), found it important to be in their undergraduate mathematics classes, it might be important for you to be in yours?

I end this chapter with quotations from three students whom I asked for advice on what to include in this book:

❖ *"If you do not understand a problem, seek help immediately, keep current in your homework assignments, do all the homework assigned, and make sure you know how to do every problem assigned."* (Calculus I student)

❖ *"Keep up daily in the course, or better yet, keep ahead of the instructor. Do as many problems in the section as you can. The ones you cannot do, find out how to do them, and then make sure that you understand them well."* (Calculus II student)

❖ *"The biggest aid in studying mathematics was to attempt the problems on my own, and after attending lecture, ask questions the following lecture, and most importantly, study with at least one other student when I needed extra help. Often in verbalizing a problem I was able to solve many of them on my own with a little assistance from a fellow student."* (Calculus II student)

Using Supplementary Materials in Mathematics

The number of available commercial supplementary materials linked to a specific lower-division mathematics textbook has increased greatly over the years. This has significantly increased the cost of taking a mathematics course for those who purchase them. Basic commercial printed materials that may be available from the publisher of your mathematics textbook include a—

- Solutions manual
- Problem workbook
- Guidebook for help in using the textbook
- Laboratory or technology manual
- Compact disk (often packaged with the textbook)

Many of these materials are probably not required by your instructor. They typically are not stocked at your college bookstore unless the mathematics department or an instructor makes a request to the bookstore. Those that you are required to purchase may be prepackaged with the textbook at the request of the mathematics department or an instructor. Many of them can be purchased online.

There are other commercial mathematics resource materials available to you that are not specifically associated with your textbook, and usually not required by your instructor. These materials include—

- Alternative textbooks
- Workbooks of problems with solutions (e.g., Schaums Outline Series)
- Condensed or concise versions of somewhat similar course content (Titles of these books include phrases like these:

short course in college algebra, calculus for dummies, algebra demystified, differential equations made easy, self-help book in geometry, and self-teaching guide in trigonometry.)
- Self-help books on the process of learning mathematics (such as this one)
- Websites (Search the World Wide Web to find sites that give online tutorial help on specific mathematics courses or topics.)
- Hardware and mathematics software

Your instructor's handouts are an integral part of the course. They may include a syllabus, the instructor's notes on specific content, study guides for examinations, problem or project assignments, and solutions to assignments and tests. It is not my purpose in this chapter to discuss instructors' handouts, per se. However, the importance of your instructor's syllabus for the course, and the content of the syllabus, is discussed in both sections of chapter 12.

Three basic questions on commercial supplementary materials addressed in this chapter are these: (1) How important is it to use commercial supplementary materials? (2) What are the benefits of using them? (3) What are optimal ways to use them? These questions are addressed for—

1. Solutions manuals
2. Alternative textbooks
3. Electronic technology
4. Supplementary problem workbooks accompanying textbooks

1. Using a Solutions Manual

A *student* solutions manual for a textbook gives solutions, not just answers, to certain exercises that appear in the textbook. If it is not available in your college bookstore, you can probably order it online. There is an *instructor's* solutions manual that gives solutions to all of the exercises in the textbook. Generally, these are not available to students. Student solutions manuals give solutions to a restricted set of exercises in the textbook (e.g., the even-or odd-numbered exercises). There are at least two reasons why solutions to some exercises are left out: (1) it gives your instructor the option of assigning some of these to be turned in and graded, and (2) it is educationally unsound to have solutions to all of the exercises at your immediate disposal.

Importance of Using a Solutions Manual

Should you use a solutions manual? The answer is a qualified, "Yes." It can be a great aid to learn mathematics, but *only if responsibly used.* Purchasing a solutions manual is typically optional, and many students don't purchase it because of the cost. Your instructor may place a copy on reserve in the library, but that makes it difficult to access. Not using a solutions manual can be a mistake. It is essential to get thorough feedback on your efforts in working assignments, and using a solutions manual is one way to do this. This is not to say that obtaining feedback on your work from your classmates, instructor, and other sources is not important. To the contrary, it is more important since there is a give and take among humans allowing you to ask questions and have questions asked of you. A solutions manual cannot do this, but it does allow you to get feedback at the time you might need it.

When to Use a Solutions Manual

The opportunities to use a solutions manual include these:

If you struggle unduly with a problem, having spent an ample amount of time trying to work it with very little success, and cannot discuss it with someone, then it makes sense to consult a solutions manual. Do this if you have difficulty knowing where to start on a problem, or if you have difficulty somewhere along the way. Follow this procedure:

1. If you don't know how to start the problem, look at the beginning of the solution in the manual, but read no further (cover up what you haven't read). Set the book aside and ask yourself how you could have thought of this approach. Realize that there may be other ways to begin the exercise. Now work to complete the solution, not looking at the manual.

2. If you get stuck, read more of the solution in the manual, stopping when you have enough to proceed in your efforts to find a solution. Continue with this process until you have a solution.

3. When you have a solution, put your work and the manual aside and work the problem from beginning to end.

Following this process is far superior to just reading the solution in the manual. It helps you develop your problem-solving skills. Reading through a solution in the solutions manual, from beginning to end, or finding a solution, with *some* assistance from the manual, are two very different activities. The latter activity provides a better learning experience.

If you are able to work a problem, consult the solutions manual to see if the solution given there varies from yours. This can be a valuable learning experience. Your ability to problem solve is enhanced more by seeing a variety of solutions to the same problem, than by seeing only one solution to each of many problems. Many students believe there is only one way to solve a problem and their goal is to find that way. If you believe this, your ability to solve problems will be inhibited since you will be looking outside yourself for this unique solution. You won't believe that you can come up with a solution that is truly yours. You will ignore your problem-solving instincts that can lead to a solution. The *best* solution to a problem *is the one you create.*

If you want feedback on problems that are not assigned, then use a solutions manual. An instructor's assignment should be viewed as minimal. If you are a student who can be challenged more, you should work additional problems from the textbook that are more difficult, or will expand your knowledge. Your instructor will probably not be willing to give you feedback on them in class, but most likely will outside of class. Regardless, having access to a solutions manual is important for getting feedback on these problems. Don't restrict your choices to only those problems solved in the solutions manual. Some of the other problems in your textbook may challenge or interest you because of the problem-solving skills needed to solve them, or because they reveal interesting applications.

Critically Reading a Solutions Manual

It is important to read a solution to a problem in a *critical* way, whether the solution is in your textbook or in a solutions manual. This will make you a better problem solver. In reading critically, make comments or ask questions similar to these:

What might have prompted the solver to begin here?

Are there other ways to begin the problem?

Does this have to be the next step of the solution or is there another one?

I don't see where this aspect of the solution came from, so I better ask someone about it.

I believe I can improve on this solution. I will talk to someone about it.

This solution makes use of notation or ideas that I haven't been exposed to. I need to ask someone about this.

Pitfalls to Avoid in Using a Solutions Manual

Many, if not most students misuse solutions manuals, which affects their development as mathematics students. They would be better off not using one. Unwise uses of solutions manuals have been alluded to earlier in this section, but the topic needs to be expounded on. The major abuses of a solutions manual are accessing it too quickly, and not reading it in a critical way. When this happens, it is being used as an immediate, easy, temporary, and harmful fix. Proper use of a solutions manual requires discipline on your part.

It is tempting to begin work on an exercise or problem and then go to the solutions manual as soon as you experience difficulty. This is consulting it too quickly. Some students do not even attempt to work a problem, but instead read a solution to it in the solutions manual. To work a problem, you need to give yourself time to try various problem-solving skills, to look over problems that were solved in class or in the textbook, or to reread your notes or portions of your textbook for ideas on how to proceed.

It is better to err on the side of waiting too long to consult your solutions manual, if there is such a thing. Delay any penchant for instant gratification. For some problems, waiting as long as a few hours before using the solutions manual may not be long enough. There are times when it is beneficial to wait at least a day, continuing to work periodically on a problem, before consulting the solutions manual. When you are not working on the problem, your subconscious is. I frequently worked on a problem without much success, put it away for a day or night, and then the solution, or an idea that would lead me to the solution, came to me during the night or immediately upon awaking the next morning. Don't let consulting the solutions manual too quickly get in the way of this happening.

Unless you are very disciplined, it is unwise to have the solutions manual at your fingertips. If you are working in a library, leave it at home. If you are working at home, leave it in a place that requires some effort to retrieve it. Do what you have to do to resist using it too quickly. You are more likely to consult a solutions manual too soon if you unduly delay working an assignment. Your time to complete the assignment is shortened, which may cause you to want a quick fix.

What Mathematics Students Say about a Solutions Manual

Here are student comments on the use of solutions manuals, followed by my evaluative comments:

❖ *"The more I used the solutions manual, the less I learned (absorbed, understood, and could relate to)."* (Calculus II student repeating the course)

It is likely that this student had not learned how to use a solutions manual.

❖ *"The solutions manual is very helpful."* (Calculus I student)

Let us hope it was for this student. It depends on what "helpful" means to this student. Did the student use it wisely and hence grow in mathematical maturity and problem-solving ability, or did the student use it unwisely, accessing it too quickly and not critically reading the solutions.

❖ *"I wish there was a solutions manual available for the even-numbered problems. It seems that many key problems are the even-numbered ones. Without answers, there is always doubt."* (Calculus II student repeating the course)

This student has to be weaned away from always looking for verification outside of himself or herself. Seeking to justify your work by looking at someone else's solution or answer can be problematic. Your eventual goal is to know that you are correct by continually asking yourself questions like these: "What can I point to that justifies this step?" "Does each step make sense to me?" "Is the answer reasonable?" This student needs to build internal checks. It appears that this student still has some things to learn about the purpose of a solutions manual and how to use it.

❖ *"Attempt to do the problems without using the solutions manual. Only use this manual as a last resort."* (Calculus II student repeating the course)

Hopefully, by the phrase, "as a last resort," the student does not mean within a short time of being stymied. It appears that the student had to learn the hard way when to use the solutions manual.

❖ *"I have found great comfort in using a solutions manual. This is, however, a dangerous statement to make. I could see how a student could use this guide as a crutch, instead of a helpful aide. This guide gives detailed solutions to the odd problems in our book and has helped me to stay alive in this course. I purchased this manual this fall but did well without it for the first two terms of calculus."* (Calculus III student)

This statement appears to be made by a mature student.

❖ *The solutions manual was a big help, especially if you are like me and hate to ask for help.* (Calculus I student)

Let's hope this student learns how to ask for help. Why do you think the student is reluctant to ask for help? Asking for help is a sign of strength, not weakness. Can a solutions manual be helpful? Yes. Is it as helpful as receiving quality help from a classmate or instructor? No.

2. Using Alternative Textbooks

During my undergraduate and graduate school days in mathematics courses, and as a mathematics instructor, I always had one or more alternative textbooks at my disposal. I knew, even as a young student, the importance of *understanding* mathematics. To accomplish this to my satisfaction in college mathematics courses, I needed to read textbooks in addition to the ones assigned to my courses. There were numerous times that I had difficulty understanding what my textbooks said on specific mathematics topics. The thrill of experiencing a light bulb going on in my mind on a topic, as I was reading another mathematics textbook, was exhilarating and confidence building. I don't recall someone suggesting that I read other textbooks, or whether I sought them out of sheer desperation. But they were so important to my growth in mathematics, that I am devoting a section of this chapter to this topic so that you will also see the value in using them. However, what I have to say presupposes that you are reading your textbook, which is your main source of knowledge. Even if you think your textbook is excellent, seeing the same ideas presented in alternative textbooks is enlightening. The importance of reading your textbook and how to read it comprises the content of chapters 15 and 16.

These topics associated with reading alternative textbook are addressed:

1. Benefits of reading alternative textbooks
2. Choosing an alternative textbook that speaks to you
3. Locating alternative textbooks
4. What students say about using alternative textbooks

Benefits of Reading Alternative Textbooks

The particular mathematics textbook you are using for your course cannot adequately meet all your mathematics reading needs. No textbook is ideal for you or for anyone else. It was written for everyone, not someone. There are differences in textbooks for the same course and that makes alternative textbooks worthwhile to read. Differences exist in—

- Topics covered
- The way definitions are worded
- Examples and counterexamples
- Applications and exercises
- Statements and proofs of theorems, and the order in which the theorems appear
- Illustrations and drawings
- Symbolic form
- Historical facts and anecdotes that humanize topics
- Styles of exposition

- Focus (giving more details at the expense of seeing the big picture, or vice versa)

Some authors have a learning style similar to yours. It follows then that their textbooks will be more understandable to you. You will feel that they are speaking directly to you, but not necessarily with respect to all the content. If you have several good alternative textbooks at your disposal, you will probably be able to find one, and it won't always be the same one, that treats a particular topic more to your liking than do the others. Authors are under pressure by publishers, due to costs, to limit the number and types of examples and other content they want to include. The examples in your textbook may be inadequate for you to work your assignment. Similarly, your textbook authors' statement or proof of a theorem, or definition of a concept, may be difficult for you to understand, whereas other authors may make it more comprehensible to you.

There are additional benefits gained from reading alternative mathematics textbooks. Instructors frequently select examination or assignment problems from them. The wording style of these problems can be different from the wording style of your textbook's exercises. Reading alternative textbooks helps you focus more on the wording of exercises, not mechanically responding to verbiage that you have been exposed to in class and in your textbook. Thus, you will be able to respond better to different styles of language, regardless of where it appears, including on your examinations. I made it a point on my examinations to not always use verbiage exactly like that of the textbook's problems. Students, who were not that concerned with understanding, or did not do what was required for understanding, would complain about this. If they understood the content better they would have had no difficulty with the wording of my examination questions. It was not *that* different!

Another benefit of reading alternative textbooks is that you will be exposed to ideas related to the topics you are studying in the course that do not appear in your textbook. This can enhance your understanding of the topics, in addition to exposing you to new ideas. It was frequently the case that, when reading an alternative textbook on a topic not related to what I was seeking, I would get so absorbed in what I was reading that I would continue reading for an hour or two, going off on a tangent (no pun intended). This may not have helped much in the course I was currently enrolled in, but it sure helped me grow as a mathematics student. I was motivated to learn knowledge that I was not required to learn and that "isn't all bad." I did this without anyone's encouragement or help. It truly became my knowledge—I understood it better and retained it longer—and this was empowering to me. A problem with some students who receive grades of A in every course they take, whether a mathematics course or not, is that they do not allow themselves to learn content that will not be tested. They only study content that they think will get them their A.

Instructors like to put problems on examinations that require their students to put some ideas together in order to solve the problems. These may be problems that have little resemblance to any you worked from your textbook exercises. Your instructor either creates them or they are taken from other textbooks. You will be better prepared to work these if you take the opportunity to read other textbooks, and work exercises appearing in them.

Choosing an Alternative Textbook That Speaks to You

The alternative textbooks you use are your decision. Pick books that appeal to your learning style. If you learn best from the written word, a book with more lengthy explanations or with a writing style that suits you, is likely to be most helpful. If you are a logical person, a book that emphasizes good reasoning and making connections to other ideas will serve you well. If you are a visual or spatial learner, a book heavy on graphs, illustrations, and diagrams may be helpful. If you are an intuitive learner, you will like a book that focuses on big ideas and their relationships. If you are a sensory learner, you will like a book that uses applications to lead into the theory or has a careful and detailed development of the content. There is no need to analyze, in a scientific way, the alternative textbooks you should select. Pick a few mathematics topics that you want to know more about, and see how several of these books present the information. It won't take you long to gravitate toward one or two of these texts for you will discover that they are speaking to you. (For more information on learning styles, see section 1 of chapter 13.)

Locating Alternative Textbooks

Here are ways you can gain access to alternative mathematics textbooks:

Check them out of college or mathematics libraries. Unfortunately, you may only be able to check them out for a short period of time. If you do a lot of your mathematics study in the library, then work near the stacks where they are located. Check them out over vacations, or when you know you cannot be in the library for at least a few days.

Ask your instructor to place them on closed reserve in the college library.

Borrow them from a college mathematics instructor. College mathematics instructors have many mathematics books on their office shelves. They get them without cost from publishers of mathematics textbooks so they can examine them for possible course adoption. Many will let you check them out for the whole semester. Some may even let you keep them. Don't hesitate to ask if you can borrow one. Ask them to recommend ones that they believe are good. Over the years instructors accumulate a lot of these books. Many, if not most of them, sit on bookshelves and are eventually thrown out when revised versions are received. Allowing you to keep them has the blessing of the publishing companies if instructors have no further use for them, and they are no longer used as a textbook for a course. You do not want to infringe on publishers' sales of new books or deprive the authors of their royalties.

Buy them online, used or unused. Buying new textbooks can be expensive. Online prices for used books can be considerably lower than the price of a new book. Online sales of used books are a great way to purchase them.

Buy them at book sales in lobbies of shopping malls, libraries, etc. You can often buy alternative textbooks for a few dollars at all of these places. But to find them when you need them is problematic, so plan ahead.

Borrow or buy them from friends attending other colleges who used them as their required textbooks.

What Students Say about Using Alternative Textbooks

Here are comments from two students about alternative textbooks:

> ❖ *"Some texts describe or analyze a certain concept better than others. Don't limit yourself to one book. You may spend hours trying to figure out what one book is trying to*

*tell you about a concept, when you may understand it in
just ten minutes by reading another book."* (Calculus III
student)

❖ *"Get a different text. Regardless of how good the instructor
and lectures are, a student depends on the text for learning.
I got through this course with the aid of a library book."*
(Calculus I student)

3. Using Technology

The suitable electronic technology that was available to me as an undergraduate and graduate mathematics student, and in the early stages
of my career as a mathematics instructor, was virtually nonexistent. I
could not carry a computer with me since computers were the size of a
large room, and they were not mine to carry anyway. How times have
changed with the electronic technology that is available for your use,
including its affordability and transportability. Your ability to learn
mathematics is significantly enhanced if you and your instructors use
this technology appropriately.

Available suitable technology to use in your mathematics courses
includes graphics and symbolic calculators, and desktop, laptop, notebook, and cell phone computers. This technology makes use of word-
processing software, dynamic or motion software, symbolic
manipulation software, spreadsheet software, and databases. A number
of these applications are more appropriately used in some mathematics
courses than in others. Computers and cell phones can be used to send
and receive mail electronically, and access websites of your instructors, publishers of your textbook, and a host of others you desire to
contact. The publishers of your textbooks may have tutorial websites,
and through your instructors' websites you have the possibility of accessing assignments, additional course materials, and other learning
aids. Email and cell phone text messaging enables you to have off-site
conversations with your instructor and classmates.

As a college student in mathematics you need to be adept using
many of these applications, especially graphics calculators and other
computers. Not only will they help you learn mathematics, but they
will also prepare you to engage the technological world of today. Entering the work world unprepared to use technology will put you at a

distinct disadvantage. Your college mathematics instructors have a responsibility to prepare you for this hi-tech world. They should not, to an unjustifiable degree, restrict the use of technology in your mathematics courses.

It is easy to understand why some college mathematics instructors are skeptical of promoting the use of technology in their mathematics courses. They have had far too many students in mathematics courses who misused technology. These students then lacked requisite mathematics knowledge. Nonetheless, not promoting the use of technology in their courses is shortsighted because students can benefit greatly when instructors integrate technology in an intelligent way. That being said, do not assume that the wise use of technology is a cure-all. Instructor A's students may have a far better understanding of the course, even though he or she does not promote the use of technology, than Instructors B's students where technology is promoted and used wisely. There are many factors that make an excellent instructor. However, as great as Instructor A's course may be without prudent use of technology, it would be greatly enhanced by its use.

Objectives Attained by the Prudent Use of Technology

If you are skeptical of the importance of integrating technology into your mathematics course, or have an instructor who feels the same way, reflect on the following objectives attained by *prudent* use of technology:

1. An enhanced understanding of concepts, which is the focus of technology
2. Assistance in looking at mathematics from multiple viewpoints—arithmetic, geometrical, and algebraic
3. Allows the use of more realistic applied problems (since computation is less of an issue)
4. Allows more time to focus on the development of concepts and principles, and on understanding, modeling, reasoning, and problem solving (since less time is spent on computation)
5. Prepares students to use technology in applied mathematics courses (e.g., in engineering, natural science, and business courses)
6. Prepares students to use technology in their careers (em-

ployers expect colleges to turn out students proficient in the use of technology)

7. Enhances the formulation of conjectures and decision-making (since information from which inferences are made is easily generated)

8. Helps in organizing data so it is more easily analyzed (through the ease of constructing charts and tables)

9. Allows more visual models that aid in understanding (since graphs are easily constructed)

10. Allows more dynamic classroom presentations and discussions (through large screen computer projections)

11. Motivates students to learn (due to the positive impact that the use of technology has on their attitudes toward learning, self-confidence, and self-esteem)

12. Provides an environment in which students can write reports that synthesize and explain their conceptual understanding (through the use of computer-based word processing with symbolic and graphic capabilities)

Research Findings on the Impact of Technology in Calculus Courses

The inclusion of technology in lower-division mathematics courses has been going on in a major way for many years, and began with reform in the calculus sequence. It began with the construction and use of experimental materials that were a radical change from what currently was being used. The content was developed in a significantly different way. These materials eventually became available commercially, in so-called innovative textbooks. Perhaps you are in a mathematics course using one of these textbooks. The success of these experimental programs has influenced other textbook writers to integrate technology into their textbooks, but to a lesser extent. The spread of the reform movement in calculus began early as noted in a 1999 article referring to a 1994 Mathematical Association of America report:

> More than 500 mathematics departments at postsecondary institutions nationwide are currently implementing some level of calculus reform. These reformed courses enroll approximately 300,000 students each year, about 32% of the total na-

tional calculus enrollment. Such growth of a movement only a decade old suggests that the influence of calculus reform will continue to spread.[1]

It is not a goal of this book to give a detailed report on the successes or failures of these curricular changes in mathematics involving more widespread use of technology. The following quote will suffice, and if you want more information consult other sources:

> The preponderance of evidence from the Connecticut study—as well as the evidence cited in this section—is consistent with the conclusion that the impact of calculus reform has been positive. If the "filter-to-pump" goal is an appropriate assessment standard, then the results from Connecticut and other calculus-reform sites suggest that calculus with modern computational and pedagogical features can promote better end-of-course mastery, significantly improve the flow rate into the technical work force, and foster more gender diversity into that flow.[2]

Abuses of Technology

The likelihood of misusing technology is high to say the least. How can you abuse it? Consider these ways:

Relying on technology (calculators or computers) for virtually all your computations. For example, you should mentally be able to determine: 3^4, $27^{\frac{2}{3}}$, $\sin 30°$, $\tan 45°$, 25×5, $90-47$, decimal form of 1/8 or 5/4, $\log 1000$, $\ln e^4$, $\int 3x dx$, $\int \sec^2 x dx$, derivative of $\frac{2}{x}$, and, in your mind's eye, sketch the graphs of $-2\sin x$ and $\sin^{-1} x$. If you consistently solve simple problems like these with a scientific calculator or microcomputer, you are becoming too dependent on them. It is important to do them mentally because in most cases you have to recall the meanings of the concepts in order to perform the computations or construct the graphs. This reinforces your understanding of them and aids in their retention, which are goals of any mathematics course. Here are ways to mentally solve some of the above problems, which require you to recall concepts and principles:

1. Calculate 3^4. ($3^4 = 3^2 \times 3^2 = 9 \times 9 = 81$.)
2. Calculate $27^{\frac{2}{3}}$. ($27^{\frac{2}{3}} = (27^{\frac{1}{3}})^2 = 3^2 = 9$.)

3. Determine sin30°. (In a 30-60-90 right triangle, it is the length of the side opposite the 30 degree angle (which is 1), divided by the length of the hypotenuse (which is 2), giving 1/2 or 0.5.)

4. Determine the decimals for 5/4 and 1/8. (5/4 = 1 + 1/4, 1/4 is 0.25; hence 5/4 is 1.25. 1/8 is one half of 1/4; hence, it is one half of 0.25, which is 0.125.)

5. Find log1000. (log 1000 is the exponent you put on the base (which is 10), to get 1000; hence, the answer is 3.)

6. Evaluate $\int \sec^2 x dx$. ($\int \sec^2 x dx = \tan x + C$, since the derivative of tan x is sec² x.)

7. Graph $-2\sin x$ and $\sin^{-1} x$. (The graph of $-2\sin x$ is the graph of sin x reflected in the $x - axis$, and stretched vertically by a factor of 2. The graph of $\sin^{-1} x$ is the reflection of the graph of sin x in the line $y = x$, restricted to $[-\frac{\pi}{2}, \frac{\pi}{2}]$. In the mind's eye this can be done quickly knowing the graph of sin x on $[-\frac{\pi}{2}, \frac{\pi}{2}]$.)

8. Find the derivative of $\frac{2}{x}$. (The derivative of $\frac{2}{x}$ is two times the derivative of $\frac{1}{x}$, which is $2 \times (-\frac{1}{x^2}) = -\frac{2}{x^2}$. The derivative of $\frac{1}{x}$ should be committed to memory since it is used so frequently.)

I hope you noticed the concepts and principles that were used to work these problems. I repeat: working them without using technology reinforces understanding and enhances retention, and can save time. Are you using calculators and computers to perform computations when you should not be? If so, wean yourself away from doing this.

When writing a paper, perhaps based on laboratory work that makes use of technology, there is a natural tendency to be primarily interested in getting computational results rather than in analyzing what the results reveal. I have seen far too many students do this—time and time again. There was something significant to be learned from the laboratory project, yet these students did not step back to critically analyze what it was. They followed instructions in a perfunctory manner, hardly questioning themselves or others as they progressed through the project. In this situation, the technology used is often referred to as the "black box." That is, something is put in the black box, a crank is turned, something comes out, and any understanding that results is accidental. Enhanced understanding should be the focus of using technology, not just ease in computing or an opportunity to avoid understanding. How thoughtful are you when you integrate the use of technology into your problem assignments and laboratory reports?

Not understanding the outputs that you get from using technology. When you use a calculator to find, for example, $\cos^{-1}\dfrac{\pi}{4}$, do you know that the answer is selected from the interval $[0,\pi]$? Do you know why this is the case?

Not understanding the shortcomings of technology in the results obtained. When you use a calculator or microcomputer, you should be able to answer questions such as these: What is round-off error and truncation error and what problems can arise because of them? What is the number of significant figures in your result? What are the shortcomings of using calculator generated numbers or graphs for the specific purposes you have in mind? For example, when you use the trace function on a graphics calculator to find relative maxima or minima of a function, do you reflect on the accuracy of your results? The more you advance in mathematics knowledge, including its applications, the deeper your understanding of technology's shortcomings needs to be, and how to manage them. Take an interest in learning this.

Appropriate Places for Technology in a Course

If the use of technology in your mathematics courses accomplishes vital objectives, than certain obligations are placed on your instructor regarding its use. Your instructor needs to encourage its use in many, if not most assignments, and to evaluate whether you use it appropriately. The latter obligation not only means that you should be allowed to use it on examinations, *or a portion thereof*, but that its use there be mandatory. Students must be tested on their intelligent use of technology if it is a goal of the instructor that students use it intelligently.

My students were always allowed to use graphics calculators on my examinations, regardless of the mathematics course and whether they were unit or final examinations. Did this interfere with my evaluation of their comprehension of mathematics? Absolutely not! I could always formulate questions in such a way that technology in the hands of a student was of little or no help, especially if I required them to show their reasoning. I did not want to deny my students the kinds of examination questions that required them to use technology. These were examination questions that (1) displayed their understanding of certain concepts, applications, and principles, (2) demonstrated whether they could use technology to gain specific information, or (3) revealed whether they knew the limitations of technology.

What If Your Instructor Does Not Encourage the Use of Technology

There are upper-division mathematics courses where technology may play a lesser role than in lower-division mathematics courses (e.g., abstract algebra; topology). However, what is disconcerting to see is the judicious use of technology not encouraged in courses where its value is readily apparent (e.g., precalculus, calculus, matrix algebra, differential equations). Unfortunately, there are instructors who do not encourage its use, or strictly forbid its use on examinations or assignments. They may be well intentioned, but they are also uninformed. College mathematics instructors have a tremendous amount of freedom regarding what they do or do not include in their courses, but not as much as some of them would like you to think. For example, multiple-section courses have department-approved syllabi which, among other things, mention the sections of the textbook to cover and whether the use of technology is required. If your instructor does not support the use of technology in your course, go to the mathematics department to see if it is a departmental requirement for that course. If it is, then go to the department chairperson and register a complaint. If it is not required, then tell the chairperson and your instructor why you think it should be. In your arguments, you can appeal to the objectives that can be accomplished by the prudent use of technology that were presented earlier in this section. If you are not successful getting your instructor to willingly accept the use of technology in the course, you should use it on your assignments. However, you also have to be able to do the work your instructor expects without the use of technology. This is not an ideal situation, but one you may have to live with.

Be Skilled in Using Technology in Your Mathematics Course

It is important to be a *skillful user* of technology in your mathematics courses. For example, if you continually find yourself confused about the operation of your graphics calculator, it will interfere with your course achievement. Frequent use of trial and error to figure out how to use it is frustrating, demoralizing, time consuming, and interferes with your learning. Perhaps you think you don't have the time to learn how to use it well. In the long run, you will save time by learning how to use it efficiently. This knowledge will also serve you well in other mathematics and mathematics-related courses. Skillful use of technology is quickly becoming a non-issue in college. Many high school students are very knowledgeable about using a graphics calculator and desktop or laptop computers.

Your college mathematics instructors may not take class time to teach you how to use technology and its accompanying software. However, many mathematics instructors have a calculator or microcomputer in the classroom hooked up to a projection system. They demonstrate how this technology and accompanying software can be used. If your mathematics course has a computer laboratory component, then you can obtain help with the technology you are using from a laboratory assistant or from a laboratory partner. There are other avenues of help to pursue, but whatever you do, what you learn will quickly be forgotten without sustained practice.

Here are additional avenues to pursue to become a skillful user of technology:

Work with another student who is more experienced. This may be the best way since you have someone who can demonstrate its use, watch you use it, ask you questions, and respond to your questions.

Read through the manuals or tutorials that accompany your technology or software, and then practice what you learn. Set aside specific times to do this, and if it is possible, start doing this before your course begins.

Use your mathematics-learning center. Many mathematics departments have a mathematics-learning center that provides learning modules on using technology as well as tutorial assistance on the use of technology.

Use online support provided by the manufacturer.

Seek help from your instructor.

4. Using a Supplementary Problem Workbook

The publisher of your lower-division mathematics textbook might publish supplementary problem workbooks. You might be able to purchase them in college bookstores or online. In the main, they are comprised of extra problems for you to work and display solutions to many problems. The big question you may have is this: "Are they worth buying?" I cannot answer this for you. You have to carefully review the workbook and make the decision. However, I have more to say on this.

Most college mathematics instructors assign only a small fraction of the exercises that appear in the textbook they are using. This means that there are many other problems for you to work in your textbook, literally pages of extra problems for each section of the textbook, to say nothing of a myriad of review problems at the end of each chapter. The question you have to ask yourself is this: "If I am not working many unassigned problems in the textbook, why would I want to pay for a supplementary book that contains problems that I also will not work?" I have a strong sense that the vast majority of students do not purchase them, unless they are required to do so. One reason they don't purchase them is they don't know they exist.

A supplementary problem workbook for your textbook may devote space to discussing specific content in a way that is more understandable to you. Or, it may provide more solutions to exercises than your textbook does, or has exercises that are especially constructed to help you work more difficult exercises in your textbook. You need to know what is in the supplementary problem workbook, whether its content is similar to what is in the textbook, how valuable you think it will be for you, and the price. Two reasons that textbook publishers and authors publish these materials are to make more money, and to further students' learning. I am sure the first reason is realized, and question whether the second reason is realized for many students who purchase materials.

19

Learning Mathematics Notation

Mathematics makes heavy use of symbols. What is unique about mathematics, besides its ideas and how they can be used to solve a myriad of problems in the world, is its use of symbolic notation (call it mathematics shorthand if you will). This, perhaps, is its most distinguishing feature, and is the primary means used to write mathematics. *The extensive use of symbols is one of the characteristics that makes mathematics an abstract subject.* Symbols are given meaning by using words and other symbols, followed by examples of their use. Once you understand the meaning of a specific symbol, you have to know later on what it means without having before you the words and examples that gave it meaning. You have to mentally supply them. If you do not adequately learn the meaning of symbols when they are introduced, or practice their use, then your success in mathematics will be seriously compromised or impeded. Many students are exactly in that position. Are you?

Since the extensive use of symbols makes mathematics more abstract, why not use fewer symbols? Textbook authors and instructors could use fewer symbols in their exposition, but this comes at a price. Using fewer symbols means using more words, which in one sense results in greater clarity; however, the conveyance of many ideas in a compact space—an indispensable part of mathematics—is compromised. It is this compactness that helps you see what you are considering, which is also a clarifying feature. Paul Halmos, a noted expositor of mathematics, addresses this issue, including what he thinks you like to see in expositors:

> I advocate the use of words more than numbers in exposition
> also, and in the exposition of mathematics in particular. The

invention of subtle symbolism (for products, for series, for integrals—for every computational concept) is often a great step forward, but its use can obscure almost as much as it can abbreviate. . . . Another correlation I wish I could prove is that the ones who do like words become famous and beloved for their clear explanations whereas the ghosts of the others are muttered and sworn at by exasperated students every day.[1]

Halmos knows the value of symbols in mathematics, but realizes that their widespread or inappropriate use can increase your difficulty to understand. For sure they will obscure understanding if you don't learn their meaning. You might want to wish them away, but they are necessary to have and are here to stay. Mathematics textbooks are filled with symbols. Learn their meaning and you will minimize the obscurity that they present.

New symbols are introduced regularly in college mathematics courses. It is desirable to delay the introduction of a symbol that represents a concept until the concept is understood, but that isn't the case in college level mathematics. In college, the symbol for a concept is introduced immediately after the concept is defined (and most likely not yet understood). The sheer number of concepts that are introduced in a short period of time at the college level dictates that you need to learn them quickly, as well as their associated symbols. Almost without delay you will use these concepts to develop other content and to work exercises. The difficulty of learning the meaning of new symbols is compounded by having to simultaneously work with symbols, introduced earlier in the course or in prerequisite courses, that you may not have learned as well as you should have.

In a sense, learning mathematics, with all its symbols, is like learning a foreign language with all the problems that go with learning a second language. Fortunately, mathematics is a *self-referring* language and picks up power from this, but to benefit from this you have to refer to it often. The process of learning the symbolic language of mathematics would be improved if there were more literature available on how to teach it. This literature would be especially valuable to pre-college mathematics teachers since learning how to work with mathematical notation is a prerequisite for college mathematics courses. This chapter shows you how to learn and work with mathematics notation.

Being adept in working with mathematics notation is important.

However, its importance to you probably increases in direct proportion to your involvement with mathematics—in college and in your career. Understanding mathematics notation helps you—

- Read mathematics with understanding
- Communicate mathematics, orally or in writing, to yourself and others
- Make connections within mathematics
- Reflect on mathematics
- Formulate generalizations
- Think mathematically
- Solve problems
- Enter information into a calculator or computer
- Perform computations
- Retain what you have learned

My concern in writing this chapter is that I am *understating* the importance of gaining adeptness or fluency in working with mathematics notation.

1. Choosing a Symbol for a Concept

In college mathematics the symbolization of a concept is introduced immediately after the concept is defined. Textbook authors or instructors introduce the symbol by using words such as these: "It is *denoted* or *named* by... (the symbol is then given). The definition of the concept gives meaning to the symbol that represents the concept. You will not necessarily understand the concept just because you read its definition. Understanding the concept often requires seeing examples of the concept as well as instances when the concept does not apply (called non-examples). It also requires working with the concept in other ways. The choice of a specific symbol to represent a specific concept is not a mindless activity or an accidental act. Mathematicians often choose a symbol for a concept to reflect, in some fashion, the meaning of the concept, which is gained from the developmental work that led to the concept. This gives rise to two important reasons for you to understand *derivations* of concepts: You are better able to (1) select the appropriate symbols to use to accomplish specific tasks, and (2) you are better equipped to make adaptations to basic symbols that will allow you to accomplish related, but different tasks. The four examples now given will help clarify what has been said. However, examples 2, 3, and 4 will be of minimal help if you have not had calculus.

Example 1. At some point in time, mathematicians wanted a symbol to represent the product of n factors of a number a. That is, they wanted a shorthand expression for this product: $a \cdot a \cdot a \cdot \ldots \cdot a$. It made sense to them to have an a and an n in the symbol they would construct because these two entities convey key information about this product (i.e., the number being multiplied and the number of times it is being multiplied). You know that the invented symbol is a^n, which contains an a and an n.

Example 2. In introductory calculus, the symbol denoting the area of the region bounded by the graph of the non-negative function $y = f(x)$, the vertical lines $x - a$ and $x = b$, and the $x - axis$ is $\int_a^b f(x)dx$. Without getting technical, this symbol represents the sum of the areas of an infinite number of rectangles with height $f(x_i^*)$ and width Δx_i, where x_i^* is an arbitrary point belonging to $[x_i, x_{i+1}]$ and $\Delta x_i = x_{i+1} - x_i$. The area of each of these rectangles has the form $f(x_i^*)\Delta x_i$, which bears a close resemblance to the symbol $f(x)dx$, which is part of $\int_a^b f(x)dx$. This is not accidental; $f(x)dx$ was chosen to reflect $f(x_i^*)\Delta x_i$. It is not clear why the symbol $\int$ was chosen. Some historians say that it is an enhanced S for the first letter of the word sum (recall that the areas of rectangles are being summed). The numbers a and b are placed on $\int$ to indicate that the area of the region sought, is between the lines $x = a$ and $x = b$. Mathematical symbols are not chosen mindlessly.

Example 3 (Adapting a Basic Symbol). Suppose you want to find the area bounded by the graphs of $y = f(x)$ and $y = g(x)$, and by the vertical lines $x = a$ and $x = b$, where $f(x) \geq g(x)$ for every $x \in [a,b]$. Your rectangles will have height $f(x_i^*) - g(x_i^*)$, width Δx_i, and area $(f(x_i^*) - g(x_i^*))\Delta x_i$. The area you seek is the sum of an infinite number of these rectangles and is represented by $\int_a^b (f(x) - g(x))dx$. This symbol is an *adaptation* of the basic area symbol $\int_a^b f(x)dx$. You were able to construct this adapted symbol based on your understanding of the development that led to the basic area symbol. The key adaptation made is to realize that $f(x_i^*) - g(x_i^*)$ represents the height of a rectangle in example 3 and $f(x_i^*)$ represents the height of the rectangle in example 2.

Example 4. The symbol used to find the volume of a solid of revolution, obtained by revolving a region bounded by the graph of the non-negative function $y = f(x)$, the vertical lines $x = a$ and $x = b$, and the x-axis, is $\int_a^b \pi(f(x))^2 dx$. In developing this symbol, a cross section of the solid is determined by cutting the solid with a vertical plane through the point $(x_i^*, f(x_i^*))$, where $x_i^* \in [x_{i+1}, x_i]$. This gives a circular disk with radius $f(x_i^*)$. The area of this disk is $\pi \times radius\ square = \pi(f(x_i^*))^2$. A cylinder is formed with this disk as its base and height $x_{i+1} - x_i = \Delta x_i$. The volume of this cylinder is the area of its base times its height, which is $\pi(f(x_i^*))^2 \Delta x_i$. We add up an infinite number of these volumes, one for each point x_i^* chosen in $[x_{i+1}, x_i]$, which gives the volume of the solid of revolution. Those of you who have had calculus know that the volume is $\lim_{n \to \infty} \sum_{i=1}^{n} \pi(f(x_i^*))^2 \Delta x_i$. Notice how closely $\pi(f(x_i^*))^2 \Delta x_i$ resembles $\pi(f(x))^2\ dx$, and that this latter symbol appears as part of the symbol for the volume of the solid of revolution, namely, $\int_a^b \pi(f(x))^2 dx$.

Don't be alarmed if you are a non-calculus student and are confused about what has just been discussed. Calculus students may struggle with what was said since most details were left out. Come back to this section after you have been exposed to these topics in calculus. Seeing how a symbolic expression is constructed from a development of a concept will help you select the correct symbol or formula when you need to apply the concept, as well as use the correct formula. You now know, through reading example 4, that $\pi((f(x))^2\ dx$ *refers* to a volume.

2. Fluency Objectives for Learning Mathematics Notation

You need to achieve specific objectives in order to be fluent in the symbolic language, or symbolic forms, of mathematics. Eleven fluency objectives are discussed in this section. You should be able to do the following, orally or in writing:

1. Given a symbolic form, translate it literally.
2. Given a literal translation of a symbolic form, give the symbolic form.
3. Given a symbolic form, give a literal interpretation.
4. Given a literal interpretation of a symbolic form, give the symbolic form.
5. Given a symbolic form, interpret it geometrically.
6. Given a geometric interpretation of a symbolic form, give the symbolic form.
7. Given a symbolic form having a name or title, give its name or title.
8. Given the name or title of a concept or property, give its symbolic form.
9. Given a symbolic form, write equivalent symbolic forms.
10. Given a symbolic form, state one (or more) of its uses or applications.
11. Given a use or application of a symbolic form, give the symbolic form.

Examples of Fluency Objectives

We look at each of these objectives through examples. Hopefully, enough examples of symbolic forms are provided for each objective so that you will know what to do with other symbolic forms you encounter. For each new symbolic form that is introduced in a unit of study in your course, check to see if you have satisfied each of these 11 fluency objectives (if they apply):

Objective 1:   **Given a symbolic form, translate it literally (orally or in writing).**

Objective 2:   **Given a literal translation of a symbolic form (orally or in writing), give the symbolic form.**

A literal *translation* of a symbolic form adheres strictly to the basic meaning of the symbolic form without further elaboration or interpretation. View it as a *direct* translation of the symbolic form. There is often more than one literal translation of a symbolic form. *When reading a literal translation, a comma indicates that you should pause in your reading.* The following are examples of objectives 1 and 2. Read from left-to-right for objective 1, and from right-to-left for objective 2.

To determine if you can accomplish objectives 1 and 2 for these examples, cover the appropriate side, read the other side, and then see if you can state the information you covered (or its equivalent).

Symbolic Form	**Literal Translation**
x^3	"x cube," "x raised to the third power," or "x to the power of 3"
$(a + b)^2$	"a plus b the quantity square," or "the square of the quantity $a + b$"
$a + b^2$	"$a + b$ square"
$-a^2$	"additive inverse of a square"
$(-a)^2$	"additive inverse of a the quantity square," "the square of the additive inverse of a," or "additive inverse of a, square" (recall that a comma represents a pause)
$\dfrac{2x - 5}{x}$	"the quantity two x minus five, divided by x"
$2x - \dfrac{5}{x}$	"two x minus five divided by x"
$(a^m)^n = a^{mn}$	"a to the mth power to the nth power equals a to the mnth power"
$g(2x) + 3$	"g of 2x, plus three (once again, a comma represents a pause)
$g(2x + 3)$	"g evaluated at, two x plus three" or "g of, two x plus three"
$f \circ f^{-1}(x)$	"f circle f inverse of x," "f composed with f inverse evaluated at x," or "f of f inverse of x"
$\sin^2 x$	"sine square of x" or "sine of x, square"
$\sin x^2$	"sine of x square"
$\lim\limits_{x \to 3} g(x) = 5$	"limit of g of x, as x approaches three, is five"
$\dfrac{dy}{dx} = \dfrac{dy}{du}\dfrac{du}{dx}$	"dy,dx equals dy,du times du,dx"
$\Delta x = -2$	"delta x equals negative two"
$\int g(x)dx$	"indefinite integral of g with respect to x"
$\displaystyle\int_c^d h(x)dx$	"definite integral of h with respect to x from c to d"

Objective 3: **Given a symbolic form, give a literal interpretation (orally or in writing).**

Objective 4: **Given a literal interpretation of a symbolic form (orally or in writing), give the symbolic form.**

A literal *interpretation* goes beyond a literal translation. That is, it is an *explanation of the meaning* of the symbolic form. The distinction between a literal translation and a literal interpretation blurs at times. The more you understand the meaning of a specific symbolic form, the more likely it is you will think of the meaning when you read the form. Hence, for specific symbolic forms, don't dwell on the semantics of whether a statement is a literal translation or a literal interpretation. There can be differences in the language used in giving a literal interpretation of a specific symbolic form. You are reminded once again that, in reading a symbolic form, a comma denotes a pause.

To determine if you can accomplish objectives 3 and 4 for these examples, cover the appropriate side, read the other side, and then see if you can state the information you covered (or its equivalent).

Symbolic Form	**Literal Translation**
$\frac{3}{4}x - 7 = x$	" a number is seven less than three-fourths the number"
$(x + 1) + (x + 3) + (x + 5)$ x an even number	"the sum of three consecutive odd whole numbers"
$\frac{nx}{ny} = \frac{x}{y}$	"common non-zero factors in the numerator and denominator of a fraction can be cancelled"
$a + b = b + a$	"the order of addition of two numbers can be changed without affecting the sum"
$(-1)a = -a$	"multiplying a number by negative one gives the additive inverse of the number"
$a^m a^n = a^{m+n}$	"in multiplying two exponential expressions with the same base, add the exponents"
If $ab = 0$, then $a = 0$ or $b = 0$	"if the product of two numbers is 0, than at least one of them is 0"

$\sqrt{a}$	"the non-negative number whose square is a"
$(\tan x)^{-1} = \dfrac{1}{\tan x}$	"the multiplicative inverse of $\tan x$ is the reciprocal of $\tan x$"
$f \circ f^{-1}(x) = x$	"the composition of a function and its inverse applied to x is x"
$\sin 2x = 2\sin x \cos x$	"the sine of a real number equals two times the sine of half the number times the cosine of half the number" [This interpretation gives rise to these symbolic forms, among others: $\sin 6x = 2\sin 3x \cos 3x$; $\sin \dfrac{x}{2} = 2\sin \dfrac{x}{4} \cos \dfrac{x}{4}$. This shows the power of specific interpretations of symbolic forms.]
$y = kx$, k a constant	"y is proportional to x"
$\Delta x = -2$	"the change in x is negative two"
$\dfrac{d}{dx}[af(x)] = a\dfrac{d}{dx}f(x)$	"the derivative of a constant times a function is the constant times the derivative of the function"

Objective 5:   **Given a symbolic form, interpret it geometrically (orally or in writing).**

Objective 6:   **Given a geometric interpretation of a symbolic form (orally or in writing), give the symbolic form.**

Even though geometric interpretations are literal interpretations, objectives 5 and 6 were created to single out their importance to your mathematics understanding and problem solving. The wonderful invention of the rectangular coordinate system, which married algebra and geometry, allows us to *picture and relate* algebraic entities in a compact space. It is often said that a picture is worth a thousand words (and I will add to this, "and many symbols").

To determine if you can accomplish objectives 5 and 6 for these examples, cover the appropriate side, read the other side, and then see if you can state the information you covered (or its equivalent).

Symbolic Form	**Geometric Interpretation**
$\dfrac{-b \pm \sqrt{b^2 - 4ac}}{2a}$	"x-intercepts of the graph of $y = ax^2 = bx + c$"
$f(-1) = 2$	"graph of f passes through the point $(-1,2)$"
$g(x) = g(x - 4)$, for every x	"graph of g is unchanged when shifted 4 units to the right"
$f(-x) = -f(x)$ for every x	"graph of f is symmetric to the origin"
$f'(x) > 0$, $x \in (a,b)$	"graph of f is rising over (a,b)," or, "slope of the tangent line is positive at each point $(x, f(x))$ of the graph of f, for $x \in (a,b)$"
$g''(x) < 0$, $x \in (a,b)$	"graph of g is concave down over (a,b)"
$\displaystyle\int_{-1}^{4} x^2\,dx$	"area of the region bounded by the graph of x^2, the x-axis, and the lines $x = -1$ and $x = 2$"
$\displaystyle\int_{a}^{b} \sqrt{1 + (\tfrac{dy}{dx})^2}\,dx$, $a \le b$	"length of the curve $y = f(x)$, from $(a, f(a))$ to $(b, f(b))$"

Objective 7: Given a symbolic form having a name or title, give the name or title.

Objective 8: Given the name or title of a concept or property, give its symbolic form.

The purpose of achieving these two objectives is evident if you imagine what it would be like if you had no names or titles associated with important symbolic forms. How would you speak or think about mathematics? Suppose you did *not* have the title "definition of derivative" associated with the symbolic form, $\displaystyle\lim_{h \to 0} \frac{f(x+h) - f(x)}{h}$, yet you

wanted to tell a classmate to find the derivative of $\sin x$ using this form. Your communication would be cumbersome, to say the least. Suppose you wanted to look up the symbolic form associated with the "Trapezoidal Rule," but this name didn't exist. You would not be aided by your textbook's index. Specific names are easier to remember than specific symbolic forms, and they often convey the use or uses for the symbolic form. *But they are of little use to you if you don't learn them.*

To determine if you can accomplish objectives 7 and 8 for these examples, cover the appropriate side, read the other side, and then see if you can state the information you covered (or its equivalent).

Symbolic Form	**Name or Title**
$\ln x$	"Natural Logarithm of x"
$y = \sin x, \ x \in [\dfrac{-\pi}{2}, \dfrac{\pi}{2}]$	"Restricted Sine Function"
$\dfrac{a^m}{a^n} = a^{m-n}$	"Rule for Dividing Exponential Expressions with the Same Base
$f(x) = ax^2 + bx + c,$ $a \neq 0$	"General Quadratic Function"
$1 + \tan^2 x = \sec^2 x$ $\sin^2 x + \cos^2 x = 1$ $1 + \cot^2 x = \csc^2 x$	"Pythagorean Trigonometric Identities"
$f(-x) = f(x)$, for every x	"Even Function"
$\dfrac{\Delta y}{\Delta x}$	"Difference Quotient"
$\displaystyle\int f(x)dx$	"Family of Antiderivatives of F"
$\displaystyle\int_a^b f(x)dx = F(b) - F(a)$, if f cont. on $[a.b]$, and $F' = f$	"Fundamental Theorem of Calculus"
$\dfrac{d}{dx}f(g(x)) = f'(g(x))g'(x)$	"Chain Rule" or "Rule for Derivative of a Composite Function"
$L(x) = f(a) + f'(a)(x - a)$	"Tangent Line Approximation of f at a" or "Linearization of f at a"

Objective 9:   Given a symbolic form, write or derive equivalent symbolic forms.

Two symbolic forms are equivalent if one can be substituted for the other. However, they may serve different purposes and that is why you may choose one form over another. For example, in finding $\dfrac{2}{3} + \dfrac{5}{4}$, you might want to use the form $\dfrac{8}{12}$ in place of $\dfrac{2}{3}$, and $\dfrac{15}{12}$ in place of $\dfrac{5}{4}$. To approximate $\sqrt{2}$ to a specific degree of accuracy, its infinite decimal form 1.4142135... is useful to use. To mentally calculate $8^{\frac{2}{3}}$, you would use the equivalent form $(8^{\frac{1}{3}})^2$, which indicates that you are to find the cube root of 8, and square it. To find the equation of a line given the point (3,2) on the line and having slope –3, you would use the symbolic form $y - y_1 = m(x - x_1)$ instead of the equivalent form $y = mx + b$.

To determine if the point $(2, \frac{\pi}{4})$ lies on a specific polar equation, you might use the equivalent form $(-2, \frac{-3\pi}{4})$. It may be simpler to find the derivative of $f(x) = 2x^2$ at $x = 3$ by using the form $\lim_{x \to 3} \frac{f(x) - f(3)}{x - 3}$, than it is to use the equivalent form, $\lim_{h \to 0} \frac{f(x+h) - f(x)}{h}$.

There is almost no end to the times in algebra, trigonometry, pre-calculus, calculus, etc., when you will want or need to use an equivalent symbolic form. You can easily derive many equivalent forms; others you have to remember.

Examples of objective 9 follow.

Symbolic Form	**Equivalent Symbolic Form**
$\dfrac{3}{4}$	$\dfrac{6}{8}$, 0.75, 75%, 3 ÷ 4
$\sqrt{3}$	$1.73205008...$, $3^{\frac{1}{2}}$, $\dfrac{3}{\sqrt{3}}$, 9^{-2}, $\dfrac{1}{9^2}$
$y = \log_b x$	$b^y = x$
$y = \sin^{-1} x$	$\sin y = x$
$\sin^2 x$	$(\sin x)^2$
$a^{\frac{m}{n}}$	$(a^m)^{\frac{1}{n}}$, $(a^{\frac{1}{n}})^m$, $\sqrt[n]{a^m}$, $(\sqrt[n]{a})^m$
$\log \dfrac{a}{b}$, $a,b > 0$	$\log a - \log b$
$\dfrac{d(\frac{df}{dx})}{dx}$	$\dfrac{d^2 f}{dx^2}$, $[f'(x)]'$, $f''(x)$, $f^{(2)}(x)$
$\displaystyle\int_a^b f(x)dx$	$-\displaystyle\int_b^a f(x)dx$

Objective 10: Given a symbolic form, state one or more of its uses or applications.

Objective 11: Given a use or application of a symbolic form, give the symbolic form.

One of your goals has to be to learn uses for many symbolic forms that are introduced in your mathematics courses. You will be expected to apply these forms as you progress through the courses. Hence, your study techniques have to take this into consideration. A common error of students is to use the wrong symbolic form to perform a specific task. For example, they will use an area formula to find a volume.

To determine if you can accomplish objectives 10 and 11 for these examples, cover the appropriate side, read the other side, and then see if you can state the information you covered (or its equivalent).

Symbolic Form	**Uses or Applications**
$y = a(x - h)^2 + k$	"to find the vertex of a parabola," "to find the maximum or minimum of a parabola," or "to sketch a graph of the parabola"
$\log_b x = \dfrac{\log_a x}{\log_a b}$	"to evaluate a logarithm in one base by using logarithms in another base"
$\dfrac{1}{2} r^2 \theta$	"to find the area of a sector of a circle"
$c^2 = a^2 + b^2 - 2ab\cos C$	"to find the third side of a triangle when given two sides and the included angle," or "to calculate an angle of a triangle when given the three sides"
$\lim\limits_{x \to \infty} g(x)$ and $\lim\limits_{x \to -\infty} g(x)$	"to find horizontal asymptotes for a curve"
$\dfrac{ds}{dt}$, where s(t) is the position of a particle at time t	"to find velocity of a particle at time t"
$\dfrac{d^2 v}{dt^2}$, where $v(t)$ is the velocity of a particle at time t	"to find acceleration of a particle at time t"

$\dfrac{d^2 f}{dx^2}$ "to find concavity of a function," "to find points of inflection of a function," "to determine relative maxima or minima of a function," or "to find the acceleration of a particle"

df "to find an approximation of Δf"

$\displaystyle\int_a^b |v(t)|dt$ "to find the total distance traveled by a particle over the time interval $[a,b]$"

$\dfrac{1}{b-a}\displaystyle\int_a^b f(x)dx$ "to find the average value of f from $x = a$ to $x = b$"

$\displaystyle\int_a^b \sqrt{1+(\dfrac{dy}{dx})^2}\, dx$ "to find arc length of curve $f(x)$ from $x = a$ to $x = b$"

After reading the examples of the 11 fluency objectives, you should see a need to achieve them. Textbook authors and instructors communicate through the use of these objectives, and it is how you have to communicate with yourself and with others. Additional suggestions on how to do this are the topic of the last section of this chapter.

3. Common Errors Using Symbolic Notation

It is prudent at this time to give examples of common errors with symbolic notation that many college students make in their problem assignments and on examinations. Appropriate study of the fluency objectives, coupled with frequent exposure to symbolic notation in your assignments, including diligent attention to all your responsibilities as a mathematics student, will help minimize or eliminate these errors.

Distributing Symbols Across Other Symbols

One of the most common errors is distributing a symbol over other symbols when it is not appropriate to do so. More precisely, you might assume that, no matter what function f is, $f(a + b) = f(a) + f(b)$. This is not true, for example, for the square root function, $f(x) = \sqrt{x} : f(a + b) = \sqrt{a+b}$), which does not equal $\sqrt{a} + \sqrt{b}$, which is $f(a) + f(b)$. What makes it easier to make this error is frequent exposure to distributive properties of other operations (where this can be done). For example, it is true for the Distributive Property of Multiplication Over Addition: $a(b + c) = ab + ac$ for real numbers a,b and c. Hence, you may be prone to incorrectly use a pseudo distributive property on a symbolic form that resembles $a(b + c)$.

Examples of distribution errors

$-(ab) = (-a)(-b)$

For $a = 2$ and $b = -3$, $-(ab) = -[2(-3)] = -(-6) = 6$, and $(-a)(-b) = (-2)(-(-3)) = (-2)(3) = -6$. Thus, $-[2(-3)] \neq (-2)(-(-3))$. It is true that $-(ab) = (-a)(b) = a(-b)$

$(p + q)^2 = p^2 + q^2$

For $p = 3$ and $q = 5$, $(p + q)^2 = (3 + 5)^2 = 8^2 = 64$, and $p^2 + q^2 = 3^2 + 5^2 = 9 + 25 = 34$. Thus, $(3 + 5)^2 \neq 3^2 + 5^2$. It is true that $(p + q)^2 = p^2 + 2pq + q^2$.

$\sqrt{a+b} = \sqrt{a} + \sqrt{b}$

For $a = 9$ and $b = 16$, $\sqrt{9+16} = \sqrt{25} = 5$, and $\sqrt{9} + \sqrt{16} = 3 + 4 = 7$. Thus, $\sqrt{9+16} \neq \sqrt{9} + \sqrt{16}$. It is true that $\sqrt{ab} = \sqrt{a}\sqrt{b}$, for $a, b \geq 0$.

$\log(a + b) = \log(a) + \log(b)$

Try a few values on your calculator to see that this is not true.

$\log a - \log b = \log(a - b)$

It is true that $\log a - \log b = \log\dfrac{a}{b}$, for $a, b > 0$.

$\sin(x + y) = \sin x + \sin y$

It is true that $\sin(x + y) = \sin x \cos y + \cos x \sin y$.

$\dfrac{df(x)g(x)}{dx} = \dfrac{df(x)}{dx}\dfrac{dg(x)}{dx}$

It is true that $\dfrac{df(x)g(x)}{dx} = f(x)\dfrac{dg(x)}{dx} + g(x)\dfrac{df(x)}{dx}$.

These examples teach us that *we have to hesitate* using a form of distributivity on a symbolic form in our efforts to derive an equivalent symbolic form. Some symbolic forms do not distribute as you think they might. You need to ask yourself these questions before distributing: "What theorem or rule tells me I will obtain an equivalent symbolic form? How sure am I of this?"

Moving a Number or Variable to the Front of an Indicated Function or Operation

A common error students make working with symbolic notation is moving a number or variable to the front of an indicated function or operation. Sometimes you can do this; other times not. When you can do it, it is because a property allows it. Unfortunately, some students think it can be done for all situations that appear similar. Here are some examples showing whether it is okay or not okay to do this.

Symbol	**Equivalent Form?**
$x(2)z$	$2xz$ (Yes)
$\sin 2x$	$2\sin x$ (No)
$\log 3x^2$	$3\log x^2$ (No)
$\int \frac{3}{4}\cos^2 x$	$\frac{3}{4}\int \cos^2 x dx$ (Yes)
$\int x\cos^2 x dx$	$x\int \cos^2 x dx$ (No—you can only take out a constant factor)
$\lim_{x \to -4} 7\sin x$	$7\lim_{x \to -4} \sin x$ (Yes)
$\lim_{x \to 8} x^2 \cos x$	$x^2 \lim_{x \to 8} \cos x$ (No—you can only take out a constant factor)
$\dfrac{d4\sqrt{x}}{dx}$	$4\dfrac{d\sqrt{x}}{dx}$ (Yes)
$\dfrac{dx^3 \sin x}{dx}$	$\sin x \dfrac{dx^3}{dx}$ (No—you can only take out a constant factor)

To avoid this type of mistake, learn well the rules and theorems that allow you to move values to the front of a function or operation, including the restrictions for doing this. Then, when you see a symbolic form that resembles the types just shown, take the time to determine whether you can or cannot move a specific value. Ask yourself this question: "What is the rule or theorem that tells me I will obtain an equivalent symbolic form?" Avoiding mistakes is mainly a matter of *mind over first instinct*.

Incorrect Symbols

To write mathematics correctly, symbols have to be written *in the form they appear* when you have been introduced to them. It is not correct to write a symbol in an abbreviated or incomplete form. When students write a symbol incorrectly they either don't know the correct symbol, or they are using their own abbreviation of the symbol for convenience sake.

Here are examples of symbols that are correctly written, followed by an incorrect writing of the symbols:

Correct Symbol	**Incorrect Symbol**
$\cos y$	cos ($\cos y$ is a function value—you are finding the cosine *of* y)
$f(x) = \sqrt{1 + x}$	$f = \sqrt{1 + x}$ (f names a function; $f(x)$ and $\sqrt{1 + x}$ name the same function *value*)
$\lim\limits_{y \to 5} y^2 = 25$	$\lim y^2 = 25$; $\lim = 25$; $\lim\limits_{y \to 5} = 25$; $y^2 = 25$ at $y \to 5$
$\dfrac{d \tan^2 x}{dx} = 2 \tan x \sec^2 x$	$\dfrac{d \tan^2}{dx} = 2 \tan \sec^2$ (these functions need to be applied to *x*—we need to have function *values*)
$\displaystyle\int x \sec x\, dx$	$\displaystyle\int x \sec x$ (dx must be written)
$\displaystyle\int_2^3 x^2 dx = \dfrac{x^3}{3}\Big|_2^3 = \dfrac{19}{3}$	$\displaystyle\int_2^3 x^2 dx = \dfrac{x^3}{3} = \dfrac{19}{3}$

Entering Symbolic Notation Incorrectly in a Calculator

Symbols are often keyed incorrectly into a calculator, especially when their grouping is not that evident. For example, to evaluate $\dfrac{3\pi}{-1 + \sqrt{2}}$, you have to view it as $3\pi \div (-1 + \sqrt{2})$, and when entering $\sqrt{\pi - \dfrac{3}{2}}$, you have to view it as $\sqrt{\left(\pi - \dfrac{3}{2}\right)}$.

Placing Equal Signs between Equations and Adding on to Sentences

This is a typical error when a student solves an equation such as $3x - 8 = 4x - 5$:

$$3x - 8 = 4x - 5 = -x = 3 = x = -3.$$

Here is a correct way to write this:

$$3x - 8 = 4x - 5 \Rightarrow -x = 3 \Rightarrow x = -3.$$

To simplify an expression such as $x - 4x + 12x + 9x$, some students write:

$$x - 4x = -3x + 12x = 9x + 9x = 18x.$$

Notice that these students are incorrectly saying that $-3x = 9x = 18x$ (this is only true for $x = 0$). Writing it either of these two ways is acceptable:

$$x - 4x + 12x + 9x = -3x + 12x + 9x = 9x + 9x = 18x.$$
$$x - 4x = -3x; \quad -3x + 12x = -9x; \quad 9x + 9x = 18x.$$

The first way is preferable.

4. More on Learning Symbolic Notation

Learning symbolic notation is not difficult but it does require your attention when it is introduced, and your ongoing consideration throughout a unit and course. You will be expected to apply it intelligently to your homework, in class, and when taking examinations. Unlike your everyday language, which you use all the time in conversation, the opportunities to converse in the language of mathematics symbolism are more limited, and for many students it is usually restricted to the mathematics classroom (and even there the opportunity is minimal). But there are ways to gain more practice, which are now discussed.

Work with a Partner

Working with a partner provides an excellent opportunity for you to learn the symbolic language of mathematics better. Here are some activities the two of you can do:

1. Have your partner read a symbolic expression or sentence to you, and you write it symbolically. This helps your part-

ner verbalize symbols and helps you write them based on this verbalization. Check with each other to see if the reading and symbolization were done correctly. Reverse your roles.

2. Have your partner read portions of a textbook section to you. Then discuss the reading, including how well your partner translated and interpreted symbolic expressions and sentences. Reverse your roles. It is also beneficial to read mathematics aloud when you are by yourself.

3. You cannot help but practice the symbolic language of mathematics when you and your partner discuss assignments. However, this practice can be enhanced if you are aware of the necessity of doing it during this discussion.

Use Note Cards

For individual study, take 3×5 cards and place a symbolic expression or sentence on one side and one or more literal translations, literal interpretations, geometric interpretations, names or titles, equivalent forms, or uses, on the other side. In your study, look at either side of the card, and see if you can vocalize or write what is on the other side. A card can be constructed soon after a symbolic expression or sentence is introduced. Other uses and equivalent forms of the symbolic form can be added to the card as they arise in the course.

Use Mathematical Chants or Mantras

In education, a chant (or mantra) is an expression or idea that is frequently repeated, often without thinking about it. Chants relating to mathematics are used by students, as well as by professional mathematicians, in their efforts to recall the literal interpretations of specific symbolic expressions and sentences. Once you learn a chant, just repeat it to recall the knowledge you need. In beginning algebra, I suspect these chants were in your repertoire of mantras:

Squaring a binominal

"To square a binomial, square the first term, add to this twice the product of the two terms and the square of the last term." Symbolically, $(a + b)^2 = a^2 + 2ab + b^2$.

Finding the roots of the quadratic equation, using the quadratic formula.

"The roots of a quadratic equation are minus b plus or minus the square root of b square minus four ac, all over two a." Symbolically, they are $\dfrac{-b \pm \sqrt{b^2 - 4ac}}{2a}$.

Here are chants for two precalculus rules:

Theorem on the sum of two continuous functions

"The sum of continuous functions is continuous." Symbolically: If

$$\lim_{x \to a} f(x) = f(a) \text{ and } \lim_{x \to a} g(x) = g(a), \text{ then } \lim_{x \to a}(f + g)(x) = (f + g)(a).$$

Finding the logarithm of the quotient of two numbers

"The logarithm of a quotient is the logarithm of the numerator minus the logarithm of the denominator." Symbolically, $\ln\dfrac{x}{y} = \ln x - \ln y$.

Caution. One danger with using mathematical chants is that your chant may *leave out the conditions* under which the chant is true. For instance, the property $\ln\dfrac{x}{y} = \ln x - \ln y$, is true if and only if x and y are greater than 0. (Note that $\ln\dfrac{x}{y}$ exists for negative x and negative y, but $\ln x$ and $\ln y$ do not.) Your chant can be more inclusive by including the conditions under which the chant is true. You don't need to do this as long as you know that you need to check for the conditions under which your chant is true.

I suspect that most calculus students have chants for these two procedures:

Finding the limit of the sum of two functions

"The limit of the sum of two functions is the sum of the limits of the functions." Symbolically,

$$\lim_{x \to a}[f(x) + g(x)] = \lim_{x \to a} f(x) + \lim_{x \to a} g(x).$$

Finding the derivative of the quotient of two functions

"The derivative of a quotient is the denominator times the derivative of the numerator, minus the numerator times the derivative of the denominator, all over the denominator square." Symbolically,

$$\frac{d\dfrac{f(x)}{g(x)}}{dx} = \frac{g(x)\dfrac{df(x)}{dx} - f(x)\dfrac{dg(x)}{dx}}{[g(x)]^2}.$$

Read Right-to-Left as Well as Left-to-Right

Normally, we read left-to-right, including the reading of symbolic form. Consequently, it may be problematic for you, as it is for many students, when given the right side of a symbolic sentence, to come up with the left side. For example, after reading $\log \dfrac{a}{b} = \log a - \log b$ numerous times from left-to-right, you may have little trouble stating the right side $\log a - \log b$, when given the left side. But you may have difficulty coming up with $\log\dfrac{a}{b}$ when given $\log a - \log b$. Students often write $\log a - \log b$ incorrectly as $\log(a-b)$ or $\dfrac{\log a}{\log b}$.

The truth of an equation does not change with a right-to-left reading, but it is viewed from a different orientation. It is important for you to read equations right-to-left as well as left-to-right so when the right side of the sentence appears in your work, you are able to give the left side. It should be as easy for you to write $\dfrac{a^m}{b^m} = (\dfrac{a}{b})^m$, as it is to write $(\dfrac{a}{b})^m = \dfrac{a^m}{b^m}$, or as easy to write $\log_b P^n = n \log_b P$, as it is to write $n \log_b P = \log_b P^n$.

Work Lots of Problems and Read Lots of Mathematics

It is appropriate to end this section by encouraging you to be heavily engaged in the mathematics of your course by reading it, speaking it, and applying it to solve many problems. More than anything, this is the best way to become more fluent in the language of mathematics symbolism. In addition, work with a partner, use note cards and chants, read symbolic sentences from left-to-right as well as right-to-left, and work to achieve the fluency objectives stated in section 2.

20

Technical Aspects of Writing Mathematics

This chapter is closely related to section 2 of chapter 14 on writing *about* mathematics. You are encouraged to view that section and this chapter as a package. Because each of these topics has its unique focus, they are separated to minimize confusion. It is difficult, at times, to make a clear distinction between writing *about* mathematics and writing mathematics. In writing about mathematics, you are engaged in writing mathematics, but the focus is less on specialized techniques, such as manipulating symbols and displaying your work, and more on giving explanations using words, sentences, and paragraphs. The focus in writing about mathematics is on *writing to learn mathematics*. Section 2 of chapter 14 addresses the benefits of writing about mathematics, the opportunities available to do it, and identifies generic categories of exercises on writing about mathematics, including examples for each category. This chapter on writing mathematics focuses on the technical aspects of writing mathematics, with special emphasis on integrating words and mathematical notation. Writing mathematics involves more than showing your computations.

Writing mathematics has its challenges. Perhaps the major reason for this is that we do not understand, as we should, the mathematics we are writing about. Until you find the right words to express your thoughts, you have to question your understanding. Fortunately, with practice, your writing of mathematics will improve, and consequently, so will your understanding. Writing helps clarify your thoughts since a necessary condition for good writing is clear thinking and organization of ideas.

The degree to which you may have to communicate mathematics

in your current or future workplace varies. Rarely is it the case that you will not be expected to do so. Surely, working mathematicians will do more communicating in mathematics, but the fact that you take mathematics in college indicates that you will know more mathematics than many of your future coworkers. You need to be prepared to communicate mathematics to them, in speaking and in writing, as well as to other coworkers who will have as much or more mathematics knowledge than you. If you communicate mathematics well your status and job security within your future place of employment will be enhanced, and you will also be more secure in your mathematics and technical courses while in college. An essential component of receiving a good grade in your mathematics course is your ability to communicate mathematics. The benefits that can result from knowing mathematics are minimized if you cannot communicate that knowledge to others.

Guidelines are given in this chapter for writing mathematically correct statements, arranging them in a clear and logical way, and then displaying them using complete sentences, paragraphs, and connective words, to create a meaningful flow of ideas.

The sections of the chapter are titled:

1. Issues to Address Before and After Writing
2. Elements of Good Mathematics Writing
3. Awarding Partial Credit to Solutions of Problems
4. Revisiting a Completed Assignment

1. Issues to Address Before and After Writing

Before giving specific guidelines on writing mathematics, there are several general issues that you need to consider, which will help you write mathematics.

Know Your Audience

A good writer writes for his or her audience. The audience you are writing for in your mathematics class is *not* your instructor or an assistant who may be grading your writing assignment. If you write for them, you are apt to leave out supporting statements since you believe they can piece your ideas together. Hence, you are not showing whether you can supply these missing statements and communicate them in a clear manner. Writing for an audience with equal or less knowledge than these individuals forces you to be more lucid in your thoughts, use fewer symbols, and be more organized in your presentation. An instructor, who expects you to write mathematics well, will not assume that you are writing for him or her. He or she assumes you are writing for someone whose mathematics knowledge is comparable to yours. A typical classmate meets this criterion. Do you write for a typical classmate?

Read Good Mathematics Writing

Many good writers do a lot of reading. The assumption here is that by reading they are being exposed to good writing, which makes them better writers, whether or not they are purposely analyzing the technicalities of the writing. This is analogous to growing up with parents, friends, and relatives who speak well. Your ability to converse is greatly influenced by listening to those who speak well. It is reasonable to conclude that reading mathematics, primarily from your textbook, will help you write mathematics, especially if you analyze the technicalities of the writing that you read. Some textbook authors or publishers violate grammatical rules, now and then (e.g., they make punctuation errors). A good knowledge of grammar and advice in this chapter should help you recognize good writing in your textbook.

Have Your Writing Critiqued

You can improve your writing of mathematics through self-critiquing, but you may need additional help. Who can help you? There are numerous sources: your instructor, classmates, tutors in a campus mathematics laboratory, campus learning and writing centers, and private tutors. Your instructor can help by making comments on your graded writing assignments and by looking at samples of your writing during his or her office hours. You can only hope that your instructor views the critiquing of your writing of mathematics as his or her responsibility. A classmate with good mathematics writing skills can help by reading some of your solutions to problems and making suggestions for improvement. Rewrite your solutions after a critique of them, and then have them critiqued again by this person. Return the favor. Do this periodically during the semester with this classmate or with others.

You may need more help than your instructor has time to give. He or she may refer you to your campus writing center if you have a general writing problem. If you recognize that you have this problem, don't wait to be referred—just go. Student tutors, who are hired by your mathematics or campus learning center, are available to help you with your writing of mathematics; however, they may not be good writers of mathematics. Perhaps you need a private tutor. One advantage of using a private tutor is that you receive more attention, but it needs to be someone who is skilled in writing mathematics. If you are serious about communicating mathematics well, which will help you learn it better, get the writing help you need. For all of the sources that were presented here, *you* have to initiate seeking help from them.

Understand Statements of Problems

Your excellent skills in writing mathematics will be of little help to you in writing a solution to a problem if you don't understand the statement of the problem. Mathematics problems are frequently misread or misinterpreted. You may have a well-written solution to a problem but it may not be to the problem you were asked to solve. How can this be avoided? One way is to realize how easy it is to misread problems. This will cause you to be more careful in reading them. It helps to rewrite the statement of a problem in your own words, and then check to see whether your rewrite is in sync with the key elements of the problem (i.e., check to see that it contains what you are given and what you are to find). See section 1 of chapter 22 on understanding statements of problems.

2. Elements of Good Mathematics Writing

Use Correct Grammar

In writing mathematics you don't need to learn a new set of grammatical rules. The rules of grammar you use in your general writing apply to writing mathematics. You need to write complete sentences using correct tenses, spelling, punctuation, and abbreviations. Your sentences need to be organized in a logical way. Paragraphs should be used whenever appropriate, beginning on a new and (most likely) indented line. Imagine the difficulty reading a page or more of mathematics where new ideas are introduced without using multiple paragraphs.

You use good grammar to help the reader understand what you are communicating. We now review some rules of grammar and apply them to writing mathematics.

Use correct punctuation

You may argue that writing mathematics has to be substantially different from writing non-mathematics since mathematics uses formulas and equations whereas non-mathematics doesn't. Your argument would be in vain for formulas and equations are contained in complete sentences and standard rules of punctuation apply. For example, formulas and equations that end sentences are followed by a period, and those that don't may or may not be followed by a comma. Following are pairs of statements; the first statement in a pair is written with correct punctuation and the second statement is written with incorrect punctuation:

1. The coordinates of the center of the circle are
$$x = \frac{-1+9}{2} = 4 \text{ and } y = \frac{-6+2}{2} = -2.$$
The coordinates of the center of the circle are
$$x = \frac{-1+9}{2} = 4 \text{ and } y = \frac{-6+2}{2} = -2$$

There needs to be a period at the end of the second statement.

2. To obtain the x-intercept, we set $y = 0$ in the equation $y = (x - 3)^2 - 14$ to obtain
$$(x - 3)^2 - 14 = 0$$
$$(x - 3)^2 = 14$$
$$(x - 3) = \pm \sqrt{14}$$
$$x = 3 \pm \sqrt{14}.$$

To obtain the x-intercept, we set $y = 0$ in the equation $y = (x - 3)^2 - 14$ to obtain
$$(x - 3)^2 - 14 = 0,$$
$$(x - 3)^2 = 14,$$
$$(x - 3) = \pm \sqrt{14},$$
$$x = 3 \pm \sqrt{14}$$

Delete the commas in the first three statements since it is convention not to write them when a vertical format is used. Also, a period needs to be inserted at the end. Here is an appropriate way to write the sentence in horizontal format: To obtain the x-intercept, we set $y = 0$ in the equation to obtain $(x - 3)^2 = 14 \Rightarrow x - 3 = \pm \sqrt{14} \Rightarrow x = 3 \pm \sqrt{14}$. Note that "$\Rightarrow$" is read, "which implies that." This implication symbol is used because the second and third equations are *derived* from the one preceding it. It is not appropriate to have a comma in place of "$\Rightarrow$."

3. The derivative of $x^{\frac{3}{2}} + \dfrac{1}{x}$ is $\dfrac{3}{2}x^{\frac{1}{2}} - \dfrac{1}{x^2}$, which equals $\dfrac{3x^{\frac{5}{2}} - 2}{2x^2}$.

   The derivative of $x^{\frac{3}{2}} + \dfrac{1}{x}$ is $\dfrac{3}{2}x^{\frac{1}{2}} - \dfrac{1}{x^2}$ which equals $\dfrac{3x^{\frac{5}{2}} - 2}{2x^2}$.

There needs to be a comma preceding "which" in the second statement.

4. Substituting 3 for x in $\sin 6x + \tan x$ gives $\sin 18 + \tan 3$.
    Substituting 3 for x in $\sin 6x + \tan x$, gives $\sin 18 + \tan 3$.

Delete the comma in the second statement.

5. The equations $2x - 5 = 8$ and $9 - x^2 - x = \sqrt[3]{7}$ are linear and quadratic, respectively.
    The equations $2x - 5 = 8$, and $9 - x^2 - x = \sqrt[3]{7}$, are linear and quadratic, respectively.

Delete the first two commas in the second statement.

6. To find the horizontal asymptotes of $h(x) = \dfrac{2x^2 - 1}{1 - x^2}$, we find the limit of $h(x)$ as x increases or decreases without bound:

$$\lim_{x \to \infty} \frac{2x^2 - 1}{1 - x^2} = \lim_{x \to \infty} \frac{2 - \dfrac{1}{x^2}}{\dfrac{1}{x^2} - 1}$$

$$= \frac{2 - 0}{0 - 1}$$

$$= \frac{2}{-1}$$

$$= -2.$$

To find the horizontal asymptotes of $h(x) = \dfrac{2x^2 - 1}{1 - x^2}$, we find the limit of $h(x)$ as x increases or decreases without bound:

$$\lim_{x \to \infty} \frac{2x^2 - 1}{1 - x^2} = \lim_{x \to \infty} \frac{2 - \dfrac{1}{x^2}}{\dfrac{1}{x^2} - 1}, = \frac{2 - 0}{0 - 1}, = \frac{2}{-1}, = -2.$$

It makes no sense to have commas in this horizontal chain; delete them.

7. The three Trigonometric Pythagorean Identities are:

$$\sin^2\theta + \cos^2\theta = 1$$
$$1 + \tan^2\theta = \sec^2\theta$$
$$1 + \cot^2\theta = \csc^2\theta$$

The three Trigonometric Pythagorean Identities are $\sin^2\theta + \cos^2\theta = 1 \; 1 + \tan^2\theta = \sec^2\theta$ and $1 + \cot^2\theta = \csc^2\theta$.

You need commas in the horizontal format after the first and second identities

Mathematics expressions need to be part of sentences

Since you should always have complete sentences when writing mathematics, it is inappropriate, for example, to have the expression $(x + 2y)^3$ appear by itself. When you write mathematics you are telling a story, just as your everyday conversations tell stories. Would you talk to a friend about the career you are seeking, and then drop in a word or phrase. You would get a strange look from your friend. For example, suppose Mary is telling Tom this story: "I am pursuing a career in accounting. I am good at it and am being encouraged by my father who owns a very successful accounting firm. Accounting courses. I am currently enrolled in two accounting courses, and . . ." Do you see the inappropriate phrase in this conversation? Determine what is improper about this writing: "To find the inverse of a one-to-one function $y = f(x)$, interchange x and y and then solve for y in terms of x. $f^{-1}(x)$. Make sure you first determine if f is one-to-one." The expression $f^{-1}(x)$ does not belong here. Your sentences must be complete and they must contribute to the mathematics story you are telling. Sentences that add nothing to the solution of a problem should not be included when you are writing the solution. Get rid of clutter.

Use acceptable abbreviations

Abbreviations are normally not used in good writing, and it is always inappropriate to use unacceptable abbreviations. The following are statements using *unacceptable* abbreviations:

1. The number midway bet p and q is $\dfrac{p+q}{2}$. ("bet" is an un-acceptable abbreviation of "between.")

2. The der of $\sin x$ is $\cos x$. ("der" is an unacceptable abbreviation of "derivative.")

3. The graph of function g is symmetric to the origin b/c it is an odd function. ("b/c" is an unacceptable abbreviation of "because.")

4. The tan of $\dfrac{\pi}{4}$ is 1. ("tan" is an unacceptable abbreviation of "tangent" when tangent is not followed by an argument. It should read, "The tangent of $\dfrac{\pi}{4}$ is 1." However, it is acceptable to write $\tan\dfrac{\pi}{4}$ is 1 (this abbreviation has been formally accepted).

Examples of unacceptable abbreviations of *symbols* appear later in this section.

Use first person in your writing

It is appropriate to use the first person in your writing. You can use "I," "we," or "us." For example, you can write, "I then substituted x^3 for ...," or, "We then proceeded to solve the trigonometric equation ...," or, "Let us now turn to the proof on ..." Using the first person makes your writing more personal. Using "I" identifies your writing with you, and using "we" or "us" indicates that you and the reader are in this together. The preference is to use "we," not "I." It is not absolutely necessary to use the first person in your writing.

Write Concisely and Clearly

Strike an appropriate balance of words and symbols

You need to be focused when you write mathematics. Don't be too long-winded or make your writing unnecessarily complicated. The beauty of mathematics is its extensive reliance on symbols, which allows you to be concise in your writing. The use of symbols allows the reader to take in a lot of information because it appears in a compact space and is easily seen. A heavy use of symbols can impede understanding since the level of abstraction is increased. For example, a particular symbol may contain other symbols, each of which has its specific meaning. If you struggle with the meaning behind symbols, either because you forgot them or didn't adequately learn them, you can see that a heavy use of symbols can obscure meaning. On the other hand, too many words can also impede understanding since words take a lot of space. The reader will have difficulty remembering the thoughts conveyed by these words when they are spread over a page or two. *You need to have an appropriate balance between the use of symbols and words*, which is determined to some extent by your audience.

Three solutions to a problem are now given, where each solution uses a different blend of words and symbols. Assume that you wrote each solution for a hypothetical classmate in precalculus who has struggled with the concepts and computations needed to work the problem.

Problem. Verify for function $y = f(x) = 3x + 1$, that composing it with its inverse results in x; that is, verify that $f \circ f^{-1}(x) = f^{-1} \circ f^{-1}(x) = x$. (It is assumed that your classmate had prior work on finding the inverse of a function and the composition of two functions.)

Solution (primarily using symbols): $y = f(x) = 3x + 1 \Rightarrow x = 3y + 1 \Rightarrow y = \dfrac{x-1}{3} = f^{-1}(x)$. Therefore,

$$f \circ f^{-1}(x) = f(f^{-1}(x)) = 3\left(\frac{x-1}{3}\right) + 1 = x - 1 + 1 = x$$

and

$$f^{-1} \circ f(x) = f^{-1}(f(x)) = \frac{(3x+1)-1}{3} = x.$$

Solution (primarily using words): Consider $y = f(x) = 3x + 1$. To find the inverse of the given function interchange the variables in the function and then solve for y in terms of x. To find $f \circ f^{-1}(x)$, replace x in $f(x)$ by $f^{-1}(x)$ and simplify, yielding $f \circ f^{-1}(x) = x$. To find $f^{-1} \circ f(x)$, replace x in $f^{-1}(x)$ by $f(x)$ and simplify, yielding $f^{-1} \circ f(x) = x$.

Solution (using a balance of words and symbols): Consider $y = f(x) = 3x + 1$. To find $f^{-1}(x)$, replace x by y and y by x giving $x = 3y + 1$. Solving for y yields $y = \dfrac{x-1}{3}$, which is $f^{-1}(x)$. To find $f \circ f^{-1}(x)$, which is $f(f^{-1}(x))$, substitute $\dfrac{x-1}{3}$ for x in $f(x)$, yielding $f \circ f^{-1}(x) =$

$f(f^{-1}(x)) = 3\,(\dfrac{x-1}{3}) + 1 = x - 1 + 1 = x$. Hence, $f \circ f^{-1}(x) = x$. To find

$f^{-1} \circ f(x)$, which is $f^{-1}(f(x))$, substitute $3x + 1$ for x in $f^{-1}(x)$, yielding

$f^{-1} \circ f(x) = f^{-1}(f(x)) = \dfrac{(3x+1)-1}{3} = x.$

Your hypothetical classmate would probably struggle with the first two solutions. The first solution does not have enough words to help remind your classmate what things mean, how they are related, and where and how to make substitutions. The second solution does not show the computational work involving symbols. Your hypothetical classmate has a better chance to understand by reading the third solution, which makes up for the pitfalls of the first two solutions. Which of the three solutions would you normally have written?

Use key words and phrases
Even though the goal is to be concise in writing mathematics, you are nonetheless writing a story and you need to make the story understandable and interesting. I once had an excellent student who turned in ten pages of writing when two pages would have sufficed. The ten pages were beautifully and correctly written, but eight of them were totally unnecessary. It is amazing how brief your writing can be, yet still say all that is needed and say it well. Avoid making extraneous remarks and taking detours. Avoid including simple computations that the reader can easily supply. To be both concise and interesting, you need to use key words and phrases (interspersed with symbols), which indicate that something is to follow (e.g., a reason, conclusion, reminder, assumption, comparison, similarity, detail, rewrite, observation, example, application, or addition). There are many of these words and phrases, including these:

we see that	indicates that	we need only to
using this fact	compared with	furthermore
for the reason	we can use	more specifically
in view of the fact	we can deduce	using the result
this means	recall the	this becomes
observe that	more specifically	we can write this
first notice that	likewise	we know that

we start with	it can be shown that	assuming that
this shows that	this gives	this suggests that
letting	yielding	it follows that
on the other hand	for example	therefore
instead of	we conclude	in order to
according to	by choosing	similarly

If you have discovered that you seldom use words and phrases like these in your mathematics writing, then you need to work to incorporate them in your future writing. It will help your readers understand what you are conveying and support your efforts *to tell a story* that captures their interests (which should be your goal).

Write for Ease of Reading

You want readers of your mathematics writing to be motivated to read at the outset of their reading, and continue to be motivated as they read. Hence, you want to make it interesting and easy to read. Besides using an appropriate balance of words and symbols, leave ample white space, write legibly, have a clean copy (no crossing out), highlight important items, and organize and display your work so the flow of ideas is logical and easy to discern. Guidelines for writing for ease of reading are now given.

Write an introduction to a project paper

If you are writing a paper on a mathematics project, a mathematics laboratory, or a more-than-routine daily problem assignment, an introduction will assist the reader. In the introduction, inform the readers what the paper is about and what they will find as they read. This necessitates stating the situation in your own words, giving its significance, as well as pointing out what is presented in various sections of the paper. The introduction is not the place to give details—they are left for the body of the paper. An introduction will not only make your paper more readable, but it will also help *you* see the larger picture being painted by the details. You are writing for yourself as well as for the reader.

Display your work from left-to-right and from top-to-bottom

The page layout for writing mathematics is the same as the page layout for non-mathematics or non-technical writing: You start on the left of the page, move to the right, and move back to the left for the next line of your work. Sometimes I saw two sets of computations, each in a column format, placed side by side. The writer most likely did this to conserve space. This is borderline acceptable since it causes problems for readers who find it hard to change from reading left-to-right and line-by-line. If you cannot avoid this, perhaps boxing off each set of computations, or inserting a vertical line between the two sets, will help. This example shows the confusion that can result by a two-column display (normal left-to-right reading is disrupted):

$$\sin^2 x + \cos^2 x = 1 \qquad\qquad \sin^2 x + \cos^2 x = 1$$

$$\frac{\sin^2 x}{\cos^2 x} + \frac{\cos^2 x}{\cos^2 x} = \frac{1}{\cos^2 x} \qquad \frac{\sin^2 x}{\sin^2 x} + \frac{\cos^2 x}{\sin^2 x} = \frac{1}{\sin^2 x}$$

$$\tan^2 + 1 = \sec^2 x \qquad\qquad 1 + \cot^2 x = \csc^2 x$$

I had students who had no plan for the layout of their work. The worse case scenario was when they employed a combination of right-to-left, left-to-right, up-to-down, and down-to-up, as part of a solution to one problem. The credit I gave them on their solution was a reflection of this mishmash. If for some reason you have to violate the acceptable order of writing mathematics (for instance, if you decide to add extra writing to your solution on an examination, but don't have time to show it in an acceptable way), by all means insert arrows, lines, or words in the appropriate place to direct the reader to your addition. Is your writing of mathematics displayed well?

Organizing and displaying calculations and equations

It is easier, at times, to read a sequence of calculations if each calculation is placed on a separate line, as shown in example 2, rather than displaying them horizontally as shown in example 1.

Example 1. $x^3 - 4x^2 = 0 \Rightarrow x^2(x - 4) = 0 \Rightarrow x^2 = 0$ or $x - 4 = 0 \Rightarrow x$
$= 0$ or $x = 4$.

Example 2.

$$x^3 - 4x^2 = 0$$
$$x^2(x - 4) = 0$$
$$x^2 = 0 \text{ or } x - 4 = 0$$
$$x = 0 \text{ or } x = 4.$$

Which display did you find easier to read? Notice the amount of white space surrounding the computations in example 2.

Place important formulas on a separate line

To point out an important formula, it is desirable to have it stand alone on a line, as shown in example 1. Contrast this with example 2, where the formula does not stand alone.

Example 1. Using the Fundamental Theorem of Calculus,

$$\int_a^b f(x)dx = F(b) - F(a), \text{ where } F'(x) = f(x),$$

we find that $\int_0^{\frac{\pi}{2}} \sin x\,dx = -\cos x\Big|_0^{\frac{\pi}{2}} = -\cos\frac{\pi}{2} - (-\cos 0) = 0 + 1 = 1.$

Hence, the area of the....

Example 2. Using the Fundamental Theorem of Calculus, $\int_a^b f(x)dx = F(b) - F(a)$, where $F'(x) = f(x)$, we find that

$\int_0^{\frac{\pi}{2}} \sin x\,dx = -\cos x\Big|_0^{\frac{\pi}{2}} = -\cos\frac{\pi}{2} - (-\cos 0) = 0 + 1 = 1.$ Hence, the area of the....

Reference and label prior statements that you use

To help the reader understand your justification for a statement, it is desirable to refer the reader to previous statements you made that provide the justification. This is often done in proofs of theorems, but it is also done to justify computations, especially if the justifications are not evident. All that needs to be done is to label the earlier statements you will use as your justifications (e.g., use labels (1) and (2)), and precede the statement you wish to justify by writing: "It follows from (1) and (2) that..." Or, immediately after making the statement you wish to justify, write, "by (1) and (2)." This is just another way to promote understanding and ease of reading.

Title, label, and describe important diagrams, graphs, and tables

Diagrams, graphs, and tables are used to support the message of your writing. They need to be neatly drawn, clearly labeled, and inserted in your paper in appropriate spots. You want your readers to easily locate them, know why you have them (i.e., what you want them to learn or see, and how they relate to your development or argument), and where to look *within them* to find the desired information. Introduce them in your prose, along with directions to locate them. Three examples of such writing follow:

Example 1. The graphs of the function f, its inverse f^{-1}, and the line $y = x$, are displayed in Figure 3. The graph of f^{-1} can be obtained by reflecting the graph of f in the line $y = x$ as shown in Figure 4. Notice in Figure 3 that the point $(3,2)$ is on the graph of f, and the point $(2,3)$ is on the graph of f^{-1}. This is not surprising since we know from past work that if $f(3) = 2$, then $f^{-1}(2) = 3$.

Example 2. The values of $\sin(x + \frac{\pi}{2})$ and $\cos x$ (for $x = 0$, $\frac{\pi}{2}$, π, $\frac{3}{2}\pi$, 2π), appear in the first two columns of Table 5, respectively. Notice that the values of $\sin(x + \frac{\pi}{2})$ and $\cos x$ are the same for these value of x. This is an indication that perhaps $\sin(x + \frac{\pi}{2}) = \cos x$ for all x.

Example 3. A cylindrical tank with height h and radius r is shown in Figure 1. Notice that the tank can be generated by revolving the rectangle shown in the figure about the segment of length h. We are now going to develop a formula for finding the volume of solids of revolution such as this one.

It is important for you to know the key items you need to address that will improve your mathematics writing. I remind you again to *look at how textbook authors write mathematics.* That will reinforce what you have learned here.

Write Precisely

The discussion here, with one exception, involves the proper use of mathematics notation; that is, the grammar of symbolic language of mathematics. It is difficult for me to decide which needs more improvement in students' writing of mathematics: English grammar or the grammar of mathematics notation. Suffice it to say that a good understanding of both is needed to write mathematics well, and you may need to improve in both areas.

Define symbols, terms, and formulas

Being imprecise or vague in writing the meanings of symbols and terms, or failing to introduce formulas is not an option. You need to let the reader know precisely what they represent.

Defining symbols. In your solution to a problem, define any letter you use that represents a variable or constant that has not been defined in the statement of the problem. You would be surprised how many students use a symbol in their writing that they have not defined. They know what it represents but they have not told the reader.

If you used p to represent the perimeter of a geometric figure, don't assume the reader knows this just because it is the first letter of the word, "perimeter." If you want it to represent the perimeter, tell your reader. Many of your readers will assume that is what you mean, but that is not writing mathematics precisely. A symbol is not sufficiently introduced by just labeling a part of a graph or diagram with it. You have to define it in your writing. Here are examples of how symbols can be introduced:

1. x and y are the length and width of the rectangle, respectively.
2. Let b be the diameter of the balloon.
3. F is the force exerted on one side of the tank by the water.
4. The water level of the pool at time t is denoted by $p(t)$.
5. $A(x_i)$ represents the area of a cross section of the solid at $x = x_i$.
6. The y-coordinate of the center of mass of the solid of revolution P is denoted $\bar{y}$.
7. Let $h(x) = \dfrac{\sin x}{x}$, for $x > 0$.
8. Let θ be the radian measure of the central angle AOC of circle O as shown in Figure 4.
9. Let C be the curve with parametric equations $x = 3t^2 + 4$, $y = 4t - 2$, and $z = 2t^3$, where t is a real number.
10. Let a be the fixed distance of the boat from the shore.

When you have completed your paper look it over to see if you have defined all of your symbols.

Defining terms. A *term*, as used here, is a word or combination of words that has specific mathematics meaning. There is no need to re-define terms in your writing if you know the reader understands them and has not forgotten them (hence, the importance of knowing your audience). For sure you need to define terms that are new to your audience or terms that can be interpreted in more than one way. Here are examples of definitions of terms that appear in writings:

1. The area of a Norman window, one in the shape of a rectangle surmounted by a semi-circle, ….
2. The area of the region between two concentric circles (that is, circles with the same center), can be found by….
3. We are seeking an answer that comes from the range of the inverse sine function (that is, from the interval $[\frac{-\pi}{2}, \frac{\pi}{2}])$.
4. To determine the intervals where the function f is nondecreasing (that is, the intervals I, where, for x_1, x_2 belonging to I, $x_1 \le x_2$ implies $f(x_1) \le f(x_2)$), we take ….

Introducing formulas. If you are using a specific formula to solve a problem, you need to introduce the formula in your writing, and include the purpose for including the formula. For example,

1. To find the area of the region we seek, we use $\int_a^b \pi[f(x)^2 dx$.

2. To find the length of the arc we seek, we use $\int_a^b \sqrt{1 + [f'(x)]^2} dx$.

3. To solve for $\cos x$ in terms of $\sin x$, we use the Pythagorean Identity $\sin^2 x + \cos^2 x = 1$.

4. To find the derivative of $f(x) = 3x^2$, we use
$$\lim_{\Delta x \to 0} \frac{f(x + \Delta x) - f(x)}{\Delta x}.$$

5. To find the radian measure of the central angle θ, we use $\theta = \frac{s}{r}$.

6. To represent $\log_a x$ in terms of logarithms to base b, we use
$$\log_a x = \frac{\log_b x}{\log_b a}.$$

7. To find the vertical asymptotes of function f, we find $\lim_{x \to \infty} f(x)$ and $\lim_{x \to -\infty} f(x)$.

If you use a non-standard formula that you derived, then you are obligated to show your derivation. You cannot expect the reader to blindly accept a formula that is completely foreign to him or her.

Use mathematics notation or symbols correctly
Since the writing of mathematics is replete with mathematics notation, you need to learn it before you can correctly write it. Learning mathematics notation is the content of chapter 19, and section 3 of that chap-

ter addresses some common errors made with symbolic form. You are not writing precisely if you are miswriting or misusing symbols. I have seen errors like these:

Writing $\int x \sec x \, dx$ as $\int x \sec x$ (inappropriate writing of the symbol)

Writing $\int x \sec x \, dx$ as $x \int \sec x \, dx$ (they are not equal)

Writing $\cos y$ as cos (inappropriate writing of the symbol)

Writing $-(ab)$ as $(-a)(-b)$ (they are not equal)

Writing $\log 3x^2$ as $3 \log x^2$ (they are not equal)

Writing $g(x) = \tan x = g'(x) = \sec^2 x$ (misuse of the equals sign)

Writing $7x - 5 = 4 = 7x = 9 = x = \dfrac{9}{7}$ (misuse of the equals sign)

Incorrect writing: Since m is an even number, m can be written as $2m$ (*Correct writing*: Since m is an even number, m can be written as $2n$ for some nonnegative whole number n.)

A common mistake in introducing a variable is to use the variable as the name of an entity or group of entities. For example, "Let x be the boys in the classroom." It should be, "Let x be the *number* of boys in the classroom."

The word "it" is *generally* off-limits when you write mathematics. Normally, good writing should not include a statement such as: "We now go through the computations to find it." What is meant by "it?" Are you referring to a specific distance, volume, number of pounds, solution to a specific equation, graph of a specific function, etc? In precise writing it must always be clear what you are referring to. As the writer, you may know what "it" refers to, but the reader also needs to know. In your solution to a problem there may be many different entities present and you need to make clear the one you are referencing at any given time. There is nothing vague about this statement: "We now progress to finding the weight of the tank" (as opposed to saying, "We now progress to finding it"). However, there are times when it is obvious what "it" refers to.

When introducing a letter in a particular discourse, say p, it can only represent one value throughout the discourse. For example, if p is introduced as the distance of a boat from the shore, then it represents this whenever it appears in your discourse. It cannot represent this and something else. This is elementary advice, but violations do occur. When you have completed your work, look back over the symbols you used to ensure that you have not used a symbol to represent two different entities.

You cannot substitute an uppercase letter for a lowercase letter since they are viewed as different letters. That is, if p is the distance of the boat from the shore, then don't assume that you can also use P to represent this distance.

When you write mathematics you will frequently refer to functions and you have to be precise in how you do it. The stars of mathematics are functions. Remove functions from mathematics and you have no mathematics. A function is a one-to-one assignment between two sets of values such that to each value in one of the sets (called the *domain* of the function), a unique value is assigned from the other set (called the *range* of the function). The particular assignment is often given by a rule, and the function name is often a letter, with or without superscripts or subscripts (such as f, h, m, r, A, F, P_n, g', $f^{(4)}$, T_n).

Consider this example: Let function f be given by the rule, $f(x) = \dfrac{x}{2}$, with domain set being the real numbers. Notice that the function has been named f, which is proper. The value $f(x)$ is not the name of the function; hence, precisely speaking, it is not proper to say, "The function $f(x)$." If you want to refer to $f(x)$, you can write, "The value $f(x)$," or just say, "$f(x)$." That being said, because so many students, and even instructors, say, "the function $f(x)$," this usage is considered acceptable even though imprecise. Even though you have this liberty, you do not have the liberty to write $f = \dfrac{x}{2}$ (a common error), instead of

$$f(x) = \dfrac{x}{2}$$

Precise writing requires proper use of "the," "a," and "an". For example, it is better to write, "The solution to $x^2 - 4x + 4 = 0$ is 2," than it is to write, "A solution to $x^2 - 4x + 4 = 0$ is 2." The reason for this is that there is only one solution, namely 2, and using "The" indicates this. The second formulation is not incorrect; it just doesn't give as much information as the first formulation—it is not as precise. It is incorrect to write, "The solution to $4x^2 = 9$ is $\dfrac{3}{2}$," since there are two solutions. But it is correct to write, "A solution to $4x^2 = 9$ is $\dfrac{3}{2}$." As a last example, it is incorrect to write, "The antiderivative of $3x^2$ is $\dfrac{x^3}{3}$," since there are an infinite number of antiderivatives. What can be written is, "An antiderivative of $3x^2$ is $\dfrac{x^3}{3}$."

Writing numerical answers

For those problems having numerical answers, there are guidelines for writing your answers.

Write exact answers unless there are reasons it is more appropriate to approximate them (e.g., if you are told to approximate them). If the answer is $\sqrt[3]{7}$, write that, not an approximation (such as 1.91). Similarly, if the answers for various problems are log 15, $-4e$, $\dfrac{2}{3}$, $\sin 4 \sin 4$, $\sqrt{2}$, $\dfrac{\pi}{3}$, and $\ln 2 \tan \dfrac{\pi}{8}$, then write those forms, not approximations of them. If an answer is $x = \sqrt{5}$ and you need to approximate it to the nearest hundredth, then write $x \approx 1.41$ (which is read, "x is approximately equal to 1.41"). If you write an answer as $x = 24.2$, then you are saying that this is the exact answer. If it is an approximate answer, write $x \approx 24.2$. That being said, many instructors will accept an answer written as $y = 2.34$ even though it may be an approximation. The thought here is that, to the nearest hundredth, it equals 2.34.

It is convention in mathematics to leave answers in fraction form as opposed to mixed form. If your answer is $\dfrac{26}{11}$, write it that way and not as $2\dfrac{4}{11}$. Yes, this advice may go counter to the instruction you received in secondary school. One reason to write the answer as $\dfrac{26}{11}$ is that this is a more convenient form to use for any additional calculations that may need to take place; otherwise $2\dfrac{4}{11}$ would have to be changed back into $\dfrac{26}{11}$ when performing these calculations. If no further calculations will be done and you want to give the reader a better feel for the size of the number, then $2\dfrac{4}{11}$ is the preferred form to give.

If you are rounding off an answer, do it correctly; that is, round up if the next digit is at least 5; otherwise, round down. Always remember that calculators approximate if the answer has more decimal places than the calculator displays.

Inform the reader of your answer. There are a variety of ways to do this, including:

- "The answer we seek is 100," or "The answer is 100."
- Write "answer" immediately after the answer.
- Underline or circle the answer (if nothing else is underlined or circled). This is not recommended for precise writing, but instructors accept it when writing of necessity is done hurriedly (e.g., on an in-class test).
- Sometimes students leave two different answers to a problem because they worked the problem two different ways, not knowing which solution is incorrect. It is inappropriate to leave more than one answer hoping the grader will pick out the correct one. Let the reader know that you cannot decide on the correct solution.

Good writing requires precise writing.

3. Awarding Partial Credit to Solutions of Problems

In this section two solutions are given to a precalculus problem. Partial credit is awarded to each solution based primarily on how well the solution communicates the thought process of the writer. The two solutions for each exercise communicate the thought process of the writer very differently. For each solution, the mathematics is correct.

Exercise. Find the maximum area of a rectangle that can be constructed with a perimeter of 40 meters.

Solution 1. Let the dimensions of the rectangle be x and y, and denote the perimeter by P. Hence,

$$P = 2x + 2y.$$

Given that the perimeter is 40, we know

$$P = 2x + 2y = 40.$$

Since we want to find the maximum area of the rectangle, we need an equation for the area. Letting A represent the area, we have

$$A = xy.$$

A can be expressed as a function of the variable *x* by solving for *y* in terms of *x* in the equation $2x + 2y = 40$, and then making a substitution.

$$2x + 2y = 40$$
$$2y = 40 - 2x$$
$$y = 20 - x.$$

Substituting $20 - x$ for *y* in the equation $A = xy$, we get

$$A = x(20 - x).$$

This is a quadratic equation in the variable *x*; hence its graph is a parabola. We complete the square in *x* so that we can determine the co-ordinates of the vertex of the parabola:

$$A = x(20 - x)$$
$$= 20x - x^2$$
$$= -(x^2 - 20x)$$
$$= -(x^2 - 20x + 100 - 100)$$
$$= (x^2 - 20x + 100) + 100$$
$$= -(x - 10)^2 + 100.$$

The vertex of the parabola is (10,100). Since the parabola is directed downward, we know that its maximum value occurs at the *y* value of the vertex, namely, 100. Therefore, the maximum area of the rectangle is 100 square meters.

Commentary on Solution 1: This solution communicates the thought process of the writer well and is easy to read. All variables are defined and the equations and different parts of the solution are well spaced. On a 10-point scale, I would give this solution 10 points.

Solution 2. $2x + 2y = 40$. $A = x(20 - x) = -x^2 + 20 = -(x^2 - 20) = -(x - 10)^2 + 100$. Answer is 100.

Commentary on Solution 2. There are no mathematical errors in this problem. Some computational steps are left out. The grader of this problem can be somewhat assured that the writer has a decent grasp of the problem. However, explanatory words are few, so the grader cannot be sure that the student isn't just mechanically working parts of this problem because he or she has worked many like it. The solution is not easy to read and would be very confusing to an average or below aver-age classmate who needs more explanation. On a 10-point scale, I would give this solution 6 points (and that is being generous, especially if the student knows how I value good writing). Instructors would vary greatly on the credit they give this solution.

These two solutions demonstrate extremes in the quality of the writing. One is written extremely well and the other is written poorly. Instructors do not expect writing as thorough as that shown in example 1, because students are not adequately taught to write such a detailed and clear solution. However, instructors need to expect considerably more from their students than the writing presented in solution 2, and they need to communicate this to their students. Work hard to improve your writing of mathematics.

4. Revisiting a Completed Assignment

It is critical to *revisit* your mathematics paper or assignment before submitting it. We have all written papers and can recall times we made changes based on rereading it. We probably made more changes in grammar than in substance. A paper with correct mathematics but poor writing can be as problematic for the reader as a paper with incorrect mathematics. Both types of papers are ineffective. Professional writers revisit their writings time and time again before submitting them for others to read. Revisit portions of your writing as you are writing. It is also helpful to set it aside for a few hours or days, time permitting, and revisit it with a fresh mind. When revisiting it, read it aloud. This involves hearing, which can pinpoint problems in your writing that your eyes might not. You may want a classmate to critique it.

A first draft of a paper or assignment, by definition, is a preliminary version of the paper; one you would not turn in. Writing well is not easy, but revising your work and having your revisions critiqued will help you improve. You need to write mathematics well, and hopefully you are supported in this by your instructors' expectations.

CHECKLIST OF WRITING GUIDELINES

It is wise to go over a checklist of guidelines before handing in your paper or assignment. This checklist of questions covers the basic content of this chapter:

1. Do I know the audience I am writing for?
2. Should a portion or all of my paper be typed?
3. Should I have my paper critiqued?
4. Do I need an introduction?
5. Does my writing progress logically, with clarity, and is it easy to read?
6. Is my writing legible?
7. Have I defined all variables, symbols, terms, and formulas?
8. Am I using mathematics notation and symbols correctly?
9. Did I state all of my assumptions?
10. Is my mathematics correct?
11. Did I use correct grammar, including spelling and punctuation?
12. Is my writing comprised of sentences and paragraphs?
13. Do I have abbreviations and are they acceptable?
14. Did I use key words and phrases to tie ideas together and to let the reader know where I am going (i.e., to help convey the story of my solution)?
15. Did I use an appropriate balance of words and symbols?
16. Does my work proceed from left-to-right and from top-to-bottom?
17. Are my equations and calculations properly organized and displayed?
18. Are my key formulas placed on separate lines?
19. Have I referenced and labeled prior statements that I used later?
20. Did I enhance my work by using diagrams, graphs, or tables?
21. Have I titled key diagrams, graphs, and tables, stated their purposes, and pointed out where to look in them to obtain specific information?
22. Have I used "the," "a," and "an" appropriately?
23. Are my numerical answers written in the proper form?
24. Did I identify my answer(s)?
25. Did I answer the questions that were asked?
26. Is my paper appealing? That is, have I used quality paper, not crossed out content, not left erasure marks, left ample white space, etc.?

Importance of Problem Solving and Attributes of a Problem Solver

This chapter and the next are on problem solving. They need to be written for a variety of reasons, including these:

- Learning mathematics well implies developing and using problem-solving skills.
- Many college mathematics students are not pleased with their problem-solving skills.
- Many college mathematics students have never been exposed to a systematic treatment of the subject of problem solving.

The sections of the chapter are as follows:

1. Describing a Problem and Describing Problem Solving
2. Necessity of Having a Good Mathematics Background
3. Solving Problems at Various Cognitive Levels of Thought
4. Problem Solving as the Focus of Mathematics Study
5. Benefits of Problem Solving
6. Attributes of Problem Solvers

If you learn and act on the information given in this and the next chapter, you will become a significantly better problem solver. Let's get started.

1. Describing a Problem and Describing Problem Solving

What Constitutes a Mathematics Problem for You

A mathematics problem for you may not be a mathematics problem for someone else. To help you determine what may or may not be a mathematics problem for someone, three exercises are given:

Exercise 1. Give a real-world problem whose solution is the product of two negative integers.

If a middle school student has no difficulty working this exercise (some won't), requiring very little thought on his or her part, then it is not a problem for this student. If a college student in calculus finds this to be a challenging exercise (some will), then it is a problem for this student. Did you solve the problem?

Solution to Exercise 1. The temperature has been steadily decreasing 4 degrees per hour. What was the temperature 3 hours ago compared to now? Let three hours ago be represented by -3 and let -4 represent a decrease of four degrees per hour. Since $(-3)(-4) = 12$, the temperature 3 hours ago was 12 degrees higher than now.

Exercise 2. What are the roots of $3x^2 + 5x - 9 = 0$? (You are not allowed to use technology to solve this exercise.)

Solution to Exercise 2. A typical second year high school algebra student should have no difficulty answering this question. He or she would recognize this as a quadratic equation and apply the quadratic formula (or complete the square to find the roots):

$$x = \frac{-5 \pm \sqrt{5^2 - 4(3)(-9)}}{2(3)}$$

$$= \frac{-5 \pm \sqrt{133}}{6}.$$

Is this a problem for this student? No, it is a *routine exercise* for him or her and can be solved in a mechanical way, using very little thought. It is a problem for someone who has little or no idea what to do after reading the problem.

Exercise 3. Find the points on the parabola $y^2 = x$ that are closest to the point (2,0).

If you have worked a few problems like this, with the only changes being the point and equation that are given, then it is most likely not a problem for you. All you need to do is apply the same procedure or scheme. It does require some thinking to solve it, but far less thinking than what is required if you have not been exposed before to a problem of this type. A calculus solution to the problem is now given.

Solution to Exercise 3 (using calculus). The shortest distance from the point (2,0) to the upper branch of the parabola, is the length of the segment from (2,0) perpendicular to a tangent line of the curve at a specific point. Represent the point we seek by (a,b). The upper branch of the parabola has equation $y = \sqrt{x}$, and the slope of the tangent line

to it at (a,b) is $y' = \dfrac{1}{2\sqrt{a}} = m$. We also know that the slope of this tan-

gent line is the negative reciprocal of the slope of the line through (a,b)

and (2,0), which is $-\dfrac{1}{\dfrac{b-0}{a-2}} = \dfrac{2-a}{b}$. Since $b = \sqrt{a}$, this slope is $\dfrac{2-a}{\sqrt{a}}$.

Setting the two expressions for the slope equal, we get $\dfrac{2-a}{\sqrt{a}} = \dfrac{1}{2\sqrt{a}}$,

yielding $a = \dfrac{3}{2} = 1.5$. Hence, $b = \sqrt{1.5}$. Since the two branches of the parabola are symmetric to the x-axis, the answers we seek are the points $(1.5, \sqrt{1.5})$,and $(1.5, -\sqrt{1.5})$.

The thinking used to solve exercise 3 used calculus and is definitely non-trivial. If you have not had calculus, you will not understand the solution, and the exercise would assuredly have been a problem for you if you had to solve it using calculus. It most likely is a problem for any student who has the necessary calculus background if this was the first time the student had seen this type of problem, and was expected to use calculus to solve it.

The three examples just given illustrate what may or may not be problems for someone. They are not a problem for you if you quickly know how to solve them. They are a problem for you if you struggle knowing what to do to solve them. Advice is given in this chapter and the next on what to do in the "face of a problem." That is, advice is given on what you can do to solve a problem that you cannot currently solve. You look at what you are given and what you are asked to find, you recall what you know that relates directly to these two things, and you then relate this knowledge in a logical order to arrive at the solution. In problem solving you have to think, and then think some more, trying this and that, until you are successful. Fortunately, thinking feeds on itself—the more you think the better thinker you become.

You are helped to become a better problem solver if your mathematics instructor gives you problems to solve that are not similar to problems you have solved, or have seen solved. A problem you can solve immediately is not much of a problem for you; nonetheless, *there is value* in being assigned some of these problems. Your instructor does not know what may be a problem for you, and neither do you until you read the problem. Once you know that you can work it, it ceases to be a problem for you. However, the problems that are most valuable for you to work, especially if you want to increase your problem-solving ability, are the problems you cannot solve. You are now in a position to learn something by working on them.

Most of us like to be successful and the sooner the better, so we feel good about the problems we are able to solve quickly, and want to do more of them. On the other hand, most of us feel badly about the ones we cannot solve, or at least cannot solve quickly. We cannot wait to leave them unfinished, most likely never coming back to them. That is human nature but it sure isn't the way to success. Let the following anonymous quote be your chant: "The only problems worth working are those you cannot do." It is also necessary to work exercises that you have little difficulty completing, since this reinforces and sharpens your existing problem-solving skills, and helps you retain past knowledge. You need to work an *appropriate* number of these exercises, as well as problems that are more challenging.

When Are You Problem Solving?

You are problem solving when you are solving a problem where a path to the solution is not clear to you (i.e., it is a challenge for you to solve the problem). But problem solving involves more: You also have to *accept the challenge* of solving the problem. Alan Schoenfeld, a mathematics and education professor at the University of California, Berkley, and a leader in problem solving, says this about problems:

> Roughly half of our students see calculus as their last mathematics course. They complete their calculus studies with the

impression that they know some very sophisticated and high-powered mathematics. They can find the maxima of complicated functions, determine exponential decay, compute the volumes of surfaces of revolution, and so on. But the fact is that these students know barely anything at all. The only reason they can perform with any degree of competency on their final exams is that the problems on the exams are nearly carbon copies of the problems they have seen before; the students are not being asked to think, but merely to apply well-rehearsed schemata for specific kinds of tasks.[1]

Don't misinterpret Schoenfeld. He is not saying that many of the problems students have to solve in their calculus studies should not be assigned. He is implying that to become a proficient problem solver you have to solve other types of problems.

2. Necessity of Having a Good Mathematics Background

Problem solving does not take place in a vacuum. When you problem solve you are putting ideas together that will lead to the solution. Hence, you need ideas to put together. The more knowledge you have, the more insight you will have in how to proceed with a problem. In problem solving you are relating information, and before you can do that you have to understand the information. You need to know specific concepts and principles, and have computational proficiency. In general, the more advanced the mathematics course, the more prerequisite ideas you need to solve problems in that course. A necessary requirement to being a successful problem solver is to have a repertoire of knowledge. This should motivate you to learn mathematics well. Imagine being a Calculus III student who has to solve a challenging problem. You may have to use knowledge from Calculus III, Calculus II, Calculus I, Precalculus, Intermediate Algebra, Geometry, Introductory Algebra, or even Arithmetic. This is a pretty daunting thought, wouldn't you say? This certainly argues for having the prerequisites for a course and for retaining knowledge. Problem solving is thinking and you need something to think about. You need to approach your mathematics study with the mindset that you will learn it well and work to retain it. Repeat this often to yourself: "Problem solving does not take place in a vacuum."

Having a good mathematics background is one of many important ingredients to becoming a better problem solver. Other ingredients are developed in later sections of this chapter. To emphatically make the point that a good mathematics background is not enough to become a good mathematics problem solver, I encourage you to read the classic problems that follow (origin unknown), which only require knowledge of arithmetic and perhaps some middle school number theory. Spend a little time thinking how you might attack them.

1. There are 17 people in a room and each person shakes hands exactly once with everybody else. How many handshakes have taken place? How many handshakes would there be if there were people in the room, $n \geq 2$?
2. How many different rectangles are there on an 8 by 8 checkerboard?
3. In how many zeros does the product of the first 100 natural numbers end?
4. Nine officials are positioned along a straight section of a marathon route. At what location should they meet to confer so that the total distance they travel is as small as possible? (It is not necessarily the case that the officials are located equidistant from each other.)
5. There is a row of 862 lockers, some closed and some open. A student runs down the row and opens every locker. A second student runs down the row and, beginning with the second locker, shuts ever other locker. A third student runs down the row and, beginning with the third locker, changes the state of every third locker (i.e., the student opens those that are closed and closes those that are open). The process continues until 862 students have run down the row. What is the state of the 862nd locker? The 722nd locker? The 529th locker?

Do these problems appear challenging to you? Many good college mathematics students struggle with them. Based upon the time you spent on these problems, are there problem-solving skills you need to learn? Say, "Yes." Please don't let any difficulty you may have working these problems keep you from reading the remainder of the chapter. They were carefully chosen to reveal to you that problems can be a challenge even though not much mathematical knowledge is needed to solve them. Specific problem solving skills are needed to solve them, and they are discussed in chapter 22, along with many other problem-solving skills. (These problems appear again in section 2 of chapter 22. Answers to some of them can be found at the end of that chapter.)

3. Solving Problems at Various Cognitive Levels of Thought

There are problems in mathematics and in other disciplines that are at different cognitive levels of thought. That is, they require different thought processes. Four cognitive levels of thought have been discussed earlier in the book: computation, comprehension, application, and analysis. (A description of each of these cognitive levels, and examples of problems at each level, are covered in section 1 of chapter 17.) The more thought processes you have used, the more equipped you are as a problem solver. Hence, working many problems at various cognitive levels of thought is important. The cognitive levels of thought that seem to be avoided (i.e., instructors not assigning problems at these levels, or students deciding not to work problems at these levels), are the upper cognitive levels of application and analysis. You must work problems at these two levels; otherwise, your problem-solving ability will not improve as it should. Look forward to working problems at these cognitive levels, as challenging as some of them may be.

4. Problem Solving as the Focus of Mathematics Study

Relative to all that goes on in a mathematics course, how important is it for you to solve problems? It has to be the *focus* of your mathematics study. You want mathematics to serve you in the future and it will primarily be through problem solving. What does it mean to say that problem solving is the focus of mathematics study? It means that all you study in a mathematics course, including concepts, principles, and computational techniques, is done with the thought that this will help you solve problems, and you act on this thought by solving problems, including those in a real world context. Mathematics work must point to solving problems.

Problem solving should be the heart of the mathematics curriculum—at all levels. The grades 9–12 standard for problem solving of the National Council of Teacher of Mathematics (NCTM) follows:

Instructional programs from prekindergarten through grade 12 should enable all students to—

- Build new mathematical knowledge through problem solving.
- Solve problems that arise in mathematics and in other contexts.
- Apply and adapt a variety of appropriate strategies to solve problems.
- Monitor and reflect on the process of mathematical problem solving.

NCTM goes on further to say:

To develop such abilities [becoming a mathematical problem solver], students need to work on problems that may take hours, days, and even weeks to solve. Although some may be relatively simple exercises to be accomplished independently, others should involve small groups or an entire class working cooperatively. Some problems also should be open-ended with no right answer, and others need to be formulated.[2]

Problem solving must also be the heart of the college mathematics curriculum as attested to by the U.S. Conference Board of the Mathematical Sciences. This Board regards "problem solving as the basic mathematical activity." Would an observer of your mathematics study come away knowing that problem solving is the focus of your study?

It has been the theme throughout this book that you are more apt to follow the advice given if you know the benefits that accrue from doing so. The benefits of problem solving follow.

5. Benefits of Problem Solving

The benefits of being a good problem solver are many. At the end of this section I hope you will say to yourself: "How could I not invest the time it takes to become a better problem solver when I see the significant benefits I will derive from it?"

Problem Solving Teaches You How to Think

Since problem solving involves thinking, improving your problem-solving skills implies improving your thinking skills. Can you think of a more valuable benefit? We are confronted with thinking situations almost continuously—in school, in a job, and during everyday life. Some thinking situations involve mathematics or other technical subjects, and others do not. Problem solving is beneficial in disciplines other than mathematics, but mathematics provides the best area to develop problem-solving skills that can be used in other disciplines. Schoenfeld's remark supports this: "Mathematics is a discipline of clear and logical analysis that offers us tools to describe, abstract, and deal with the world (and later, world of ideas) in a coherent and intelligent fashion."[3] There is no other discipline quite like mathematics for enhancing your problem-solving ability, and, concomitantly, your ability to think. Who's not for thinking?

Problem Solving Is Valued in the Work World

You either have a career or are pursuing one. Problem solving is valued in any career. This is the 21st century and human robots are not much in demand in the work world. Individuals with problem-solving or thinking skills have a definite advantage over those who don't, and recruiters look for these skills. Individuals who have these skills are in short supply, and are cherished.

This is a good time to mention that *mathematics models* are used to solve real-world problems. A mathematics model is an abstract representation (or idealized version) of the real-world entities involved. For example, a poker chip is represented by a circular disc, an electrical wire by a line segment, a can of green beans by a cylinder, and so on. The accuracy of the solution to a real-world problem, through the use of a mathematics model, depends on how well the mathematics entities represent the real-world entities. A humorous story involving the well-known mathematician W. Feller and his wife is told by Paul Halmos:

> W. Feller enjoyed telling about the time when he and his wife tried to move a large circular table into the dining room. They

pushed and turned and tugged it, but they could not get it through the door. Annoyed and tired, Feller took time out to diagram the situation, and, having found a workable mathematical model, he proved that what they were trying to do couldn't be done. In the meantime, as he was constructing his proof, Clara kept tilting and pulling the table; by the time the proof was finished, the table was in the dining room.[4]

Problem Solving Motivates You and Helps You Learn Mathematics

One great advantage of focusing on problem solving is that mathematics will become more exciting to you. Why? Because now you are engaged in searching for ideas and determining how they relate as you work to find a solution to a problem. What motivates you to keep working on a problem is the anticipation that the solution will eventually evolve. There is nothing like solving a problem when you had no idea how to begin your work on the problem. You will experience an exhilarating sense of accomplishment and unbridled enthusiasm when you say to yourself, "I did it!" If you have never experienced this joy, then you probably think you can never be a problem solver and will work in a self-defeating manner that fulfills your prophecy. You will most likely give up on finding a solution when you sense the solution is not within your grasp. Success at problem solving increases motivation to learn the other content of your mathematics courses. You have experienced first-hand how important this content is to problem solving. But you will also discover that problem-solving skills help you learn mathematics content. You will apply problem-solving skills to develop concepts, construct theories, generalize, and abstract. So there you have it—a two-way street: Problem solving helps you learn mathematics content, and learning mathematics content helps you become a better problem solver. That's a win-win situation.

6. Attributes of Problem Solvers

This section addresses some key perceptions, attitudes, and virtues of problem solvers. Not every problem solver has all these attributes or has them to the same degree. If you don't have them, get them, for in most cases it is nothing more than making up your mind to take them on.

Problem Solvers Trust Their Ability to Learn Mathematics

Do you think you were behind the door when mathematical ability was passed out? That is, do you feel you were not blessed with what some people call a mathematical mind? Can you see how this belief can keep you from developing as a problem solver? If you think that you don't have a mathematics mind, then when you look at a problem and don't immediately know how to solve it, you get discouraged and say, "Why bother, I cannot change something that is innate."

Yes, there is something called mathematics aptitude—a natural ability to understand mathematics and apply it, and you may or may not have less of this than others. However, you have some mathematics aptitude and you have to believe that it can be more fully developed. We cannot all be equals as problems solvers, but each one of us can get better. I taught a mathematics course in problem solving for many years and the improvements I saw in students' problem-solving abilities, as they progressed through the course, were impressive. Exposure to this chapter and the next is not the same as taking a course in problem solving, but it can help you improve as a problem solver.

Problem Solvers Do Not Compare Their Problem-Solving Ability with That of Their Instructors

Many students compare their problem-solving prowess to that of their instructors, coming out a distant second in their minds. Inherent in this comparison are beliefs they hold that will inhibit their growth in problem solving. Most often they see instructors work problems with ease causing them to say: "I wish I was as smart as my instructor. There is no way I can be as good. Mathematics is so easy for him or her." They believe their instructor is innately blessed with this wonderful problem-solving talent, and they are not. They need to realize that instructors have a lot of mathematical experiences, which contribute greatly to their perceived problem- solving ability.

Let's take a closer look at not-to-atypical Calculus I instructors who have been teaching this course for at least several years. Their preparation is extensive and includes the following:

- They took this course as students, as well as several other lower-division calculus courses that followed it, and one or more upper-division courses in advanced calculus. This alone would make it easy for them to solve most problems in elementary calculus that students have to solve. These problems hardly qualify as problems for them.

- They worked hard and obtained one or more graduate degrees in mathematics, qualifying them to teach Calculus I, among other mathematics courses. In preparation for teaching the course, they determined the problems they would assign their students, and made sure they could work them before arriving in class. They may not assign problems that caused them considerable difficulty. If any of them are not routine for them, but are important to assign, they will spend time out of class working toward a solution. At times they will struggle trying to reach solutions, but they finally do, perhaps with the help of an instructor solutions manual.

- After teaching the course several times, the problems they assign become second nature to them and they can easily respond to any question from the class. Their presentations of solutions are works of arts, having a beautiful logic and flow to them.

Imagine being exposed for the first time to a non-routine problem in calculus. Imagine also struggling in your attempts to arrive at a solution, and then watching your instructor solve it with dispatch. Should you be impressed with his or her ability? Should you be expected to solve the problem in a similar fashion? The answer to both questions is, "Absolutely not." Discounting your instructor's vast formal preparation and teaching experience, you may not be that different from him or her. Believe it, and stay away from making comparisons.

Problem Solvers Know That Solving Problems Is Difficult and Takes Time

Why do so many students have difficulty with problem solving? This is easy to answer. By definition, when working a problem you are in a difficult situation—you are in unfamiliar territory without a road map. You may contribute to the difficulty but that does not mean that problem solving is not inherently difficult. You are expected to find a path to the solution and that isn't easy. It will be less difficult if you know more about problem solving. Unfortunately, many students think that it should not be difficult, which lessens their resolve to spend the time that is needed. It is critical to realize upfront that you must set aside an appropriate amount of time in your mathematics course to solve problems. Are you overextended because you are carrying too many credits, working too many hours on a job, etc.? The best advice in the world on problem solving, coupled with having excellent problem-solving skills, will be of little use to you if you do not have an appropriate amount of time to spend solving problems. Solving a problem is a non-trivial task—it takes time and you must have it. If you don't have the time, take steps to get it, which most likely means you will have to reprioritize things in your life.

Problem Solvers Know That Perseverance Is the Key to Solving Problems

If I were asked to give you one problem-solving attribute to develop, it would be perseverance. What does it mean to persevere in finding a solution to a problem? It means that you will not give up on the problem. It means having the attitude: "I am not able to solve this problem now, and maybe not this evening or tomorrow, but I will eventually solve it. I may need a hint or two, but I know the solution is within my grasp and I will unearth it." Persistence does not mean that you will never take a break from working on a problem. Putting a problem aside for a while is a wise thing to do and is the subject of the next subsection.

The students who impressed me the most in the mathematics problem-solving course I taught, were the students who never seemed to give up on a problem. At times they would turn in pages and pages of data they had generated. Their analysis of this data helped them discover a pattern or another way to approach the problem, which led them to a solution. Some of the problems could have been solved in a short paragraph or two, but they didn't discover those solutions. There was no doubt that they spent many hours over several days working on a problem. These students were always among my best problem solvers. They were the only ones who were able to solve some of the more difficult problems. They were successful because they persevered—not because they had great problem-solving aptitude.

Problem Solvers Know That You Need to Set Problems Aside for Awhile

Some of you may find it hard to believe that your best ideas for solving problems can come to you when you are not intentionally engaged in finding a solution. How wonderful it is to get ideas that lead to a solution of a problem when you are not consciously seeking them. Putting a problem to rest for a while can cause viable avenues of attack to surface. The problem-solving literature mentions this often, and it has happened numerous times to many of my colleagues and to me. I have suddenly awakened during the night with a solution to a problem or an idea for attacking a problem. It was pretty incredible. Obviously my subconscious was working on the problem, increasing my understanding of ideas I had earlier, and allowing new ideas to surface. This could not have happened without taking a break and giving my subconscious permission to work on the problem. My subconscious was aided by (1) my diligence in working on it earlier, and by (2) my persistence—telling myself that "I can't get it now but I will later."

Taking breaks from problem solving can occur because of normal interruptions such as fulfilling an appointment, going to a job, working on assignments for other courses, or having to eat. But beyond these, you need to make conscious decisions to take breaks from problem solving, whether it be for 20 minutes, several hours, overnight, or one or more days. Obvious signs that you need to take a break include losing your focus and continually rehashing the same thing over and over. Allow your subconscious to work for you. The famous author and problem solver Isaac Asimov said: **"When I feel difficulty coming on, I switch to another book I'm writing. When I get back to the problem, my unconscious has solved it."** Similarly, another famous author, John Steinbeck, had this to say: "It is a common experience that a problem difficult at night is resolved in the morning after the committee of sleep has worked on it."

Problem Solvers Learn More from Receiving Hints for Attacking Problems than from Seeing Solutions

At times you need to see solutions to problems, and there are ways to read through a solution that will help you become a better problem solver. (Advice is given in section 1 of chapter 18 on how to use a solutions manual for your textbook.) What is more beneficial to you, in the long run, is not to read through solutions as soon as you are stymied, but to receive hints for attacking problems that you are unduly struggling with. You may still have to give considerable thought to the problem, but hints allow you to progress toward a solution. A good instructor will often give you a hint rather than show you a solution. Unfortunately, you are given complete solutions much more frequently than you are given hints. Classmates and instructors love to tell you how to work a problem, but most often that is not to your benefit. What you need to ask them is this: "I don't want to see a solution, but can you give me a hint so I can keep working?" If you are able to progress further by using the hint, but eventually get stopped again, then ask for another hint. Your eventual goal in problem solving is to rely less and less on hints. When asking for a hint, ask for one that is less direct (that is, ask for one that requires more thinking on your part to come up with a next step). You may need to ask for a more direct hint later. Return the favor when your classmates ask you for help—give them hints, not solutions. In the final analysis, you are in the driver's seat; that is, you are responsible for your education, not your classmates or instructors. Only solicit and take from them what is most beneficial to you.

Problem Solvers Expect to Be Confused When Solving Problems

I have found that one of the best ways to eliminate anxiety in my life is to avoid setting myself up for failure. One way I do this is by being more realistic in my expectations. We often get bent out of shape when our expectations are not met, and more often than not it is because our expectations are unrealistic. It is unrealistic to expect problem solving to be a walk in the park. You need to tell yourself upfront that you will most likely be confused, and that you will spend a considerable amount of time solving some problems. That is the nature of problem solving. It is much more difficult to become frustrated, disappointed, and anxious when you know what can happen, and it does happen. With these realistic expectations, you are more apt to accept them with equanimity when they occur. Until you can tolerate working in an uncertain environment such as problem solving, you will exhibit certain behaviors that will keep you from being a successful problem solver. These behaviors include telling yourself you cannot solve it, asking someone for major hints or a solution too quickly, or permanently stopping work on the problem. Herbert Hoover captures the essence of this paragraph when he said: "When I decided to go into politics I weighed the cost: I would get criticism. But I went ahead. So when the virulent criticism came I wasn't surprised. I was better able to handle it."

Problem Solvers Know That the Time Spent Solving Problems Is not Wasted, Even if a Solution Is Not Found

It is anxiety provoking to spend a lot of time on a problem, make little progress toward a solution, and then feel that the time spent was a total waste. You are operating on the false expectation that time spent on a problem without solving it is wasted time. How wrong you are. Do you believe you have wasted your time if you spend a half hour, an hour, or more time than that on a problem, yet no solution is in sight? Many students believe this. This belief can cause you to abandon your problem-solving efforts.

You often hear that the journey is more important than the destination, and that is true when it comes to problem solving. Your efforts in problem solving prepare your mind to see or accept a solution. Unplanned learning takes place when you are working on problems. You are led in directions that will surprise you. You review and apply concepts and principles, learning them better and relating them in different ways. You are engaged in assimilating them in a way that makes them usable. This happens in stages, not all at once. A solution is only the final step in problem solving. Even if you don't reach a solution, what you have learned can most likely be applied to solving other problems. When a solution comes it usually comes suddenly, and comes from the exploration you have done. This analogy by Jacob Riis applies here: "When nothing seems to help, I go and look at a stonecutter hammering away at his rock perhaps a hundred times without as much as a crack showing in it. Yet at the hundred and first blow it will split in two, and I know it was not that blow that did it, but all that had gone before." The smallest progress you make is significant. The secret is to *continue* to do something, small as it may be. Spending time trying to solve a problem is not wasted time.

Problem Solvers Have Faith in the Originality of Their Thoughts, and Know That There Are Different Ways to Solve a Problem

We often talk about the uniqueness of finger and voiceprints, but not too often do we talk about the uniqueness of one's thoughts in solving problems. Suppose you and another student are separately working on the same problem. Will your solutions be different? Yes, since no two individuals think exactly alike, no two solutions are exactly alike. Your solution consists of what you write *and* what you think. You have a way of processing information unlike anyone else. If you think you should be like others, you will not focus on what *you* can bring to the problem.

What hinders many students in their problem-solving efforts is a false belief that there is only one path to the solution and they have to find it. This thought stifles creativity and hinders success. These students suppress, perhaps subconsciously, strategies that do not seem to fit their preconceived notion of the path to the solution. Perhaps much of their previous instruction in mathematics reinforced this notion. Instructors can help their students more by showing several solutions to a problem, rather than showing one solution to each of several problems.

I continued to be amazed by the different solutions to the same problem that students in my problem-solving course gave. I would have had great difficulty discovering some of them. They would present their solutions on a blackboard or overhead, and many of their classmates would marvel at the uniqueness of the solutions. Students' confidence in their ability to problem solve was boosted when other students, often those with greater problem-solving ability, made comments to them such as, "I would never have thought of that. What an interesting solution." You have the power and ability to solve problems. You may need more problem-solving experience and more knowledge about mathematics to fully realize the problem-solving potential that is within you.

You may find it instructive at this time to see different solutions to a rather simple problem. Work this problem before reading the given solutions (please, no peeking at the solutions):

Problem. A fence is to be placed around a square field using 84 posts. If each side has the same number of posts, how many posts are on a side?

Solution 1.

Collect data, place it in a table, and look for a pattern.

# of posts at your disposal	# on a side
4	2
8	3
12	4
16	5
20	6

The pattern I see is this: Divide the number of posts by 4, and then add one to this. Hence, for 84 posts, $84 \div 4 = 21$, and $21 + 1 = 22$. *Answer:* 22 posts on a side.

Solution 2. I will make an educated guess at the answer, check it, and then revise my guess to come up with another guess. I continue this until I solve the problem. Here's my educated guess: *Divide 84 by 4 (the number of sides), giving 21 on a side.* That leaves 19 in the middle (since two are at the ends). *Check:* 19 times 4 equals 76, plus the 4 corners, gives 80. I am short 4 posts. Place one more in the middle of each side, giving 20 posts in the middle of each side. *Answer:* 20 middle posts on a side + 2 end posts on a side = 22 posts on a side.

Solution 3. I will work the problem using smaller numbers (similar to solution 1), and then infer from this work what to do with the 84 posts. Suppose there were only 4 posts total. This leaves one at each corner, none between the corners. Suppose there were 8 posts. This leaves 1 at each corner, and 4 left which means that there is 1 on each side between two corners. I now see what to do with 84 posts. I subtract the 4 corner posts from 84 giving 80, which has to be the number of posts not on a corner. Since there are 4 sides, 80 divided by 4 yields 20 on a side, which are between two consecutive corners. *Answer:* 20 + 2 = 22 posts on a side.

Solution 4. Let x be the number of posts on a side. $4x = 84$ implies there are 21 on a side. But this assumes that each corner post has been counted once, but in reality, when you add the number of posts on each of the 4 sides, each corner post has been counted twice. Hence you need to have 4 more posts to give you a total of 84. This means that you need 1 more on each side. *Answer:* 21 + 1 = 22 on a side.

Solution 5. Act it out. Take your 84 posts and begin to place them on a side. Begin by placing one on each corner. That leaves 80 posts left to distribute. 80 divided by 4 means you need 20 on a side, interior to the corner posts on that side. *Answer:* 20 + 2 = 22 on a side.

All of these solutions to the problem are similar, but not the same. You may find some of these five solutions easy to follow; others not. You can be sure that a student who solves the problem using one of these five solutions, or who devises a solution different from these five, found that solution to be most natural to him or her. Do not stifle your problem-solving creativity by falsely thinking that there is only one solution to a problem and you have to find it. Be encouraged and excited about the unique problem-solving strategies that you have. Allow your problem-solving instincts to surface.

Problem Solvers Know There Is No Best Way to Solve a Problem

As you now know, there can be a variety of solutions to a problem, some more elegant than others. In your quest to become a better problem solver don't get caught up in evaluating the merits of your solution or the solutions of others. You may think that your instructor's solution to a problem, a classmate's solution, or the solution in your textbook is the *best* way to solve a problem. You may think they are better than yours. You will quickly dismiss this thought if you realize that different instructors and textbook authors have different solutions to the same problem. They are expressing their own individuality and creativity, as you should.

You can learn something from analyzing your classmates' solutions (and they from analyzing yours), so don't ignore them. But the moral of the story is this: *The best solution to a problem is the one that works for you.* That is, the solution you develop is the best one, rather than the solution someone else develops. Keep this in mind and you will value your solution, which will motivate you to rely more on your problem-solving instincts in the future.

Problem Solvers Know That Going Down Wrong Paths Is a Necessary Part of Problem Solving

If you think a problem solver's work is a neat and tidy process, think again. You will make plenty of errors and take paths that do not lead to the solution. It can't be any other way, for a path to the solution is not known at the outset. You and your instructors have to tolerate this untidy search process. Making errors is inevitable and should not be viewed as failure. If you do not go down any wrong paths, you are most likely not problem solving—you are performing a rote procedure. Alan Schoenfeld describes how a problem solver works:

> He may not fully understand the problem, and may simply "explore" it for a while until he feels comfortable with it. He will probably try to "match" it to familiar problems, in the hope it can be transformed into a (nearly) schema-driven solution. He will bring up a variety of plausible things: related

facts, related problems, tentative approaches, etc. All of these will have to be juggled and balanced. He may make an attempt at solving it in a particular way, and then back off. He may try two or three things for a couple of minutes and then decide what to pursue. In the midst of pursuing one direction he may back off and say "that's harder than it should be" and try something else. Or, after the comment, he may continue in the same direction. With luck, after some aborted attempts, he will solve the problem.[5]

He comments further:

Does that make him, at least in that domain, a bad problem solver? I think not. In all likelihood someone proficient in that domain (i.e., someone who knows the right schema) could produce a solution that puts his to shame. But that isn't the point at all. The question is: How effectively did the problem solver utilize the resources at his disposal?[6]

Problem Solvers Know That the Solution to a Mathematics Problem Will Seem Trivial Once It Is Revealed

You have probably seen a solution to a problem that you could not solve and made a comment similar to this: "That is so simple. Why couldn't I have come up with that solution?" It is not fair to evaluate the difficulty of a problem by how easy it is to understand the solution. There are some very difficult problems whose solutions are very short and easy to understand. What makes a problem difficult for you is the difficulty you have devising a solution, not understanding it once it has been found. Chiding yourself for not being able to solve a problem (or for taking too long to solve it), whose solution is easy to understand, can only make you think that you are a poor problem solver. Don't feel alone if you think that way, for Thomas Edison, the great inventor, asked, "Why didn't we think of this sooner?" I can answer that: "Because we didn't."

Problem Solvers Know the Value of Working Independently, and with Others

You can work on problems independently and with others. Working independently is absolutely critical. No amount of work with others, including your instructor, a classmate, study group, or tutor, can replace individual work. That being said, it is highly beneficial, at appropriate times, to work with others to solve problems. Working with others supports the individual work that you do. Your *initial* work on a problem assignment should be done independently so you can—

- Take the time you need to explore.
- Schedule appropriate breaks so your subconscious can work.
- Use the opportunity to consult written resources.
- Identify the problems that are causing the most difficulty.
- Pose questions you need to ask others to proceed further.
- Prepare to discuss ideas with others.

A good scenario to follow, repeated as needed, is to work alone, then work with others, and then work alone again.

When you work with an individual or a group, you will improve your problem-solving ability by asking and answering questions, and observing how others solve problems. You can jointly brainstorm strategies to attack a problem, decide on the approaches most likely to lead to a solution, and then determine if they lead to a solution. You will be amazed at what you learn from all the conversation that takes place during this process. Solving problems is fun and exciting, and including others makes it even more so.

You are encouraged to read the following content, which gives the benefits of working with others to solve problems and improve your thinking, and methods for doing so:

- Section 3 of chapter 9 (on thinking in mathematics)
- Section 4 of chapter 14 (on discussing mathematics)
- Chapter 23 (on seeking assistance from your instructor, study group, mathematics department tutor, office of academic support, or personal tutor)

An important goal in life, and it is a lifetime pursuit, is to know who you are. This requires being with others (i.e., being in community). You need others to know who you are as a problem solver and where you can be as a problem solver. Oliver Goldsmith said: "People seldom improve when they have no model other than themselves to copy." In summary, you must work problems independently; however, it is wise to also get support by working with others.

Problem Solvers Know the Importance of Communicating Their Solutions to Problems, Orally and in Writing

Along with being able to solve problems, the ability to communicate your solutions to problems, orally and in writing, is important. A primary means of developing oral communication is by discussing mathematics with others. A primary means of developing written communication is by writing mathematics, and writing about mathematics. You have to practice writing solutions to problems so others can understand them. This is not easy to do. You have to assume that the reader has no idea how to solve the problem; hence, you have to present your thinking in a clear and logical way, letting the reader know how you solved the problem. Writing about mathematics and technical aspects of writing mathematics are discussed in section 2 of chapter 14, and chapter 20, respectively.

Problem Solvers Ask Questions and Analyze Solutions

A critical skill for a problem solver is a questioning mind. How can you improve your problem-solving skills without continuously questioning yourself and others? You need to ask questions as you read your textbook, attend class, and work with a tutor, classmates, and your instructor. You have to question, question, and question some more, and be dedicated to finding answers to your questions. Do you do this? If you don't do this, why not?

Analyzing solutions can give you insight into how problems are attacked by others. You have plenty of solutions to look at, including those in the textbook, solutions manual, and class notes. You can ask yourself questions such as these: What prompted this thought? Would I have come up with this thought? What might have prompted me to come up with this thought? Pay attention to how problems are solved. Also ask yourself questions about problems you have solved. That is, look back over your solution to a problem and ask yourself if you could have taken another path to the solution, or could have streamlined what you did. If I had to use only one word to describe how to learn, it would be "question."

Problem Solvers Are Always on the Lookout for Generic Methods to Use to Attack Problems

It is good to have a repertoire of generic problem-solving skills from which you can select skills to attack a problem. Some of them you will pick up from reading your textbook, reading alternative textbooks, your instructor, your classmates, or a tutor. Most likely you will have to go beyond these sources to have a suitable repertoire, including reading books or articles that specifically address the topic of problem solving. Use the exploratory resources you have to locate these skills, including the Internet. Many generic skills for solving problems are addressed in the next chapter.

In summary, it is a good idea to periodically review the following list of attributes of a problem solver to see how you match up. If you find that you do not have a particular attribute, work to attain it. If there is a will there is a way.

LIST OF PROBLEM SOLVING ATTRIBUTES

1. Problem solvers trust their ability to learn mathematics.
2. Problem solvers do not compare their problem solving abilities with those of their instructors.
3. Problem solvers expect to be confused when solving problems; hence, minimizing their anxiety.
4. Problem solvers know problem solving is difficult and takes time.
5. Problem solvers know that perseverance is the key to problem solving.
6. Problem solvers know that you need to set problems aside for a while.
7. Problems solvers learn more from receiving hints for attacking problems than from seeing solutions.
8. Problem solvers know that the time spent problem solving is not wasted, even if a solution is not found.
9. Problem solvers have faith in the originality of their thoughts, and know that there are different ways to solve a problem.
10. Problem solvers know there is no best way to solve a problem.
11. Problems solvers know that going down wrong paths is a necessary part of problem solving.

12. Problem solvers know that the solutions to mathematics problems will seem trivial once they are revealed.
13. Problems solvers know the value of working independently, and with others.
14. Problem solvers know the importance of communicating their solutions to problems, orally and in writing.
15. Problem solvers ask questions and analyze solutions.
16. Problem solvers are always on the lookout for generic methods to use to attack problems.

22

The Process of Problem Solving and Development of Problem-Solving Skills

There is a process to problem solving and that is good news. It is not a haphazard romp through a morass of ideas that, with complete luck, can be put together to arrive at a solution. There is a body of knowledge on problem solving to which you might not have been exposed. In this chapter a problem-solving process is discussed, along with development of key problem-solving heuristics or skills. This knowledge will help you become a better problem solver, especially if you have or develop the attributes of a problem solver that were presented in section 6 of chapter 21.

The problem-solving process can be outlined in four steps, which have been described by George Polya (1887–1985) in his 1945 book titled, *How to Solve It.* This book has sold over one million copies and has stood the test of time. A second edition was published in 1988 by Princeton University Press[1] and is still available. Polya was a famous Hungarian mathematician and excellent problem solver with an interesting biography. He immigrated to the United States in 1940. It was his belief that the skill of problem solving could be taught. I also have this belief.

Polya's four steps of the problem-solving process are the titles of the sections in this chapter:

1. Understand the Problem
2. Devise a Plan
3. Carry Out the Plan
4. Look Back

Polya's framework for problem solving can be used with many types of problems. Considerable time in section 2 is devoted to describing specific problem-solving skills (or heuristics), and examples of their applications.

1. Understand the Problem

The first step of the problem solving process, *understand the problem,* is a vital one. You would probably say that this is so obvious it doesn't need to be stated. However, evidence shows that many students are not able to solve a problem because they do not understand the statement of the problem. I suspect most of them think they understand the problem. You are not successful in solving a problem if you misunderstand the problem and end up solving a different one. Credence for this concern can be found in this statement by Alan Schoenfeld, who reported on this experiment with calculus students at the University of California, Berkeley:

> Fifty-eight protocols were obtained from randomly selected calculus students who were asked to rewrite problem statements "more understandably." Of these, 5 simply rewrote the problem verbatim. The 53 remaining rewrites tell a sorry tale: 5 (9.4%) included information which directly contradicted the input, and 11 (20.4%) included information that was so confused as to be unintelligible; 2 (4%) made both kind of errors. This information is the more striking since two-thirds of these students were to write simple declarative sentences, if possible, to make their task simpler. Thus, before they would normally have put pen to paper, a quarter of the 53 students had already seriously garbled or completely misinterpreted the problem statement.[2]

Language can be tricky. Here are seven simply stated problems that illustrate the importance of reading carefully or not making unwarranted assumptions:

Problem 1. An automobile buff had 12 antique cars. All but 7 cars were non-operational. How many cars were operational?

Problem 2. Helen took 5 chickens from a flock of 20 chickens. How many chickens did Helen have?

Problem 3. Frank's two coins totaled 26 cents. One is not a quarter. What are the two coins?

Problem 4. How many cubic feet of dirt is in a hole that is 20 feet long, 12 feet wide, and 3 feet deep?

Problem 5. You have an unlimited supply of 4 cent and 9 cent postage stamps. What is the greatest amount of postage you can't make by using these stamps? (Many students have great difficulty understanding the question that is asked. Do you?)

Problem 6. What is the maximum number of regions into which 15 chords can divide a circle? (Many students make an unwarranted assumption about this problem. Did you?)

Problem 7. A mother and her daughter were in a car accident in front of the daughter's school. The daughter's first grade teacher was the first upon the scene and said, "That is my daughter." How can this be? (Many students make an unwarranted assumption about this problem. Did you?)

Work Problems 1–7.

Answers or hints to selected problems appearing in a section of this chapter are given at the end of that section. The focus of this chapter is not on the answers; rather, it is on deriving the solutions that yield the answers. Typically, an answer is given for a problem if the answer does not interfere with deriving a solution to the problem. *Only look at the answer to a problem if you want to check an answer you have derived.*

You May Think That Not Enough Information Is Given

Some problems you are asked to solve seem to be missing information in the statement of the problem. If you struggle with solving the problem you might think you should have been given more information in the problem statement. That may be true, but not necessarily so. The problem may require you to use the information you are given in a way that allows you to derive additional information that leads you to a solution. Here are examples of two *classic* problems in which adequate information is given to solve them, but that doesn't seem to be the case when an attempt is made to solve them:

Problem 8. A woman was 3/8 of the way across a bridge when she heard the Wabash Cannonball Express approaching the bridge at 60 miles per hour. She quickly calculated that she could just save herself by running to either end of the bridge at top speed. How fast could she run?

Problem 9. A commuter ordinarily reaches the railroad station nearest his home at 5 p.m. His wife meets him there and drives him home. One day he unexpectedly arrived at the station at 4 p.m. and instead of waiting, he started for home on foot. After a certain length of time he met his wife on her way to the station and they traveled the rest of the way home in the car. They arrived home 10 minutes earlier than usual. At what time did his wife pick him up?

Can you solve these two problems (no advanced mathematics needed)? Don't lose hope if you cannot solve them *at this time*. They are referred to again in section 2 where more problem-solving skills are presented.

I had numerous students over my many years of teaching who were not able to work a problem because they misinterpreted the problem. Suggestions are now given that will help you understand problems.

Take the Time You Need to Understand the Problem

Many of us undertake a job without adequately preparing for it. It is no different when it comes to searching for a solution to a problem. You must understand a problem before you can solve it. For your sake, please don't hurry this stage of solving problems. The urgency students feel to find a solution to a problem is no more evident then observing how quickly they begin writing after receiving their examinations. They see other students writing and feel they better get writing. They would have made more efficient use of their time if they had spent more time understanding the problem. Furthermore, understanding a problem helps illuminate a solution to it. Gilbert Chesterton said, "It isn't that they can't see the solution, it's that they can't see the problem," and Stanley Arnold said, "Every problem contains the seeds of its own solution."

Sometimes work on a problem has to begin before a complete understanding of the problem is possible. You might think you understand it, but as you begin working on a solution you discover that you don't understand the problem; hence, you have to spend more time reflecting on the statement of the problem.

Strategies to Help You Understand a Problem

In your efforts to understand a problem you can do the following:

Ask yourself questions about the problem. A secret to understanding a problem is to ask yourself questions about the problem, and get answers to them. Your questions may include these: "What information is given that I can use to work the problem?" "What are the constraints that I must abide by?" "What am I being asked to find?" "What are the implied assumptions?" "What words, concepts, or principles mentioned in the problem do I need to understand better?"

Outline the statement of the problem. List each thing that you are given, and each thing that you are to show, find, or determine. Use short sentences to do this—one item to a sentence. This breaks the problem into parts, which helps you see all of the parts and to focus on each one. This is especially helpful when problem statements are complex.

Problem 10. Two vertical poles of heights 10 meters and 40 meters are 50 meters apart on level ground. A wire goes from the top of each pole to the foot of the other. Find the height in meters of the point of intersection of these wires.

Example of an outline of the statement of the problem.

Given:

Vertical pole A on level ground of height 10 meters
Vertical pole B on level ground of height 40 meters
Poles A and B are 50 meters apart
A wire goes from the top of pole A to the foot of pole B
A wire goes from the top of pole B to the foot of pole A
The two wires intersect in a point

Find:

The height in meters of the point of intersection of the wires.

State the problem in your own words. Give an oral or written description of the problem in your own words. You want the problem to be part of you and until you can state it using your own words, it won't be. If you struggle doing this, then you know you need to understand the problem better. You might want a classmate to react to your restatements of problems, at least until you get better at giving an account of problems in your own words. Here is an example of a restatement of problem 10:

On level ground are two vertical poles of heights 10 and 40 meters. The poles are 50 meters apart. Wires that go from the top of one pole to the bottom of the other intersect in a point. How far is this point off the ground?

Exercise. Read the original statements of problems 9 and 10, and give *your* restatements of them.

Write a brief description of each variable that appears in the problem statement. For example, write y is the number of 4-cent stamps, or V is the volume of the sphere. I was always surprised when I asked a student what one of his or her variables represented and the student could not tell me. Many students who understand what their variables represent, do not describe them when they are writing solutions to problems that are turned in and graded. Is the grader of the solution supposed to supply this information? No!

Draw diagrams and label them to help you picture the problem. This technique is very beneficial to understanding a problem, because it allows you to display the key ideas of the problem in a small space, thus making it easier to see all of them and hold them in your mind. It may surprise you to learn how reluctant many undergraduate college students are to picture problems. At the outset of each problem-solving course I taught, I formed small in-class groups of students and gave them a non-routine problem to work. I recall times when the problem they were to solve begged for a picture to be drawn. After ten minutes and no solution was found, I asked each group if they had made a drawing of the problem. Not one student had made a drawing. (This was a teachable moment.) I frequently saw this same behavior early on in calculus-track courses when I gave problems to students to solve in class. How can you expect to see a solution if you do not literally see the problem? Seeing the problem is a facet of understanding the problem.

Why are students hesitant to draw a diagram of a problem? Probably because constructing a diagram was not emphasized enough in previous mathematics classes. This should be instinctive behavior by the time students reach college. Not only were my students reluctant to make a diagram, but they often didn't write anything down. Continuing to stare at the statement of a problem is not very helpful to understanding and solving the problem. If you are able to understand a problem without using a picture or diagram, constructing one will help you find a solution.

In constructing a diagram, label the various quantities found in the problem, either with letters, or a word or two. Your letters will communicate better if you use suggestive ones. For example: use h for the height of the building, d for the distance of the boat from the dock, r

for the radius of the circle, and A for the area of the parking lot. There are times you cannot use a suggestive letter (e.g., the letter may already be used to represent a different entity). Examples of words or phrases to use on a diagram are: "shore," "base," "road," "wall," "starting place," "path of the car," "airport runway," and "Tower A." An added benefit of a labeled diagram is that it aids in picturing *connections* between the quantities identified, which aids in understanding and solving the problem.

Exercise. Draw a diagram and label it to help you picture this problem: Consider $y = f(x) = |2x|$. Find the area of the polygon with vertices $(-1, f(-1))$, $(-1, 0)$, $(-3, 0)$, and $(-3, f(-3))$.

Seek help to understand a problem. A problem might be ambiguously worded (e.g., more than one interpretation is plausible), or you just don't understand it—for whatever reason. Seek help from others if you have spent considerable time trying to understand it without success. If there is no opportunity to do this before an assignment is due, and more than one interpretation is possible, state the interpretation you have decided to use and note the others.

Know the meanings of non-technical words. An aspect of knowing what you are being asked to do in a problem requires an understanding of non-technical words such as *verify, illustrate, derive, demonstrate, deduce, formulate, and transform.* If you are confused about the meaning of these types of words, consult your instructor or classmates.

Have a good knowledge of prerequisite mathematics. Instructors are often criticized by students for the confusing way in which they have written examination exercises. These are unfair criticisms when the students making them do not have the prerequisite knowledge to understand them. The topic of having prerequisite knowledge comes up often in this book. As I have stated before, you will succeed in your mathematics course if you have the prerequisite knowledge, appropriate attitudes, and a viable method of studying. This cannot be repeated often enough.

ANSWERS OR HINTS TO SELECTED EXERCISES
IN SECTION 1

(1) 7 **(2)** 5 **(3)** a quarter and a penny **(4)** 0 **(5)** 23¢ (any amount of postage can be made after 23¢) **(6)** 121 (the unwarranted assumption is that the chords have a common point) **(7)** the teacher is the daughter's father **(8)** 15 mph **(9)** how much time is saved driving in one direction? **(10)** 8 (use similar triangles to obtain a solution)

2. Devise a Plan

Once you have satisfied yourself that you understand the problem, you need to devise a plan to attack the problem. The statement of the problem holds clues for what may work. These clues, along with your mathematics knowledge and your problem-solving experience, will be used to devise your plan of attack. The more clues you can find, the more prerequisite mathematics knowledge you possess, and the more experienced you are as a problem solver, the easier it will be for you to discern a viable path to follow.

Before we look at specific tools to use in devising a plan, there is a need to discuss a topic called *heuristics*, or problem-solving skills. A heuristic is a general procedure used to arrive at a solution to a problem. For example, "Look for a pattern" is a heuristic, as is "Make a drawing." Many heuristics, such as these two examples, are content-free; that is, they can be used to solve problems in many disciplines. Heuristics that are not content-free are restricted to specific disciplines, to subject areas within a specific discipline, or to specific professions. If a specific heuristic is part of your plan to solve a problem, you cannot assume that you will be able to solve the problem using it. If it doesn't help you solve the problem, then you have to identify another heuristic (i.e., problem-solving skill) that seems feasible to use.

Heuristics are powerful problem-solving tools, and to be a good problem solver you need to have an ample repertoire of them. The primary goal of this section is to focus on key heuristics or problem-solving skills that will help you in many college mathematics courses. These will also be of value to you in many areas of your life. Only so much can be done with them in this book since this is not a book on problem solving. Many of the problem-solving skills frequently used in calculus-track courses can be picked up incidentally, or you already possess them; however, to learn others you will need formal exposure to them. If you want more information on problem-solving skills than what appears in this book, read books and articles on problem solving.

Go online and you will be directed to articles and books on problem solving.

You often need to use a combination of problem-solving skills to solve a problem. The more skills you know, the easier it is to find a path that leads to the solution of the problem. There is more to being an excellent problem solver than having a large collection of problem-solving skills. Other important attributes include an excellent knowledge of mathematics, proper attitudes, perseverance, willingness to question, confidence, patience, tolerance for frustration, seeking help when needed, and problem-solving experience. You can develop all these attributes.

Problem-Solving Skills or Heuristics

Devising a plan of attack to solve a problem requires (1) understanding the problem, followed by (2) reviewing your repertoire of problem-solving skills or heuristics to determine those you want to apply to the problem. Twelve important problem-solving skills are now presented, with examples of problems that are solved using them.

1. Analysis

You will always use the skill of analysis to solve problems, whether or not it is combined with other problem-solving skills. It is used to identify the essential parts of a problem that are necessary to use in order to solve the problem, and requires the use of logical reasoning to see how these parts can be put together to find a solution. You have to make comments to yourself about what you need at any point, and ask yourself, and then answer, a series of meaningful questions that lead you to a solution. Analysis is a necessary skill for all problem solvers and is the key skill used to solve mathematics problems, college level or otherwise.

Example (simple problem using the skill of analysis). A square of side x and circle of radius 4 have the same area. Find the ratio of the perimeter of the square to the circumference of the circle.

Analysis. We need to find the areas of the square and circle. Since they are equal, we set them equal and solve for x, whose value we need to determine the perimeter of the square. We then find the perimeter of the square and circumference of the circle, and determine the desired ratio. (Work this problem and see if you get the answer $\dfrac{2}{\sqrt{\pi}}$.)

Following are two more examples of problems that primarily use the skill of analysis to arrive at a solution. (The solutions will not be meaningful to you if you do not have the prerequisite content knowledge.) These problems are typical of the *non-routine* problems in your college mathematics courses that make use of the concepts and principles of the course. Most likely the concepts and principles that appear in a section of the textbook you are using—which precede a problem set for the section—are used to solve many of the problems in that problem set. Solving the majority of these problems does not require overly complex reasoning; rather, it requires a follow-your-nose (or common sense) type of reasoning or analysis.

Problem 1 (from precalculus). Find the value of b such that the distance from the origin to the vertex of the parabola, with equation $f(x) = x^2 + bx + 1$, is as small as can be.

Devising a Plan (analyze the problem). The seeds of a solution to this problem are quite evident. Since we are being asked to minimize a distance, we need to express this distance as a function of the variable b. Since the two points mentioned in the problem are the origin and the vertex of the parabola, we need to find the coordinates of the vertex of the parabola. We then can use the distance formula to find the length of this segment. We can find the coordinates of the vertex of the parabola if we write its equation in standard form. Once we have the distance equation, we can analyze the equation to determine a value of b that will make this distance a minimum. (It is readily apparent from this analysis that analysis cannot take place in a vacuum; that is, you have to have knowledge to analyze. This sure argues for having a good supply of mathematics knowledge if you want to become a good problem solver.)

Solution to Problem 1. We begin by completing the square for the equation of the parabola:

$$f(x) = x^2 + bx + 1$$
$$= x^2 + bx + \frac{b^2}{4} - \frac{b^2}{4} + 1$$
$$= (x + \frac{b}{2})^2 + 1 - \frac{b^2}{4}$$
$$= (x + \frac{b}{2})^2 + \frac{4 - b^2}{4}.$$

Hence, the coordinates of the vertex are $(-\frac{b}{2}, \frac{4-b^2}{4})$. Let z denote the distance from $(0,0)$ to $(-\frac{b}{2}, \frac{4-b^2}{4})$. Using the distance formula, we get

$$z^2 = (-\frac{b}{2} - 0)^2 + (\frac{4-b^2}{4} - 0)^2$$
$$= \frac{b^4 - 4b^2 + 16}{16}.$$

The value of z that minimizes z^2 will also minimize z, so we dispense with getting an expression for z. The numerator can be rewritten as $(b^2 - 2)^2 + 12$. This was done so b occurs in an expression that is squared. (Do you know why this is needed?) Hence,

$$z^2 = \frac{(b^2 - 2)^2 + 12}{16}.$$

Clearly, the smallest value for z^2 occurs when $(b^2 - 2)^2$ is the smallest, and that is when $b^2 - 2 = 0$, or $b = \pm \sqrt{2}$.

Commentary. Logical reasoning or logical thinking was used to solve the problem. A sequence of steps was taken suggested by the statement of the problem. The problem just solved is a formidable problem for many students in a calculus-track course. You need to develop the type of logical thinking required to solve it. It is the predominate type of thinking required to solve problems in calculus-track courses. If you currently are weak in this skill, know that you can develop it if you practice it diligently and wisely. However, you will be stymied in your efforts to apply the skill of analysis to solve problems if you don't have substantial requisite knowledge, as this problem required. I repeat what I say several times in this book: Problem solving does not take place in a vacuum.

Problem 2 (from calculus). Show that for a quadratic function defined on $[p, q]$ the conclusion of The Mean Value Theorem is satisfied by only one point belonging to the interval (p, q).

The Mean Value Theorem (MVT)

Let f be a function continuous on the closed interval $[a, b]$ and differentiable on the open interval (a, b). Then there is a number c in (a, b) such that $f'(c) = \frac{f(b) - f(a)}{b - a}$.

Devising a Plan (i.e., analyzing the problem). The seeds of a solution to this problem are in the statement of the problem. Since we are to show something about an arbitrary quadratic function defined on $[p,q]$ we need to represent such a function. Secondly, since we need to verify something about the conclusion of The Mean Value Theorem (MVT) for this function, we need to (1) apply this theorem to this function (first checking to see that it satisfies the hypothesis of the theorem), (2) look at what we get for the conclusion, and then (3) determine how this can be used to give us what we are to show. Now it's just a matter of using the mathematics we know to see if this plan of attack answers the question (i.e., the conclusion of The Mean Value Theorem is satisfied by only one point on the interval).

Solution to Problem 2. Let the general quadratic function be $f(x) = ax^2 + bx + d$. This function satisfies the hypothesis of the MVT since all quadratic functions are continuous on a closed interval and differential on an open interval. To apply the MVT, we need to have the derivative of f, which is $f'(x) = 2ax + b$. Applying the MVT, we get the existence of a c belonging to (p,q) such that $f'(c) = \dfrac{f(q) - f(p)}{q - p}$. Since $f'(c) = 2ac + b$, we have

$$2ac + b = \frac{f(q) - f(p)}{q - p}$$

$$= \frac{aq^2 + bq + d - (ap^2 + bp + d)}{q - p}$$

$$= \frac{a(q^2 - p^2) + b(q - p)}{q - p}$$

$$= a(q + p) + b.$$

We need to show that c is a unique number. Hopefully, we will see this when we simplify the above and solve for c.

$$2ac + b = a(q + p) + b$$

$$c = \frac{q + p}{2}.$$

Yes, c is a unique value belonging to (p,q) since p and q are constants. This is what we were asked to show.

Commentary. Same commentary as that for problem 1. Reread that commentary.

2. Sketch a diagram

The problem-solving skill of "sketch a diagram" was discussed in a previous subsection of section 1 on understanding the problem. You may want to reread that discussion. Sketching a diagram of the problem will help you devise a plan to solve the problem. You may not have needed to sketch a diagram to identify the key parts of the two problems that were just stated and solved, but each of those problems begged for a diagram. Diagramming a problem can help you see a path to the solution because you can see, in a small space, the key parts and how they are related. You cannot deny that a picture is worth a thousand words. I always sketch a diagram of a problem if it lends itself to one. It is the *first* problem-solving skill that I try to employ. Sketch a diagram for each of these problems, devise a plan for solving them, and then carry out your plan:

Problem 3. A steel band is fitted tightly around the equator. The band is removed and cut, and an additional 5 feet is added. The band now fits more loosely than it did before. Assuming that it is equally distanced from the ground all around the earth, how high off the earth is this larger band?

Problem 4. A goat is tied with a 4-meter rope to an outside corner of a 3-meter high wall that has the shape of an equilateral triangle of side 3.5 meters. What is the area of the lawn that the goat can graze?

Problem 3 does not use the language of geometry in its statement, but geometric objects are used in a diagram of the problem (e.g., circles for the equator and steel band). Problem 4 uses geometric language in its statement (e.g., equilateral triangle), but in constructing a diagram you have to come up with geometric representations of a rope (a line segment) and grazing area (a portion of a plane). A diagram of a problem can also be constructed if the problem has disconnected objects that are related to each other in some way (e.g., classmates with the same birth month, cities connected by roads, handshakes among people). In those situations the objects can be represented by points and the relationships by line segments connecting them. Use diagrams in your efforts to solve these three problems:

Problem 5. There are 6 people in a room and everyone shakes hands exactly once with everyone else. How many handshakes have taken place? (Represent each person by a point, arrange the points in a circular pattern, and let a line segment between two points represent a handshake between the corresponding individuals.)

Problem 6. A basketball falls from a height of 18 feet. Each time it hits the ground it bounces up to 2/3 of the height it fell. How far will the ball have traveled when it hits the ground for the third time?

Problem 7. How long will it take a train 1/4 mile long traveling at 30 mph to go through a tunnel 1/8 mile long? (You may want to draw a series of diagrams, each one representing a key position of the train.)

Sketch a diagram for problem 1, showing an arbitrary parabola whose axis is parallel to the x-axis, and the line segment whose length you are minimizing.

3. Think of a Related Problem you have Solved

Seeing similarities in a problem you are to solve with one you have solved, should suggest to you that using the same or modified techniques to solve it is a good place to start. Perhaps you recognize the problem as related to one solved in the textbook, in class, earlier in the course, or in other courses you have taken. Regardless of where and when you were exposed to a related problem, either recall what was done to solve the earlier problem, or do some investigative work to see how the problem was solved. The more problems you have solved, the more likely it is that you will recall problems that are related to those you are currently working. We now state two problems that are similar in wording, which suggests they are related problems:

Problem 8. Prove this statement: If the first derivative of a function is greater than zero on an open interval, then the function is increasing on that interval.

Problem 9. Prove this statement: If the first derivative of a function is less than zero on an open interval, then the function is decreasing on that interval.

The main difference in the proofs of these theorems is the interchanging of "less than" and "greater than" in the statements in the proofs that contain these phrases. One of these theorems is most likely proven in a calculus textbook. If you are in a calculus course, use the proof of the theorem that is in your textbook, to construct an analogous proof of the other.

If you are asked in a precalculus or calculus course to prove a theorem involving a maximum value, try to recall if you have seen the proof of an analogous theorem involving a minimum value. It is common practice for textbook authors to prove a theorem and then have an analogous theorem as an exercise. They expect you to look back, if need be, at the theorem they proved and adapt their steps in the proof to those used in the theorem you are proving. If you need to look back, be sure you understand what you are reading. A mechanical adaptation of what you read to what you have to prove is a meaningless activity.

Suppose you had solved this problem:

Problem 10. How many different squares are there on a checkerboard? (Note. A checkerboard is 8 by 8, displaying 64 one-by-one squares; however, there are squares of different sizes on a checkerboard.)

Would solving problem 10 help you solve problem 11?

Problem 11. How many different rectangles are there on a checkerboard? (Note. A square is a rectangle, but there are rectangles that are not squares.)

It is reasonable to think that a plan similar to that for working problem 10 can be used to work problem 11. However, you would soon see (if you haven't surmised it already), that the second problem is more complicated, requiring an extension of the plan you used to solve the first problem. Work problems 10 and 11 now if that is your desire, but these problems will appear again later in the section after more problem-solving skills are discussed.

Problem 5 is related to this problem:

Problem 12. How many diagonals are there in a convex polygon of 6 sides?

Problems 5 and 12 have similar diagrams. Their answers will differ because you do not count the number of sides of the polygon in problem 12, but you have to do this for problem 5.

4. Use symmetry

There is symmetry in the proofs of problems 8 and 9 concerning increasing and decreasing functions. The steps in the proof of one of them are "reflections," with a small adaptation of the steps in the proof of the other.

Another precalculus example of symmetry involves a function f and its inverse f^{-1}. We know that their graphs are symmetric to the line $y = x$. Hence, with appropriate adaptations, you can use what you know about the graph of f to say things about the graph of f^{-1}. For example, if you know (2,3) is on the graph of f, then (3,2) is on the graph of f^{-1} (since (2,3) and (3,2) are symmetric to the line $y = x$).

A form of symmetry in calculus involves the definite integrals $\int_a^b f(x)dx$ and $\int_c^d g(y)dy$, where the x-axis is horizontal and the y-axis is vertical. Assume that if you solve $y = f(x)$ for x in terms of y, you get the function $x = g(y)$. For the function f, the independent variable is x, and for the function g, the independent variable is y. Working a problem in which you have to set up the definite integral $\int_{c}^{d} g(y)dy$ to solve the problem (e.g., to find the area of a region) becomes routine if you *mirror* what you do when setting up the definite integral $\int_{a}^{b} f(x)dx$ to solve an area problem. You are exposed to many more problems in calculus-track courses where the independent variable axis is horizontal, versus problems where the independent variable axis is vertical. The difference is in the orientation of their geometric views.

When devising a plan to solve a problem, see if there are symmetries among the entities that relate to the problem. There are many examples of symmetrical relations in college mathematics. Making use of them is a powerful problem-solving skill.

5. Study the proof of a theorem that relates to the problem

If you have some sense that a specific theorem can be used to solve a problem, study the proof of the theorem. Aspects of the proof of the theorem may hold the key to solving the problem. Justification of the theorem may help you see how the theorem can be used to solve the problem. This is shown in examples 1 and 2.

Example 1. Suppose you have two functions f and g that are continuous on the interval $[a,b]$, where $f(x) \geq g(x)$ for $x \in [a,b]$, and you are asked to develop a formula that gives the area of the region bounded above by the graph of f, below by the graph of g, and on the sides by the lines $x = a$ and $x = b$.

You want to look at the proof of the theorem that preceded this one, namely, the proof that yielded the formula for finding the area of a region bounded above by the graph of a continuous function f, below by the x-axis, and to the left and right by the lines $x = a$ and $x = b$ (call this Theorem A). The only change here is that the region for which you are now asked to develop the area formula is bounded *below* by the graph of g, whereas the region described in Theorem A is bounded below by the x-axis. If you understand the proof of Theorem A, you know that you are summing up an infinite number of areas of rectangles with width $\Delta x_i = x_{i+1} - x_i$ and height $f(x_i)$. The adaptation you need to make to solve the problem in question is to realize that the region whose area you seek is the sum of an infinite number of rectangles with the same width Δx_i, but with different heights, namely, $f(x_i) - g(x_i)$. (This problem begs for a diagram of the *ith* rectangle.) The formula you get is $\int_a^b [f(x) - g(x)]dx$, which is an adaptation of the formula $\int_a^b f(x)dx$ for the area referred to in Theorem A.

Example 2. Knowing the basics of the proof of a theorem can also help you choose the correct formula to use to solve a problem. For example, there is a calculus formula for finding the volume of a solid of revolution generated by revolving around the *x*-axis, the region between the graph of the continuous function f and the *x*-axis, from $x = a$ to $x = b$. The formula is $\int_a^b \pi[f(x)^2]dx$. It is amazing to see the number of students who know there is a formula for calculating the volume of a solid of revolution but use the wrong formula on an examination. They fail to realize that the integrand yields a clue as to the formula to use. They rely on rote memorization to give them the formula, which often fails them. A basic understanding of the proof of the theorem that yielded this formula can help avoid this error. Students need to know that a cross section of the solid at $x = x_i$ gives them a circular region with radius $f(x_i)$. This circular region sweeps out a cylinder with height $\Delta x_i = x_{i+1} - x_i$. (This problem also begs for a diagram of the rectangle and the cylinder this rectangle creates when it is revolved). Hence, the volume of this cylinder is the area of the base times the height, or $\pi[f(x_i)]^2\Delta x_i$. Dispensing with all formality from this point on, we add up an infinite number of these volumes to eventually get the volume we seek. Notice that $\pi[f(x)]^2dx$ resembles the form of the *volume* of the *ith* cross-section cylinder, which is $\pi[f(x_i)]^2\Delta x_i$. Because of this relationship, we should know that the formula $\int_a^b \pi[f(x)^2]dx$ is a formula for finding the volume of a solid of revolution with the conditions stated in example 2.

To conclude, in reading proofs of theorems, (1) concepts of mathematics are being related (i.e., stories in mathematics are being told), (2) mathematical understanding is enhanced, and (3) problem-solving skills are advanced.

6. *Work backward*

The problem-solving skill of "work backward" can be a method of attack when you are given initial conditions and a final result, and are to find a sequence of steps that lead to this final result, or a piece of information that helps lead you to the final result. In working backward, you start with the final result and the first question you ask yourself is this: "What is a step just before this one?" Then you ask yourself, "What is a step just before that one?" You continue until you reach your initial conditions. At some point in this sequence you may get stymied and have to turn to the initial conditions and work forward, coming up with a next step, then the next step, and so on. You may be switching back and forth between working backwards and forwards. Hopefully, your actions of going backwards and forwards will result in arriving at the same step somewhere between your initial conditions and the final result.

This heuristic of "work backward" should not be new to you. This is what we use, at times to prove theorems. For example, recall in high school having to prove two triangles congruent, given certain relationships among the sides and angles of the two triangles. Working backward, you would say to yourself, "I know that the triangles are congruent if I can show that (1) three sides of one triangle are congruent to three sides of the other triangle (SSS), (2) two sides and the included angle of one triangle are congruent to two sides and the included angle of the other triangle (SAS), or (3) two angles and the included side of one triangle are congruent to two angles and the included side of the other triangle (ASA). You then proceeded to work forward, using your initial conditions, to see if you could derive one of these three conditions.

Apply the *work backward* heuristic to solve these five problems:

Problem 13. Assume that in Lake Tonka algae grows on the surface of the lake, the algae approximately triples in area every week, and it takes 48 weeks for the lake to be completely covered with algae. How many weeks does it take for Lake Tonka to be one-third covered?

Problem 14. Three individuals are playing three rounds of a card game. Each round consists of 15 hands played. There are two winners and one loser for each round. The loser of a round is the one who has lost the most number of hands in the round. If two individuals are tied for the least number of hands won after a round, they play a hand and the loser of that hand is the loser of the round. The loser of a round has to double the number of chips that each of the other two players has by giving up some of his or her own chips. Assume that Player A loses the first round, Player B the second round, and Player C the third round, and that each player has 80 chips at the end of the three rounds. How many chips did each player have at the beginning of the game?

Problem 15. Laura entered a store one day and had to pay two dollars to enter, spent one-half of her money in the store, and had to pay one dollar to leave. She did this three times, and was completely broke after leaving the store for the third time. How much money did she start with?

Problem 16. In a two-person game, an arbitrary whole number > 1 is selected (e.g., 30), and the players take turns subtracting any single-digit number greater than zero and less than or equal to the starting number (e.g., 30 − 8 = 22, 22 − 6 = 16, etc.). The player who is forced to obtain zero loses the game. Describe a strategy for winning the game. Will you always win?

Problem 17. I am thinking of a number. If you add 8 to the number, square the result, divide this result by 9, subtract 15 from it, multiply what you get by 5, then the answer is −30. What is the number?

You probably don't realize that you are using the "work backward" skill when you solve the equation $7 + x^2 = 22$. To construct this equation you start with x, square it, add 7, and get 22. To find x, you begin with the end result, namely, 22. You reverse your steps (i.e., go backward), by subtracting 7 from 22, giving you 15, and then find plus or minus the square root of 15, yielding the answer $\pm \sqrt{15}$. Solve the equation $\dfrac{x + (-6)}{5} + 7 = 16$ by working backward. I will start you off (you also talk as you move through the solution): "The last step before getting 16 was to add 7, hence I have to subtract 7 yielding 9; the step before that I divided by 5 to get 9, hence I have to..."

Set up an equation for problem 17 and solve it as you normally would.

You will have problems to solve where it will be wise for you to ask yourself this question as frequently as need be: "What do I need to show to come up with this next step?" When you do that you are working backward.

7. Solve a simpler problem

If there is a problem you can't solve, then look for an easier problem within this problem that you can solve. Some of the ways to come up with easier or simpler problems related to the more difficult problem you have to solve, include these: (1) use smaller, fewer, or less complex numbers, (2) work a special case of the problem, (3) change the dimensions of the problem, or (4) use less complex symbolic form. What you learn from solving the simpler problem can be used to solve the more complex one, perhaps with modification.

Solve this problem using the numbers given:

Problem 18. Joey has rectangular ceramic tiles with dimensions 30*mm* × 56*mm*. He wants to form a solid square by laying them out with the long side horizontal. What are the dimensions of the smallest square Joey can make?

Are you not sure where to start? *Hint:* Change the size of the tiles to 3*mm* by 4*mm*, and work on forming a square by sketching copies of them on paper, labeling the size of two adjacent sides for each tile (to help you see when you have formed a square). *(STOP! Do not read on until you have done this!)* The use of these smaller numbers and your resulting drawing should help you come up with a scheme for solving the problem for the larger numbers. Perhaps you need to work the problem again using other small numbers such as 4 and 5. Eventually you will probably realize what you need to do with the numbers 30 and 56 to solve the problem. You may know what to do with these numbers immediately after reading the problem, but this is not the case for many college students.

If your problem is to find the number of rectangles on an 8 by 8 checkerboard, you might want to first solve the simpler problem of finding the number of rectangles on a 3 by 3 checkerboard. Using smaller numbers often makes solutions accessible.

Problem 19. Find the smallest number, which when divided by each of the numbers 2, 3, 4, 5, 6, 7, 8, 9, and 10, will give in each case a remainder that is one less than the divisor.

Solve problem 19 using 2 and 3, instead of all nine numbers given. Now do it for 2, 3, and 4. Are you beginning to see a scheme for doing it for all 9 numbers? If need be, work it for 2, 3, 4, and 5. Do you now see a way to use the nine numbers to get the answer?

There are situations where too much symbolic notation can get in the way of seeing what to do to solve a problem. The goal is to *look for the simple in the complex.* For example, suppose you are asked to solve the absolute value inequality $\left| \dfrac{4+5x}{4} \right| < 5$. If you know that $|y| < a$

is equivalent to $-a < y < a$, then you can use this simpler form to solve the more complex one viewing y as $\dfrac{4+5x}{4}$. Hence, $\left|\dfrac{4+5x}{4}\right| < 5$ is equivalent to $|y| < 5$, which is equivalent to $-5 < y < 5$, which is equivalent to $-5 < \dfrac{4+5x}{4} < 5$ (once you get to this last inequality, you have to use other knowledge to complete the work).

If you are a Calculus III student, it will help you to find the center of mass of a solid three-dimensional object by knowing how to find the center of mass of a flat plate (which is a two-dimensional object), and the center of mass of a rod (which is a one-dimensional object). Many calculus theorems involving two-dimensional coordinate systems can be extended to three-dimensional coordinates systems. Hence, when working problems in three-dimensions, your understanding of how to proceed can be assisted by looking at similar problems in two dimensions, which are simpler problems. This is one of the major reasons why students in a Calculus III course are helped greatly by their knowledge of simpler, but related content from a Calculus I or Calculus II course. If you are in a calculus course, you will find it educational to look in your textbook at the developments of the center of mass of one-dimensional, two-dimensional, and three-dimensional objects. (How important is having a good background in Calculus I and Calculus II to students in Calculus III? Very!)

There are many ways for a problem to be viewed as simpler than the one you are working. It is not possible to characterize all these ways. What you have to do when working a problem is to ask: "Is there a simpler problem I can work?" It is up to you to determine one or more ways a problem can be simpler than the one you have to solve.

8. Establish subgoals

Following a path that yields a solution to a problem requires you to accomplish certain tasks along the way, called subgoals. Solving the problem may be a formidable task, but solving a subgoal might not be. This does not necessarily mean that solving a subgoal signifies that you will see a clear path to solving the problem. But solving a subgoal can lead to solving other subgoals, and eventually to a solution of the problem. When you work on subgoals you are dividing the problem into smaller problems that are simpler to work. Contrast this approach with having no subgoals and just staring at the statement of the problem. You have to get started somewhere, so look for a subgoal and proceed.

Examples of subgoals are given for problems 20, 21, and 22.

Problem 20. Prove that 2 and 5 are the only prime numbers that differ by 3.

Subgoal 1. *Look at the differences of various prime numbers.* You will discover that the differences are all even numbers except if you use the prime number 2 as one of the numbers. All the other prime numbers you used are odd.

Subgoal 2. *Prove that the difference of two odd numbers is an even number. Proof:* Any odd number is of the form $2n + 1$ for n a whole number; hence,

$$2p + 1 - (2q + 1) = 2p - 2q = 2(p - q).$$

Since $2(p - q)$ is divisible by 2, it is an even number.

Subgoal 3. You wonder if there are even primes other than 2 because the only way you can get the differences of two primes to be odd (and 3 is an odd number), is for one of the primes to be an even number. Your subgoal here is to *prove that 2 is the only even prime. Proof:* Any even number greater than 2 has at least 3 divisors, namely, 1, 2, and the number itself. Hence, it cannot be prime since a prime number, by definition, has exactly two divisors.

Putting it all together: 2 and 5 have to be the only primes that differ by 3 since all other prime numbers are odd numbers and the difference of any two odd numbers is an even number.

Problem 21. Determine a formula for finding the number of divisors of a non-prime whole number using its prime factorization. (Note, for example, that the prime factorization of 1,260 is $2^2 \times 3^2 \times 5 \times 7$ and that a divisor of a number is a factor of the number.)

A way to attack problem 21 is to work these three special case problems: (1) Find a formula for the number of divisors of numbers of the form $N = p^n$, p a prime. (2) Find the divisors of $N = 2^3 \times 3^2$ by picking out the divisors from this factorization. For example, one divisor is $2^2 \times 3$. Leave each of your divisors in its prime factorization form, look at them, and then determine how many divisors there are by working with the form $2^3 \times 3^2$. (3) Use what you learned in the first two cases to state a formula for the number of divisors of N, where $N = p_1{}^{a_1} p_2{}^{a_2} p_3{}^{a_3}$ ... $p_n{}^{a_n}$. Use your formula to find the number of divisors of 49,000.

Problem 22. Seven appointed observers at a track meet are situated along a straight portion of the track. Where should they meet along the track so that if they wished to discuss an issue, the total distance they travel is minimized?

A subgoal for problem 22 would be to assume that they are equally spaced along the track, which might not necessarily be the case. Perhaps as a subgoal you might want to start with placing a smaller number of officials along a track (say three officials), vary their positions, and then determine where they should meet.

It is almost impossible not to have subgoals in solving a problem. Most solutions to problems are comprised of various parts, which came about through working subgoals. A key aspect of solving problems is to identify subgoals. *The goal is to start doing something.*

9. Look for a pattern

Some problems are solved, or progress toward a solution is made, by recognizing a pattern. Your pattern may be arithmetical, geometrical, or algebraic. Success in recognizing patterns is dependent on experience working with them.

Problem 23. For the following sequences of numbers, describe a pattern for each in a manner similar to the descriptions that are given for selected sequences.

    a.   1, 4, 7, 10, 13, 16, 19, ... (*Pattern*: The difference of two consecutive terms is 3.)

    b.   1, 4, 9, 16, 25, 36, 49, ...

    c.   8, 7, 9, 8, 10, 9, 11, ...

    d.   1, 3, 6, 10, 15, 21, 28, ... (*Pattern*: The difference of two consecutive terms is increasing by 1 as you proceed in the sequence.)

    e,   27, 9, 3, 1, 1/3, 1/9, 1/27, ...

    f.   3, 6, 12, 24, 48, 96, 192, ... (*Pattern:* The factor 3 in each term is multiplied by a power of 2, which is increasing by a factor of 2 as you proceed in the sequence; for example, $6 = 3 \times 2^1$, $12 = 3 \times 2^2$, $24 = 3 \times 2^3$.)

    g.   5/4, 5/2, 15/4, 5, 25/4, 30/4, 35/4, ...

    h.   1, 1, 2, 3, 5, 8, 13, ...

Problem 24. Use a pattern to predict the 35*th* term and the *nth* term of each sequence in a manner similar to those for which this has been done.

a. 1, 4, 7, 10, 13, 16, 19, …, 35*th*, …, *nth* term, … (Since 3 is added to a term to get the next term, we need to keep track of the number of 3's that have been added to the first term to get the 35*th* term. We see that the second term is 4 = 1 + 3, the third term is 7 = 1 + 3 + 3, the fourth term is 10 = 1 + 3 + 3 + 3, and so on. The number of 3's added on is one less than the number of the term. Therefore, the 35*th* term is 1 + 34 × 3 = 103. The *nth* term is 1 + (n − 1) × 3. Realize that there are other patterns that may be seen. *Not everyone sees the same pattern.* Someone may have discovered that the *nth* term is 3*n* − 2, which equals 1 + (n − 1) × 3.

b. 3, 6, 12, 24, 48. 96, …, 35*th* term, …, *nth* term, …

c. 1, 3, 6, 10, 15, 21, 28, …, 35*th* term, …, *nth* term, … (first term is 1, second term is 3, third term is 6, fourth term is 10, and so on. With some trial and error, we see that each term is the number of the term times one more than the term number, divided by 2. For example, the fourth term is 4(4 + 1) ÷ 2 = 20 ÷ 2 = 10. Hence, the 35*th* term is 35(35 + 1) ÷ 2 = 1260 ÷ 2 = 630. The *nth* term is $\dfrac{n(n+1)}{2}$. Others may see that 3=1+2, 6=1+2+3, 10=1+2+3+4. And so on. Hence, the 35*th* term is 1+2+3+4+…+35 (which equals 630), and the *nth* term is 1+2+3+4+…+n, which is the sum of the first *n* whole numbers, and equals $\dfrac{n(n+1)}{2}$.

d. 3/2, 4/3, 5/4, 6/5, 7/6, …, 35*th* term, …, *nth* term, …

e. 0, 5/4, 5/2, 15/4, 5, 25/4, …, 35*th* term, …, *nth* term, …

Consider this problem:

Problem 25. Twenty teams are participating in a *single* elimination basketball tournament (i.e., one loss and you are out). How many games must be played to crown a champion?

There are a variety of ways to solve Problem 25, including looking for a pattern. Solve the problem for 2 teams, 3 teams, and 4 teams. (Some teams would receive byes if the number of teams were odd.) Analyze your data to come up with a pattern and then use it to determine how many games would be played if there were: (a) 48 teams. (b) *n* teams.

A *table* or *chart* can be used to organize your information. Use appropriate labels for your table headings. Tables are also helpful to use with other problem-solving skills such as "make an organized list" and "guess and check," which are discussed later. Use a table or chart, along with patterning, to work Problems 26 and 27.

Problem 26. There are 17 people in a room and each person shakes hands exactly once with everybody else. (a) How many handshakes have taken place? (b) How many handshakes would there be if there were *n* people in the room, *n* a natural number?

The table you use to solve problem 26 might look like this:

# of people	*# of handshakes*
2	1
.	.
.	.
.	.

Problem 27. In a double elimination tournament each player must lose two sets to be eliminated from the tournament. When you lose for the first time your next match is in the losers bracket. The final survivors of the winners and losers brackets play for the championship. (a) How many tennis sets have to be played in a *double* elimination tournament when 32 players begin the tournament? (b) How many have to be played if *n* players begin the tournament, $n \geq 2$?

Problem 28. What is the maximum number of regions into which (a) 85 chords divide a circle? (b) *n* chords divide a circle, *n* a natural number?

Problem 29. Use patterning to solve problems 10 and 11.

Work problems 30 and 31 by looking for a pattern.

Problem 30. Show how patterning can be used to conjecture the derivative of $f(x) = x^n$, for *n* a natural number greater than 0. (*Hint.* Use the definition of the derivative to find the derivatives of x, x^2, and x^3. Recall that, by definition, $f'(x) = \lim_{h \to 0} \dfrac{f(x+h) - f(x)}{h}$, and $x = x^1$.)

Problem 31. Sketch graphs of each pair of functions: (a) $f(x) = -3x^2$ and $g(x) = -3(x+1)$, (b) $f(x) = \sin 2x$ and $g(x) = \sin 2[(x + \dfrac{\pi}{3})]$, and (c) $f(x) = |3x|$ and $g(x) = |3(x+4)|$. Look for a pattern in these graphs in your efforts to state a rule that tells how the graph of $f(x)$ is related to the graph of $g(x)$, where $g(x) = f(x + a)$ for any function f and non-zero whole number a.

Problem 32. Find the total number of toothpicks needed to construct a square of dimensions (a) 52 by 52 (where a 1×1 square contains 4 toothpicks), and a square of dimensions (b) n by n, where n is a non-zero whole number. (Note, for example, that a 2×2 square is comprised of four 1×1 squares.)

Problem 33. How many angles of measure less than 180 degrees are formed by n rays with a common endpoint, where $n \geq 2$?

Problem 34. You need dot paper or a geoboard for this problem. The problem is to discover a formula for finding the areas of polygons that can be made on a geoboard (or drawn on dot paper), where the vertices of the polygon have to be pegs on the geoboard (or dots on the dot paper). The independent variables in the formula are B, the number of pegs or dots on the *boundary* of the polygon, and I, the number of pegs or dots *interior* to the polygon. To discover the formula, work these three subgoals (and more than these if needed):

Subgoal 1. Make five figures on your geoboard or dot paper that have 0 interior points and different numbers of boundary points. Use a table to record, for each figure, the number of boundary points (B), the number of interior points (I)—which is 0 for this subgoal, and the area (A) of the figure. Conjecture a formula using B, I, and A that works for this subgoal. (To find the areas of your polygons, you will have to count squares and parts of squares, sometimes directly and sometimes indirectly. It requires a little creativity on your part to find the areas of some of these figures.)

Subgoal 2. Same directions as subgoal 1, but your figures have exactly 1 interior point.

Subgoal 3. Same directions as subgoals 1 and 2, but your figures have exactly 2 interior points.

To finish the problem, use the formulas you have conjectured to come up with a formula that works for all of your polygons. Does your formula work for figures with 5 interior points? 7 interior points?

10. *Make an organized or systematic list*

A list is organized if there is a scheme around which the data is organized, accounting for all of the possibilities. Work problems 35–37, devising a scheme for each one that ensures that you are accounting for all possibilities.

Problem 35. Karen wanted to rid her pond of algae by treating it with copper sulfide. She can purchase it in 5-lb. and 7-lb. bottles, costing $3.49 and $4.20, respectively. She needs to buy 30 pounds. What should she buy to obtain at least that amount at the lowest cost? (*Hint.* Use these headings in your list: # of 5-lb. bags, # of 7-lb. bags, total # of pounds, and total cost. Use a table or chart.)

Problem 36. In how many ways can three people divide 25 pieces of candy so that each person gets at least one piece? (*Hint.* Work it for smaller numbers (start with 3 pieces of candy), and look for a pattern. Let these be the first four rows of your list:

Person 1	Person 2	Pearson 3
1	1	1
2	1	1
1	2	1
1	1	2

Problem 37. A small pasture is to be fenced off with 96 meters of new fencing along an existing fence, using the existing fence as one side of a rectangular enclosure. What are the whole number dimensions of the pasture of largest area that the new fencing will enclose? (If you have the background, also work this problem using (a) precalculus content, and (b) calculus content.)

11. Guess and check

With this skill you *guess* the answer, and *check* to see if it is the correct answer. If it is not the correct answer, continue to revise your guess until you get it. Analyze your incorrect guesses so your subsequent guesses are more educated. The process of checking your answer helps you better understand the problem and perhaps see another way to solve it. *This is an incredibly powerful problem-solving skill.* When all else fails, this is the skill that often yields a solution, especially when you are assisted by technology in the guessing and checking. Using this strategy does not always yield a correct solution, but it can help you progress toward one.

Use guess and check to solve problems 38–40.

Problem 38. There are two two-digit numbers that satisfy the following conditions: (1) the digits in one number are the same as the digits in the other number; (2) the sum of the digits in each number is 11;

and (3) the difference between the two numbers is 45. What are the two numbers?

Problem 39. Solve $x^2 - \sqrt{x} = 7$, to the nearest tenth.

Problem 40. A woman was 3/8 of the way across a bridge when she heard the Wabash Cannonball Express approaching the bridge at 60 miles per hour. She quickly calculated that she could just save herself by running to either end of the bridge at top speed. How fast could she run?

12. Make a model, act it out, or use simulation

Make a model to solve this problem.

Problem 41. A log sits on the top of three congruent cylinders on their sides, and each cylinder's base has a diameter of one meter. The cylinders revolve one revolution. How far does the log travel?

Perhaps you solved problem 41 without the use of a model, but if your answer is π think again. I suspect you will have to use a model to get the correct answer.

Act out this classic problem by drawing it on a sheet of paper:

Problem 42. A farmer has to get a fox, a goose, and a bag of corn across a river in a boat that is large enough only for him and one of these three items. If he leaves the fox alone with the goose, the fox will eat the goose. If he leaves the goose alone with the corn, the goose will eat the corn. How does he get all the items across the river?

Students typically use several different skills to solve this problem, including (1) act it out, (2) guess and check (3) make a drawing, and (4) analysis.

It appears that the skill of "think of a related problem," such as problem 42, along with other problem-solving skills, would help solve this classic problem:

Problem 43. Two fishermen, each weighing 200 pounds, and their two sons, each weighing 100 pounds, cross a river in a small boat that can only carry 200 pounds. How do they all manage to get across?

The use of simulation, with or without the use of technology, can help find the answer to a problem. Let's look at how non-computer simulation can be used for this geometrical probability problem:

Problem 44. At a county fair, a game is played by tossing a coin of diameter 4 centimeters onto a large table ruled into congruent squares. If the coin lands entirely within a square, the player wins a prize. What is the probability that a random toss of the coin will result in a win if the squares have sides of length 5 centimeters?

There is an analytical solution to problem 44, but suppose it is not accessible to you and you want to get a good approximation of the answer. You can take a large piece of paper and rule it into congruent squares of side 5 centimeters. Randomly toss, one at a time, many pennies onto your paper (do it at least 30 times). Then divide the number of times your pennies fall entirely within a square, by the number of tosses you made. This experimental probability is your approximation of the answer. The more pennies you toss, the more faith you can place in your approximation. Compare the answer you obtained experimentally with that obtained analytically, which is given at the end of the section.

Consider this classic problem:

Problem 45. A husband and wife on a shopping expedition agree to meet at a specified street corner between 4:00 p.m. and 5:00 p.m. The one who arrives first agrees to wait 15 minutes for the other, after which that person will leave to continue shopping. What is the probability that the couple will meet, assuming that their arrival times are random within the hour? (Compare your answer with that obtained analytically, which is given at the end of the section).

Problem 45 is another example of a geometrical probability problem. It has an analytical solution, but an approximation of the solution can be determined through *computer simulation*. Write a computer program instructing the computer to randomly select two values between 0 and 60, one representing the number of minutes after 3 p.m. that the husband arrives, and the other representing the number of minutes after 3 p.m. that the wife arrives. Then instruct the computer to find the absolute value of the difference of these two values and determine if it is less than 15 (thus constituting a success). Do this for a large number of pairs of randomly selected values. Finally, have the computer divide the number of successes by the number of times it performed the experiment, which is an approximation of the probability you are seeking. (Compare your answer with the answer at the end of the section that was found by analytical means.)

Do You Have Unusual Difficulty with Formal Reasoning?

I recall a student in my college advanced geometry course who was taking the course as her last required upper-division mathematics course as a mathematics concentrator. She had unusual difficulty making the simplest of proofs in the course. This was surprising since she had completed seven upper-division mathematics courses, most of which made heavy use of formal reasoning (i.e., reasoning requiring her to construct "original" proofs). Granted, all formal reasoning is not alike, and geometric reasoning has its own peculiarities. However, I sensed this student had difficulty with formal reasoning, regardless of the subject matter. Fortunately, she did well in my course primarily because she (1) worked many graded assignments, (2) had to redo "proofs" in those assignments that were not correct, (3) received *frequent* help from me where I supplied hints that allowed her to progress further in constructing original proofs, and (4) had a strong desire to succeed (as evidenced by her work ethic). All of a sudden the light came on for her. From that point on I was thrilled with her ability to make proofs, but not as much as she was!

Do you see yourself in this student, struggling with formal reasoning or an inability to construct proofs? If so, there is a lot you can do to rectify the situation, including:

- *Understand* the problem. (See section 1.)
- Employ a variety of problem-solving skills. (See the subsection of section 2 on problem-solving skills. Before you can reason formally you have to be able to reason informally, and problem-solving skills such as sketching a diagram, looking for a pattern, or using simulation are used to reason informally.)
- Acquire attributes of problem solvers. (See section 6 of chapter 21 on "Attributes of a Problem Solver.")
- Know what it means to understand mathematics and what you need to do to acquire that understanding. (See section 2 of chapter 9 on "What it Means to Understand Mathematics.")
- Have your efforts in reasoning formally critiqued by your instructor, classmates, or others. (In chapter 9, see the subsection of section 3 on verbalizing your thoughts to a partner—but, for this situation, let your thoughts be on your work in reasoning formally.)
- Don't give up!

Do I think you can learn to improve your ability to reason formally in mathematics, including the ability to prove theorems? Without doubt, and I don't even know you! However, you most likely will not be able to do so if you don't believe you can.

Problem-Solving Skills Discussed in the Section

1. Analysis
2. Sketch a diagram
3. Think of a related problem
4. Use symmetry
5. Study the proof of a theorem that relates to the problem
6. Work backward
7. Solve a simpler problem
8. Establish subgoals
9. Look for a pattern
10. Make an organized or systematic list
11. Guess and check
12. Make a model, act it out, or use simulation

Add these 12 skills to your repertoire of problem-solving skills. Commit the names of these skills to memory so you can ask yourself questions, or make statements, similar to these:

- "Is there a simpler problem I can work?"
- "I believe an organized list can help me here."
- "Do I know of a problem that is related to this?"
- "What is a subgoal I can start with?"
- "I need to generate some data and look for a pattern."

There are problem-solving skills other than these (is there ever an end to them?), but these are some of the more basic ones. Draw upon them in your attempts to solve problems. With experience solving problems using these and other skills, you will learn to recognize from the statement of a problem, which skill it might be best to start with in your attempts to solve the problem. Solving a problem most likely requires the use of multiple skills. If the use of a specific skill doesn't seem to help, try another one. *This is what you do in the face of a problem.*

Additional Problems

You are encouraged to work the following set of problems:

Problem 46. (a) What is the sum of the cubes of the first 80 natural numbers? (b) Of the first n natural numbers?

Problem 47. A set of numbers has the sum s. Each number of the set is increased by 15, then multiplied by 6, and then decreased by 11. What is the sum of the numbers in the new set in terms of s and n?

Problem 48. In how many 0's does the product of the first one hundred natural numbers end?

Problem 49. Suppose you have an unlimited supply of a ¢ postage stamps and b ¢ postage stamps, where a and b are relatively prime. Find a formula for determining the greatest amount of postage that you cannot make by using these stamps.

Problem 50. There are 8 coins and a balance scale. The coins look alike, but one is a fake and is lighter than the others. How can you determine the fake coin using two weighings on the balance scale?

Problem 51. There are 5 identical-looking coins and a balance scale. One of these coins is counterfeit and either heavier or lighter than the other 4. With only 3 weighings on the balance scale, explain how the counterfeit coin can be identified, and whether it is lighter or heavier than the others.

Problem 52. There is a row of 862 lockers, some closed and some open. A student runs down the row and opens every locker. A second student runs down the row and, beginning with the second locker, shuts every other locker. A third student runs down the row and, beginning with the third locker, changes the state of every third locker (i.e., the student opens those that are closed and closes those that are open). The process continues until 862 students have run down the row. What is the state of the 862nd locker? The 722nd locker? The 529th locker?

ANSWERS OR HINTS TO SELECTED EXERCISES IN SECTION 2

(3) $\dfrac{5}{2\pi}$ ft **(4)** $\dfrac{27\pi}{2}m^2$ **(5)** 15 **(6)** 58 ft **(7)** $\dfrac{3}{4}$ min **(10)** 204 **(11)** 1296 **(12)** 9 **(13)** 47 **(14)** A had 130, B had 70, C had 40 **(15)** $28 **(16)** leave the other person with 1, 11, 21, 31, 41, ... (you will always win if you have the first move) **(18)** $840mm \times 840mm$ **(19)** 2519 **(21)** 48 **(23)** (b) the squares of consecutive natural numbers, (c) subtract one then add 2 as you progress in the sequence, (e) subsequent term is obtained by dividing preceding term by 3, (g) numerator is increasing by 5 (note that $5 = 20/4$), (h) to obtain a term in the sequence, beginning with the third term, add the two preceding terms **(24)** (b) nth term is $3 \times 2^{n-1}$, (e) $35th$ term is 170/4, nth term is $5(n-1)/4$ **(25)** (a) 19 (b) $n-1$ **(26)** 136 **(27)**

(a) 62 or 63, (b) $2n - 1$ or $2n - 2$ **(28)** (a) 3,656 (b) $\dfrac{n^2 + n + 2}{2}$ **(32)**

(a) 5,512, (b) $2n(n + 1)$ **(33)** $\dfrac{n(n - 1)}{2}$ **(35)** two 5-lb. bags and three 7-lb.

bags **(36)** 276 **(37)** $24m \times 48m$ **(40)** 15 *mph* **(44)** 1/25 **(45)** 7/16

(46) (a) 10,497,600, (b) $(\dfrac{n(n + 1)}{2})^2$ **(48)** 24 (*hint:* what are the prime

numbers whose products give zeros?) **(49)** *Hint:* Work the problem for several pairs of postage stamps and look for a pattern **(52)** locker 862 is closed, locker 722 is closed, locker 529 is open (*Hint:* act it out and look for a pattern.)

---

3. Carry Out the Plan

Considerable information has been presented in this chapter that will help you carry out your plan for attacking a problem. Here are more tips to help you carry it out.

Write Down What Comes to Mind

When working on a problem, it is difficult to keep several thoughts in mind. This can be improved if your thoughts are written down as they occur. If a thought does not seem to lead to a solution at the time it arises, you will want to remember it in case it may bear fruit later. Write your thoughts down, as abbreviated or incomplete as they may be. Use words, pictures, or symbols—whatever helps you notice your thoughts and relate them. Professional mathematicians would be lost without scrap paper! Writing my thoughts down has been indispensable to me, regardless of whether I am solving mathematics problems or working on rewording a statement. When I struggle with a mathematics problem it often dawns on me that I have not written anything. When I write key thoughts down, I often notice soon after that I am making progress toward solving the problem. A thought can be fleeting—you have it one minute and it is gone the next. It is often said, "Out of sight, out of mind." Recording a thought on paper or a video screen, keeps it in sight.

Don't Get Stuck in Your Initial Approach to Solving a Problem

Perhaps the place students get bogged down the most in problem solving is when they are carrying out their first plan of attack. When they don't see progress toward a solution *they continue to stay with this initial approach*, modifying various aspects of it, but basically not changing it, as they try again and again to derive a solution to the problem. Somehow they have convinced themselves that this must be the way to solve the problem and all they need to do is tinker with it. They only thought of one method of attack, and are determined not to deviate from it. This is obviously a natural thing to do since so many students do it. What you have to do is resist this natural inclination and *make a lateral leap in thought*. Rather than continually reworking various parts of your initial plan of attack, there comes a time when you have to discard your initial problem-solving skill, and try a different one. You have to step back and tell yourself to do this since you are breaking a habit. Allowing yourself to change your initial method of attack will allow you to solve many more problems. Staying too long with an initial approach to a problem, that seems to be going nowhere, can be compared to digging yourself deeper and deeper into a hole.

The flip side of staying too long with your initial method of attack is to dispense with it too quickly. Perhaps you have made errors carrying out the skill or only need to make a few minor modifications to your plan. As an example, you may be employing the "look for a pattern skill," and make an error calculating one or more numbers, which will keep you from finding a pattern. Check and recheck your computations as you proceed. Even if you haven't made any errors, your initial method of attack may lead to a solution, but at the time you may not see how that is possible. Trying another method of attack does not mean that you throw away your first method of attacking the problem. Tell yourself that you might have to come back to it.

You will experience considerable psychological interference as you carry out your problem-solving plans. You have to trust your ability, accept the difficulty of the work, expect to be confused, expect to go down wrong paths, and know that your time has not been wasted if you don't arrive at a solution. You have to work through all of this, and are helped by reflecting periodically on the attributes of a problem solver that are discussed in section 6 of chapter 21.

4. Look Back

Many students do not *look back* at their solution to a problem, except perhaps in a perfunctory way. They miss out on an essential aspect of problem solving. Looking back at your solution involves much more than reading through it. It requires you to analyze each step of your solution, asking yourself appropriate questions at each step. Begin looking back as soon after completing the solution as you can. This section describes the benefits of looking back at your solution and questions to ask yourself as you look back.

It is also good to look back at solutions derived by others, including those in your textbook, solutions manual, and class notes. Many of the benefits you will accrue from looking back at your solutions also apply to looking back at solutions of others. However, it is likely that you will gain more from looking back at your solution since it is the one in which you have been most intimately involved. In analyzing your solution, recall your thoughts as you worked your way to the solution. The time it takes to do this is short in comparison to the time spent solving the problem, and will pay proportionately larger dividends.

Benefits of Looking Back, and Questions to Ask Yourself to Attain These Benefits

The benefits of looking back at your solution to a problem, and the questions to ask yourself to help attain them, include these:

Find errors in your solution. There are times you will not have solved the problem when you think you have. However, you can take steps to help ensure that you have a correct solution. Ask yourself questions such as these to help you find computational or conceptual errors:

- "Did I interpret the statement of the problem correctly and does my solution answer the question?"
- "Did I meet all conditions of the problem?"
- "Is my answer reasonable? (For example, "Is the order of magnitude of my numerical answer correct?")
- "Is my logic correct?"
- "Is each step correct?"
- "Do I have contradictions anywhere?"

- "Have I satisfied the hypotheses of the theorems I have used?"

Better understanding of your solution. You may think you have a solution that you completely understand, and don't need to spend any more time with the problem. Polya doesn't think so and I agree with him. He said:

A good teacher should understand and impress on his students the view that no problem whatever is completely exhausted. There remains always something to do; with sufficient study and penetration, we could improve any solution, and, in any case, we can always improve our understanding of the solution.[3]

Questions such as these may help you better understand your solution:

- "Can I justify the steps that need justification?"
- "Can I close my eyes and go through an outline of my solution?"
- "Is there any part of my solution that is not entirely clear to me?"
- "Do I clearly understand all the concepts and principles that I used in my solution?"

Improve the clarity of your solution. Looking back will help you better organize your statements and see how well you communicated your solutions. Questions such as these may help:

- "Do I need to reorganize my statements so the flow is better?"
- "Do I need to bring in more explanatory statements, so the reader will have an easier time following my solution?"
- "I know what each statement says, but will the reader know? If not, how can I reword them?"
- "Are my statements grammatically correct?"
- "Can I shorten, simplify, or streamline my solution without shortchanging understanding and readability?"

Discover alternative solutions that will help in future problem solving. There may be other solutions to the problem and it is a learning experience to uncover them. You are better positioned to uncover a different solution since having a solution indicates that you have a good understanding of the problem. Spending time uncovering alternative solutions will help you apply them to solve future problems that are similar. Furthermore, you may prefer an alternative solution that you uncover because it is easier to communicate to others. Asking yourself questions such as these may help you discover alternative solutions:

- "Are there any other skills in my repertoire of problem-solving skills that I can use to solve this problem?"
- "Can I change my initial approach to the problem and still arrive at a solution?"
- "As I progressed through various parts of my solution, did I have any thoughts on other ways to attack some of these parts?"

Extend the problem. Other problems can evolve out of a problem and the solutions to them may involve a minor modification of your solution to the initial problem. It is interesting and informative to look for these problems and to solve them (you are then operating as a professional mathematician). Questions such as these will help you extend a problem:

- "If I changed or deleted one of the conditions of the problem, what would the conclusion be?"
- "I wonder how much more difficult it would be to determine how many rectangles there are on a checker board as opposed to how many squares?"
- "I wonder how many games would have to be played if it were a double-elimination tournament instead of a single elimination tournament?"
- "I just proved it for two dimensions. I wonder if I can state and prove an analogous three-dimensional problem?"
- "Would the conclusion still hold if I enlarged the set of numbers I was working with from whole numbers to integers?"
- "Would this conclusion hold for all convex polygons, not just for triangles?"

This chapter and chapter 21 do not come close to presenting what you can learn in a mathematics course that is devoted to teaching problem-solving skills and applying them to solve problems. (The title of such a course might be, *Problem Solving in Mathematics*.) Nonetheless, chapters 21 and 22 cover many of the basics of problem solving. Your increased understanding that comes from reading these chapters, working the problems contained in them, and being committed to solving problems in your mathematics courses, will give you the boost you need to continue progressing as a problem solver.

23

Obtaining Assistance in Mathematics

There are many people available to help support your study of mathematics. You need to know who they are, and use those that best meet your needs. The people resources discussed here are your instructor, mathematics study group, tutors supplied by your campus (e.g., mathematics department, college, university, or student organizations), and a hired mathematics tutor. The benefits of using each of these resources are presented, along with accompanying remarks on how to access and use them.

1. Professional Responsibilities of Your Instructor, and Obtaining His or Her Assistance

Other than yourself, the most important person who affects your learning is your instructor. This should be no mystery to you, especially if you have read some of the previous chapters. Your instructor determines the quality of your mathematics course. You have the opportunity to interact with your instructor in a variety of ways, not the least of which is receiving one-on-one help. For these interactions to be productive it is important for you to understand the professional work-life of a college instructor. Instructors are busy people and experience pressures as we all do. It is clear that you can relate better to someone, whether it is a parent, sibling, friend, minister, or your mathematics instructor, if you are aware of what a person deals with in life. This knowledge, along with a glimpse at some personality characteristics of instructors, can be helpful when you need assistance from your instructors. Their professional responsibilities and personalities have a bearing on how well they will serve you.

The Professional Life of a College Mathematics Instructor

Not all college mathematics instructors have the same professional responsibilities and commitments. Their responsibilities depend on many things including whether they are—

- Employed by a community college or a four-year college or university
- Working in an undergraduate college that focuses on teaching, or in a doctoral degree granting university that focuses on research
- Adjunct professors, teaching assistants, full-time lecturers with a fixed renewable appointment, or tenure-track professors
- Teaching lower division, upper-division, or graduate courses
- Working toward tenure or already have tenure

Tenure-track faculty at most four-year colleges and universities are evaluated for tenure and promotion considerations based on teaching, service, and research. These three areas for a faculty member are reviewed by various faculty committees to determine promotions from assistant to associate professor, and from associate to full professor. Typically, tenure is granted at an institution for a faculty member when the person is promoted to associate professor. At many colleges and universities, faculty who do not receive tenure by the sixth year of employment lose their tenure-track position, and most likely their employment. You need to understand that these faculty members are under considerable pressure during their tenure-seeking years. Once tenure is achieved, continued achievement is required for promotion to full professor, but a drop-off in performance in one or more of the three areas is seen with some of them as they age. In general, non-tenure track faculty, which typically includes adjuncts, lecturers, and teaching assistants, do not have research responsibilities.

Research responsibilities

A junior or community college faculty member most likely does not have a research requirement, and four-year colleges and universities vary as to the strength of the research demands on their faculty. Research is difficult, time-consuming work, and requires minimal interruptions.

To be a successful researcher, faculty have to be protective of their time. It is virtually impossible for most faculty to get any significant research work done while on campus. Hence they may stay home two to three days per week to work on their research. This means that they are not as available to give extra help to students, may spend less time preparing for their classes than what is ideal, and limit their service commitments.

Teaching responsibilities

Teaching responsibilities of college faculty are discussed in section 2 of chapter 12. You are encouraged to read or reread that section. You will see that faculty members could devote all of their professional time to improving the quality of their teaching. They don't have this luxury because of research and service obligations. It is important for you to be realistic in the expectations you place on your instructor. All of them will not be rated as excellent instructors for a variety of reasons. One reason is that this would require them to spend more time on teaching—time they don't have. Nonetheless, at a minimum they need to be capable instructors and you hope for more than that. There are excellent researchers who are excellent teachers and for good reason. When doing research one is intellectually alive, working at the forefront of knowledge in a specific research area. This can contribute significantly to an instructor's classroom performance.

In general, community college instructors teach more hours per week (typically, from 10 to 15 hours) than their counterparts in four-year colleges and universities (typically, from 6 to 12 hours). At many colleges and universities, released time from teaching is given to faculty for various research and service responsibilities, or for teaching large classes of students. For example, a college instructor may teach only three hours per week because he or she has a large research grant; thus, being granted released time from teaching.

It is typical to see faculty at four-year colleges and universities posting from 2 to 6 office hours per week. Faculty at community or junior colleges typically post from 6 to 10 hours per week. Some schools require their faculty to post a specific minimum number of office hours. A problem arises with adjunct faculty who teach evening courses. They may have another job off-campus and are not on campus long enough to offer many office hours (they may not even have an office). Normally, they will schedule office hours just before or after class, perhaps meeting in your classroom or one nearby. Not many students want to stay around for help after an evening or late-in-the-day class, and the same can be said for instructors.

All college faculty need to be aware of the changes that take place in the teaching of mathematics. Thus, it is important for them to attend local, state, regional, or national mathematics or mathematics education conferences. These conferences may only be for a few days or can be as long as a week. An instructor's attendance at a conference often necessitates the canceling of classes, or finding substitute instructors. You will typically see a note on an instructor's door such as, "No office hours from 2/13 through 2/15."

Service responsibilities

Professional service activities of mathematics faculty within a university can be within the mathematics department, the college in which they reside, or at the university level. Some of these activities can be very time consuming for those who take leadership roles, and especially for chairpersons of committees. Professional service outside the university is typically with local, state, or national professional organizations, or could even be with an international organization. Many faculty are engaged in community service activities. They may give presentations to community groups, serve on church committees, or volunteer for community projects. It is not uncommon for an instructor to come late for a scheduled office hour, or not come at all, because a meeting has been called at short notice, or a meeting lasted longer than expected.

Why discuss the professional responsibilities of instructors?

Your instructor may appear to be hurried and harried when you go to him or her for help. The body language of your instructor may indicate that you are an imposition. And sometimes you are! But don't take it personally, since helping you is an instructor's responsibility. An instructor may be apprehensive about work he or she needs to complete after you leave. He or she may be anxious to leave campus to grade examinations, prepare for the next day's classes, or work on research. Nonetheless, you have a right to be there, and the instructor knows this, so don't hesitate to get help. It is not your responsibility to relieve an instructor of pressures and anxieties, but it is your job to understand that he or she has them.

Here are common sense guidelines on instructors' office hours:

1. Instructors, or their designates, should provide an adequate number of office hours and students have a right to use them, without being encouraged *not* to use them.

2. If possible, instructors should leave a note on their office doors if they will be late for office hours, or that office hours have been canceled.

3. With few exceptions, an office visit should not be a mini-lecture to a group of students.

4. Instructors have the right to limit their time with you in order to serve other students.

5. Students with appointments have first priority to see instructors. Otherwise, students should be taken in the order they arrive at an office, unless there are extenuating circumstances.

Receiving Assistance from Your Instructor

Don't underestimate the importance of receiving one-on-one help from your instructor during office hours. No one is in a better position to help you. It is one thing to be with your instructor in class, surrounded by many classmates, and quite another to have your instructor's undivided attention in an office setting. If you are experiencing difficulty with your health, would you want to meet with a medical doctor in a group setting or individually where you have his or her complete attention? The main goal of an instructor when you come for help should be to diagnose the reasons for your difficulty, and then prescribe a process to overcome them. Getting one-on-one help from your instructor can be the difference between success and failure in a course. However, there are a small number of instructors who perform poorly during one-on-one meetings. They may have personality traits that are difficult to deal with in an office setting. Some of these traits are mentioned later in this section.

You have to take initiative to get help from your instructor. Your instructor may not know that you are having difficulties, and even if he or she did, you will not be lassoed with a rope and dragged into the instructor's office. In an in-depth interview study of forty college sophomores reported by Harvard professor Richard J. Light, a key predictor of troubling outcomes for students was their unwillingness to seek help. He reported:

> Of the 20 students who were struggling yet were able to share their problems and to seek help from one of these many sources [professor, departmental advisor, teaching fellow, residence hall advisor], all, *without exception,* were able to work at developing strategies to improve their academic performance. But most of the 20 who were unable to share their problems remained distressingly isolated. They became caught in a downward spiral of poor grades and lack of engagement with other people at the college. It was far harder for them, struggling alone, to turn their situation around.[1]

Aside from getting your instructor's help in mathematics during an office visit, there is a good chance that you will get to know him or her better, and vice versa. Share information about yourself and show some interest in his or her personal life. Your instructor may drop hints in class about what interests him or her, and that can be a starting point for you. Your efforts here may not go anywhere, but it is worth a try. It is exhilarating to experience the warmth, joy, and value that come from establishing a friendly relationship.

Why you may hesitate to get help from your instructor

There may be many reasons why you are reluctant to get help, many of them probably due to misconceptions on your part. Regardless of your reasons, get help when you need it. It is key to your success. The following are reasons for students' hesitancy to get help:

Believe they can learn the material themselves if they work long and hard at it. The belief that we don't need others is often promoted in an individualistic nation such as the United States. You will experience mathematics content for which you will need help. This is not to say that spending significant time on something yourself isn't a necessary and commendable trait. However, there comes a time when additional time spent pays very few dividends. You don't have forever to learn what is required. If you don't understand something in a timely manner, class time will be less meaningful, assigned projects and problem sets will be more confusing, and your performance on quizzes, unit tests, and a final examination will be negatively affected.

Do not want to expose their weaknesses since instructors assign course grades. Yes, your weaknesses will be exposed since that is the purpose of your office visit, and, yes, your instructor does assign your course grade. However, your instructor will grade you on the criteria established at the outset of the course. If anything, you will be viewed more favorably because you took steps to help yourself. Most instructors want to grade high and look for excuses to do so. Trust your instructor's integrity and professionalism as you trust your physician with your health. Would you stay away from your physician because he or she might find weaknesses in your body? Some people do and they may die because of it. You can also expire as a student, so to speak.

Believe that receiving help is a sign of weakness. This relates to the previous two reasons for not receiving help from your instructor. Receiving help is a sign of strength, regardless of the difficulty you want to overcome. I don't know of one instructor who doesn't believe this. How could it possibly be viewed otherwise? You won't have to convince your instructor that it is not a weakness; most likely you will have to convince yourself. It is interesting to note that it is typically the students who struggle the most who do not make use of their instructor's office hours. These students can benefit the most if they did use them, but for the reasons just listed, they stay away until it is too late. They may show up near the end of the semester when there is no hope to rectify their poor performance. They believe that students who are not struggling as much as they are don't get help from their instructors. How wrong they are about this. It is often the better students that instructors see in their offices, and that is one reason why they are better. Good students don't feel it is a sign of weakness to get help. They do what they have to do to improve their performance. Do you make use of your instructors' office hours?

Don't want to experience the frustration, impatience, or anger of their instructors when they cannot adequately respond to their instructors' questions. You may have experienced this behavior from past mathematics instructors. An instructor who exhibits these characteristics may do so because he or she (1) has other work to do and feels pressure to get to it, (2) does not know what to do when the student doesn't seem to understand, or (3) has a personality trait that lends itself to this type of behavior. In all likelihood, instructors do not mean to behave badly, and are probably unaware of it. Don't take it personally. It is an instructor's shortcoming, not yours. Avoiding him or her is not the answer.

You can tolerate your instructor's poor behavior, or you can tell him or her how you feel and that may improve the situation. However, do not tolerate any abusive behavior. Here are two remarks made by students about the poor behavior of their instructors:

❖ *He gets mad when you go for help and don't understand something.* (Calculus I student)
❖ *Instructor becomes impatient when in individual contact with me during office hours. This doesn't always happen, but it does sometimes when I'm unable to comprehend fast or well.* (Calculus III student)

Believe that they don't have time to see their instructors for help. My response here is simple: find the time. It may mean that you have to schedule an appointment with your instructor at a non-office hour time, have a telephone meeting, or reschedule or eliminate a less important engagement. If all of your other engagements are consistently more important than seeing your instructor for assistance, then you need to reconsider your priorities.

Believe that instructors do not want to give help to their students outside of class. The vast majority of instructors want students to get assistance from them. One of the more frequent complaints I hear from instructors is comparable to this: "I am saddened that so few students make use of my office hours for assistance. I am continually inviting them to come and so many need it, but they stay away. I even make comments on their returned examinations to see me about this or that problem. They still don't come." Instructors know how valuable it is for their students to come for help, and most of them know that it makes teaching more enjoyable.

Should I make an appointment to see my instructor?

An appointment is typically not required to see your instructor for help during a stated office hour. He or she has reserved this time to help students. However, it is wise to make an appointment, and preferable to set it for the beginning of the office hour. Your instructor will appreciate you making an appointment since that will help him or her better plan for that office hour. Making an appointment helps ensure that your instructor will see you at the stated time. Otherwise, you will have to wait until he or she finishes with other students who have arrived before you, or you will have to wait until students who made appointments are finished. Arrive on time or early for your appointment; otherwise, your instructor may take another student and work with that student during part or all of your scheduled time.

Appointments to see your instructor can be made via phone, email, a visit to his or her office, or in the classroom just before or after a class session. Expecting to get help from your instructor without an appointment with less than fifteen or twenty minutes left in an office hour can be problematic. Your instructor may be getting ready for a class or to go to lunch, dinner, or home. It is common for instructors to leave their office near the end of a stated office hour if no student is present. Instructors like to schedule office hours around their class times since that is most convenient for them. Unfortunately for you, and for them, they are then more apt to feel pressure from all the work that needs to be accomplished in a very short time. But this poses no problem for the well-organized instructor.

You need to make an appointment to see your instructor if you cannot see him or her during stated office hours. If there is a critical need to see your instructor as soon as possible, you can drop by the instructor's office, but expect that he or she may not be able to assist you at that time. I had students call me while they were on campus to ask how soon they could see me on an important matter.

When should you get help, how often should you get it, and for how long?

You need to get help from your instructor as soon as you think you need it. Students seem hesitant to come in for help in the early stages of the course. It is as though they need to see the results of their first examination before they know they need extra help. This is a form of denial, since many signs of confusion will surface well before the examination. It is not too soon to see your instructor between the first and second class periods of the term if you are confused over an aspect of the lecture, what you have read in the textbook, or on an assigned exercise. What rule would you be obeying that says you should wait until more time has passed? No such rule exists except perhaps in your mind. The pace of mathematics in college is very fast, to say the least; hence, a lack of understanding can get out of hand quickly. Confusion on the first day of class can interfere with understanding the content of the second class period, and so on. For some students and some mathematics courses, confusion throughout the first one or two weeks of a term is enough to warrant dropping the course. That is, there may be little hope of catching up in the course. Don't wait to get help from your instructor. There are other sources for getting individual help, but no one is positioned better than your instructor to help you.

If you are struggling with many aspects of your course, then seeing your instructor a few times a week for a minimum of 15–20 minutes each time is recommended. This presupposes that your instructor has that much time to give you. Get help from your instructor and other individuals, and continue until your situation improves.

I had students who did badly on their first examination in my course. Encouraged by me, they came in for specific help soon after, asking how they could improve their study in the course. Unfortunately for some, they only came in once or twice prior to the next examination. When they performed badly on the second examination, I asked why they didn't come in more often. Their response was something like this: "What difference would that have made since the help I received didn't help me on the second examination?" This thinking is comparable to the thinking of a basketball player who, at the start of a season, lacks many basic skills, has only one or two short 15-minute practices with the coach before the next game, and then expects to do well in the next game. Changes in your study methods and your ability to understand evolve over time, not in a few brief meetings.

Your instructor is not your personal tutor

If you are unduly struggling with a course then you most likely will have to drop it if you don't have the prerequisites. If you have the prerequisites to an acceptable degree, you may have to hire a personal tutor since you will need considerably more one-on-one help than your instructor can provide. Instructors do not have time to serve as personal tutors. Suggestions for getting a personal tutor are discussed in section 4.

Topics of discussion during an office hour

There is almost no limit to the topics that can be discussed during your instructor's office hours. These topics include—

1. Your questions on class work, textbook readings, or assignments

2. Advice on how you can do better in the course
3. The prerequisites you need to succeed in the course
4. Concerns over mathematics or test anxiety
5. Concerns on how the class period or course is structured
6. Concerns over lack of discipline in the classroom
7. An overview of your performance in the course so far, including whether you should drop the course
8. Concerns over how your examinations or assignments are graded
9. Concerns over how you are being treated by your instructor or classmates
10. Personal problems that are interfering with your performance in the course
11. Reasons why you need to miss a class period or examination, or why you cannot hand an assignment in on time
12. Getting advice on subsequent mathematics courses to take, and when to take them
13. Getting advice on instructors to select for subsequent courses based on your learning style and their teaching styles
14. Getting advice on whether to pursue a mathematics major or minor
15. Getting advice on graduate schools to attend or on a possible career in mathematics

To get the most value out of your instructor's office hours, you need to prepare for your visit. It is so much easier for your instructor to help you if you have first tried to help yourself, and then show and discuss your efforts. The next few paragraphs give ideas on how to prepare for a visit to your instructor's office.

Preparing to get help from your instructor

It is not prudent to come to your instructor's office hours and say, "Help me, I don't understand anything in the chapter or section." This is comparable to asking the instructor to lecture once again on that content. The lecture didn't work the first time and probably wouldn't work the second time. That is not the best way to use an office hour and your instructor will be reluctant to do this. Come to the session prepared to show your instructor precisely *where you first became confused* in your reading of the textbook, in the class lecture, or on a problem, and make specific statements or ask specific questions similar to these:

- "I don't understand how this statement follows from the previous statement."
- "I started working this exercise but could not complete it. Here's my work so far."
- "This theorem makes no sense to me. It was used in this example but I don't see how."
- "I devoted considerable time to trying to work this problem. Eventually I saw how it was worked in the solutions manual, but I don't see how the solver thought of starting where she did. Can you help me with this?"

Here are other examples of initial comments you can make when you see your instructor:

- "Can you help me analyze my examination results to figure out the types of problems that I received full credit on, those that I received no credit on, and those for which I received partial credit? This will help me perform better on subsequent tests."
- "I am not sure if I am prepared for this calculus course. I took precalculus last semester and received a grade of B, but I am currently struggling with several ideas from that course. Can you quiz me on them and perhaps on other key ideas to see if I need to retake the course?"
- "There are a few students who sit near me in the back of the classroom who are making it difficult for me to concentrate in class. They have loud conversations and make snide remarks. Would you please do something about this? They know I am bothered by this, yet that hasn't stopped them."
- "I don't understand why I didn't receive full credit on this examination problem. I have gone over it many times and see nothing wrong. Would you please look at it again?"
- "I am not having any difficulty with the assignments, yet I do very poorly on examinations. Can you help me determine why?"
- "I don't see how I can possibly get the assignment to you by next Monday's due date. I have three examinations in the

next two days and am leaving Thursday evening to attend my sister's out-of-state wedding on Saturday. Can I turn it in two days later?"

Bringing one or more classmates to an office hour

Some students like to bring a friend or classmate to their instructor's office hours, especially the first time they come. This makes the session less threatening to them. That may be the case, but you will receive less help if the friend is there to receive help as well. It may also be a problem since you may not react the same way you would if you were there alone. When two or more students come in together for help, the session often becomes a mini-lecture; hence, your specific concerns will not be adequately addressed. Frequently, students will come in and say that they are struggling with the same problem. That may be true, but the reasons for their struggles may be totally different. If you want the best help from your instructor, he or she has to diagnose your problem by asking you questions, hearing your responses, and listening to your questions. It is difficult, if not impossible, to do this when simultaneously trying to respond to the student who is with you.

If you need the crutch of having one or more students accompany you to your instructor's office, work to wean yourself away from that crutch. An astute instructor will encourage you to come in alone to future visits. I cannot tell you the number of times I saw three to six students form a circle around an instructor sitting at his desk, watching and listening to his or her explanation on a particular topic. This often happens just before an examination when procrastinating students find out very late that they are confused.

Pretending to understand

When you receive individual help from your instructor, you have to be willing to ask questions until you understand. I have seen far too many students feign understanding during an office hour, and in class. I suspect they also feign understanding when working with a group of students. They don't want their instructor or classmates to think that they are slow to understand. How sensible is that? Stop feigning understanding if you are serious about learning. If your instructor asks you if you understand something, and you don't, your comment can be as simple as this: "No, I'm still confused. Can you work another example, or give me an example to work, and then help me if I get stuck?" If you are still confused after that, let your instructor know. Don't say you understand until you do. An experienced instructor should be able to diagnose why you don't understand, and then select from his or her repertoire, an idea to help you understand.

These two sentences should characterize the focus of a visit to your instructor's office: Good teaching or learning is less a matter of what your instructor says to you than what you say back to your instructor. In the same vein, watching your instructor work a problem is not as meaningful as your instructor watching you work a problem.

You are not the instructor

There are students who believe they should have the final or an equal say about how the course should be taught. For example, they come into the office implying or stating that their examination or assignment was graded unfairly. When specific solutions of problems are jointly reviewed, and the instructor states reasons for the grading, some students won't agree, even though they are not able to justify their position. They don't allow the facts to get in the way of their desire to receive more credit on the problems, implying that their view should carry as much or more weight than their instructor's. There may be times the instructor makes mistake, but that is not what I am referring to here.

I had students come to me with another student's examination paper in hand and point out that they had the same solution or response as this student, yet they received less credit. In almost all of these situations I found that the solutions or responses were not even close to being the same. Even after my explanation, some of them still had difficulty discerning how their responses were different, or they didn't want to acknowledge the differences. Thus, they didn't move from their original stance and were angry with me for my inflexibility. I am not implying that you should not try to explain your position to your instructor. On the contrary, it is important to do so if you genuinely believe you have a point. But it should be done calmly, respectfully, and with an openness to hear your instructor out. Don't display a confrontational attitude, make demands, or act as though you are the instructor.

Take notes during your office hour visit

When a student came to my office hours, we would write on paper on a shelf that slid out of a corner of my desk. Sometimes I would write and sometimes the student would write, with frequent pauses for discussion. It didn't take me long to realize that students wanted to take this paperwork with them. Hence, I became more careful organizing my work on the paper. Many instructors don't think to ask their students if they want to take this work with them. If it is meaningful for you to have this work, don't hesitate to ask your instructor for it. Take notes if your instructor is writing on a chalkboard.

Will this be on the examination?

I saw a one-frame cartoon showing a first grader standing next to a chalkboard with his head at the height of the chalk tray. He had just finished working the problem, $1 + 1 = 2$, which was displayed on the board. He then asked his teacher, "Will this be on the test?" I don't suggest that you ask your instructor this question about specific content. He or she will not, or should not, welcome that question. It indicates that you are more interested in obtaining a good test score than learning mathematics.

2. Participating in a Mathematics Study Group

Many mathematics learning goals and other worthwhile goals can be accomplished through informal group study; thus, the topic of group study needs to be addressed per se. To give you a general sense of the importance of study groups, Richard Light reported on this symptom of trouble for a college student, besides low grades:

> a student feels a sense of isolation from the rest of the college community. A handful of undergraduates may relish such isolation, but only a handful. With a bit of effort, an advisor can spot isolated students. They are not involved in any extracurricular activities. They are not members of a study group in many of their courses. And they deal with their low grades by going from classroom to dorm room, closing their door, and studying, and then studying some more, nearly always alone. If their grades don't improve as the year progresses, they don't change their behavior pattern. They just do more of the same, stay up later and later at night, or, in a few cases, simply give up on coursework.[2]

Some instructors use a portion of the class period for group study. There is no need to specifically address that here, other than to say that this is an excellent use of class time; however, it only allows for a limited amount of group study. More informal mathematics study groups are addressed here. These are groups that meet outside of class and are typically formed by students. Basically, study groups are discussion groups, and any activity that fosters discussion of mathematics is desirable. These topics are discussed in this section:

- Limitations of Group Study
- Reasons to Belong to a Study Group
- Two-Person Study Group
- Large Study Group

Limitations of Group Study

It needs to be said upfront that group study cannot replace individual study. The bulk of your mathematics study has to be done independently. You own what you discover through individual study. It is part of you, which means that you understand it better and remember it longer than if it comes from others. No one can do your learning for you, but others can support your learning in a variety of ways, and we now look at some of them.

Reasons to Belong to a Study Group

The students in your study group are your discussion partners, and when you engage them in discussion you reap many benefits. (Learning objectives accomplished through discussion of mathematics are presented in section 4 of chapter 14.) Suffice it to say that *discussion is second to none* in improving or fostering the following: attitudes, motivation, understanding and applications of concepts and generalizations, problem-solving ability, transfer of knowledge, leadership qualities, ability to work cooperatively, and increased confidence. Having multiple discussants in your study group allows for diverse discussions that enhance the attainment of all the benefits of group study. Another benefit of belonging to a small study group is that it is less threatening to ask and respond to questions than it is in class. It is unfortunate to see the vast numbers of college students who hesitate to ask questions in class because they are afraid that they will appear to be unintelligent. The elimination of this fear can be fostered by first asking questions in a small group.

An additional benefit of discussion is what you gain by helping students who are struggling with a topic. You are performing an act of kindness and nothing is better than that. As I said earlier in this book, if you want more joy in your life, help someone. An analogy would be visiting the sick. You do this to comfort them, but in all likelihood you find out that you are the one being comforted. As you are helping other students, you are also enhancing your learning. Any mathematics teacher will tell you this: "I first really learned mathematics when I had to teach it." Why is this the case? Because teachers are bombarded with all sorts of questions and to answer them they have to know what they are talking about. If a teacher cannot adequately provide students with answers to their questions, or come up with good questions for them, they continue to study the content until they can. When you help your classmates you are acting as a teacher. You will also improve your ability to communicate mathematics. Helping other students benefits you at least as much as them since you are reaping the learning benefits that come from teaching.

Finally, in a study group you can talk about aspects of your mathematics course that help promote good mental health. That is, you can share your feelings and hear the feelings of others on the assignments, quality of the examinations, grades received, instructor's use of class time, etc. Conversing with members of your small groups helps alleviate the feeling that you are all alone with these thoughts. However, the focus of your meetings should be on learning mathematics, not on airing all your complaints about the course.

Two-Person Study Group

There are many advantages of having a study partner. It is easier to find times to meet and there is a greater chance you will be in sync since you are selecting each other. Find a classmate, or a student taking the same course as you, and meet at least once or twice a week to discuss mathematics. It is important to think carefully about the student you want for a partner. You need to know if his or her knowledge is too far above or too far below yours, if he or she has adequate helping skills (including a sincere desire to give help), if he or she is reliable, and if there is good chemistry between the two of you. The best situation occurs when the partnership is mutually beneficial, and what makes it such may be different for each person. You can work together in an empty classroom, in the student union, in a dorm room, apartment, etc., and at times for convenience sake, over the telephone or through email.

How to work with your partner

Your first attempt to learn specific mathematics content has to be done independently. You don't want a partner to become a crutch. If you have set times for meeting during the week, be sure and have your initial work done beforehand. When you meet you become sounding boards for each other with the goal of helping each other help themselves. For example, if you understand the reason why the author of your textbook can go from step four to step five in a proof, give hints to your partner by asking leading questions so he or she can figure it out. When helping your partner solve a mathematics problem, look at his or her work on the problem and begin to ask questions. It is less beneficial to say, "This is how I worked the problem," or "This is what you should have done." This may need to be done every so often, but it must not be the standard mode of operation. It is good occasionally for you or your partner to write the solution of a problem that you both have solved independent of each other, explaining to your partner what you are doing at each step. (A technique for verbalizing your thoughts to a partner is given in section 3 of chapter 9.)

It is very unwise for you and your partner to do everything together, including going to your instructor on the same problem and expecting a joint explanation. You have to decide how beneficial it is to continue with your partner. If the benefit is minimal, consider finding another partner.

Large Study Group

Large study groups are defined here as having three to six members. An ideal size for a group is four. Since there will be short dialogues interspersed with whole group discussion, four people can be split into two pairs, in three different ways. With three- and five-person groups, one person is often left out of discussions. With six or more members in a group, introverted students can be intimidated and withdraw from the conversation.

A few disadvantages of a study group larger than two members include (1) finding it more difficult to meet, (2) less opportunity to be as involved when meeting, (3) less control of what is accomplished at the meeting, (4) the possibility of more members being unprepared for the meeting, (5) greater danger of being ignored, and (6) more difficulty staying on task. The advantages of a larger study group include (1) greater diversity of abilities, ideas, and helping skills; and (2) the opportunity to pair up with different members of the group during a study session to accomplish specific tasks. Most likely there is still much to be learned after group study has taken place; hence, you will need to go back to individual study. Suppose there is a study group of four members. Here is a good scenario indicating what can take place before, during, and after the group meets:

- Each member studies individually in preparation for the group study.
- The whole group is involved in the discussion when they first meet.
- Members of the group pair up at times to tackle specific tasks during the group study.
- Each member returns to individual study at some point after the group study session ends.

Forming your large study group

Some instructors form out-of-class homework teams to work on specific assignments or projects that will be graded. One set of solutions from each group is turned in to the instructor. Besides the advantages of group study that have been discussed, students on these teams can be given more difficult problems and the instructor has fewer papers to grade. Some students resent receiving a group grade for the assignment. This is especially true if there are group members who do not carry their load. Would you resent this?

If your instructor does not take initiative to form out-of-class study groups, then you can by inviting some of your classmates (or others) to join a group you are forming. At your first meeting you have to work out logistics such as—

- Exchanging phone numbers and addresses
- Appointing a coordinator
- Determining the duration, frequency, starting time, and location of the meetings

- Creating a procedure to follow if a member cannot make a meeting
- Outlining a general structure or agenda for the meetings, including a statement on individual preparation

Ask your instructor if he or she will help form out-of-class study groups. There are a variety of ways for an instructor to do this. Here is a set of *procedures for instructors* for forming out-of-class study groups (you might want to discuss them with your instructor):

1. During the second week of class help students form out-of-class study groups by listing one-hour time blocks during the week on sign-up sheets, and have the students sign the sheets as the sheets circulate around the room (the class is going on as usual). Don't do this during the first week of classes since students' schedules are changing. Most students sign up for times that are within a few hours of the class time. Unless your course is scheduled later in the day or in the evening, students will avoid mid-to-late afternoon times, as well as evening times (at least this is true for a commuter campus). *Request that students sign up for three or four different times, and rank their preferences.*

   Another option is to list times on the board and have students sign up for several times (on the board), again listing their preferences. A student can copy this information for you to use later when making assignments. The disadvantage of the latter method is that class cannot begin until this is completed.

2. When the sheets have circulated around the room, circulate them once more so that those students who signed at times when only a few or none have signed, can sign up at more popular times. A student may decide the second time the sheets are circulated to switch times for other reasons.

3. Form study groups before the next class session, taking into consideration the students' preferences and the minimum group size desired. (I tried to maintain a minimum of four students to a group, but at times had to form a few three-student groups.) Groups are formed based on the group's ability to meet for one hour during the week, although

members of a group may decide later to meet more often, for a longer meeting time, or at a different time.

4. Hand out information to your class on the study groups you formed based on the information on the sheets, which includes the names of the students comprising each group and the meeting time. Allow those students who were absent from class when the survey was taken to join any of the groups formed, as long as this doesn't make the group too large. As part of this handout, list the responsibilities of a group member (for example, come to the group meetings, call the coordinator if you cannot make a meeting, prepare for the meetings, etc.).

5. Have the group meet for about five minutes in class to decide where they will have their first meeting, and who will be the coordinator. Better yet, you could have their first meeting be an in-class meeting in which they also discuss a previous assignment or any topic that you wish addressed. This allows them to get a little familiar with each other and provides a good start for ongoing discussions. (My in-class study groups were different from my out-of-class study groups because I wanted the former groups to be more fluid.)

6. Don't insist that every student belong to a group. Be amenable to a student who wants to change to a different group. You can casually monitor these groups by periodically asking how they are functioning. You can talk to the coordinators or invite anyone in the class to come to your office and discuss how their group is working. Periodically, encourage the groups to continue meeting. If you find out that a group is not functioning well or has disbanded, you might want to find out why. Show your students that you value group study by showing interest in how they are functioning.

Be prepared to discuss

It is demoralizing for group members to meet with members of the group who haven't prepared. They provide little or no help to those who are prepared, and they themselves get little out of being there. Their lack of preparation becomes evident very quickly and it creates a bad atmosphere. You need to come to the group meetings prepared (i.e., having read the assigned sections in the textbook, worked on the assignment(s), and ready to ask questions).

Good discussion requires good listening

Discussion is enhanced if you are a good listener since you will be responding in a meaningful way to what is said, not to what was not said. Hence, be sure and hear your classmates out before responding; that is, resist interrupting either by speaking too soon or by body language that indicates acceptance or non-acceptance of what is being said. This is not easy to do. Good listeners are the exception, not the rule. (Learning mathematics by listening is discussed in section 3 of chapter 14.)

Guidelines for working in a large study group

There are some basic guidelines to follow that will help your study group be more productive. As a group member you have to help ensure that they are met. These guidelines include the following:

Stay on task. Discussion on matters not pertaining to the task at hand should be infrequent and short.

Work on the same task. All group members are to address the same issue; that is, they are to be working as a group, not as individuals or as subgroups of the group. A time may come during the meeting when subgroups are formed, perhaps with the goal of working on different tasks, but that is a *deliberate* decision of the group.

Each member of the group is responsible for the achievement of the other members of the group. The group has accomplished its task when everyone in the group understands the topic at hand. If a group member is confused, than you and others need to work to clear up the confusion. You need to lead the person to the desired understanding by questioning and supplying hints. Hopefully, the times are minimal when a group has to move on when someone is still struggling with the topic. In that situation, someone in the group can volunteer to work further with that person at another time.

Be respectful of the comments, questions, and ability of others. Avoid unacceptable behavior such as arrogance, verbal put-downs, inappropriate body language, and paying little or no attention to someone.

To help ensure that you and other group members are following the guidelines for working in a study group, review them periodically.

3. Seeking Assistance from the Department, College, and University

There is substantial opportunity to receive assistance with your studies from your department, college, or university. Know these resources and use them when appropriate. Their services are typically free. Opportunities for assistance at the departmental, college, and university levels are now discussed.

Help Provided by the Mathematics Department

Many mathematics departments have a *mathematics learning or resource center* where you can get help on most lower-division mathematics courses. Most helpers are undergraduate students, but some may be mathematics instructors who hold some of their office hours in the center. These centers are typically open many hours each week, including evenings, and help is available to everyone, campus-wide. Not only are helpers available, but also the centers may have special handouts, computer tutorials, and videos available for learning prerequisite knowledge you may be lacking. Some centers also conduct short workshops or teaching sessions where specific remedial topics are addressed.

Undergraduate student helpers are normally students who have done well in their previous mathematics courses, and are most likely recommended as helpers by mathematics instructors. The quality of their help depends on their academic backgrounds and tutorial skills. Hopefully, they will help you help yourself rather than just provide answers to your questions. Seek out the better helpers. Realize that these student helpers are not your personal tutors; the time they can spend with you is limited. Expect them to tell you, at times, that they cannot spend more time with you since other students are waiting to see them. In the meantime, they may give you a specific task to do that relates to your concern and tell you to come back when no more students are waiting. If a department does not have a mathematics-learning center per se, it most likely will have undergraduate student helpers available during given hours at specific locations on campus.

The type of assistance that the mathematics department *office staff* may be able to give you includes—

- Information on tests that you can take to determine course placement or college credit
- Information on mathematics tutorial services that are available
- Information on department designated private mathematics tutors, including hourly pay rate
- Your instructor's office hours
- Access to your instructor's department mailbox
- Forwarding something to you at the request of your instructor
- Help locating your mathematics instructor in case of an emergency
- Distribution of handouts on mathematics programs
- Referring you to a mathematics advisor
- Making an appointment for you to see the mathematics department chairperson

Help Provided by the Division or College

The division or college you are enrolled in, or the division or college where the mathematics department is housed, provide certain services for you. These include:

Meeting with the Director or Dean. If you are not satisfied with the resolution of a complaint that you took to the mathematics department chairperson, see the director or dean of the division or college in which the mathematics department is housed.

Academic advising. As far as receiving advice on mathematics issues, the more specific the question the less likely you will obtain the information you need from someone outside of the mathematics department. Specific questions about mathematics courses will most likely be better answered by consulting mathematics faculty who have an affinity for advising.

Tutorial service. There are several sources that provide tutoring. For example, a minority-engineering program in the college of engineering may provide mathematics tutoring by senior undergraduates or graduate students. Also, student organizations may provide tutorial help. Examples of these are student business, engineering, and education organizations.

Help with writing. Your Humanities Department may be responsible for a campus-writing center whose function is to help students improve their writing. Poor writing skills can affect academic perform-ance, including mathematics performance.

Grievance procedures. If you have an unresolved complaint against an instructor, you can file a formal grievance and have the com-plaint heard by a grievance committee.

Campus-Wide Help

There are many campus-wide services you can access and they cost nothing. Consult your *undergraduate announcement* for these services and other pertinent information covering many diverse topics. This an-nouncement is one of your most valued possessions as a college stu-dent. Keep it handy, look at it periodically, and know what is in it that applies to you.

Typically, most universities have campus-wide services that bear on academic performance. Included among them are the following, perhaps referred to by other names at your campus:

Campus Tutoring Center. These centers may provide free indi-vidual or group session tutoring for all students who demonstrate a need for help. Tutors are often graduate or upper-class undergraduate students. A tutor is often assigned to a student and together they deter-mine when and where they will meet. This is a wonderful opportunity to have a personal tutor at no cost. A campus tutoring center might ask mathematics department faculty to recommend student tutors the cen-ter can hire to tutor students in specific mathematics courses.

Program of Academic Support. This support is primarily de-signed for students who have been accepted to college, but have not met all the standard qualifications for acceptance. These students have records that show strong signs of college ability. Many of these stu-dents need extra support learning mathematics and this program is one means of providing the support.

Counseling Services. This service, among other things, resolves barriers to the learning process through personal counseling. Some problems that students have are deep-seated, others not. Students work out a simple plan with a counselor to minimize or eliminate a specific problem, and if that doesn't work, they look deeper into the problem and may conclude that there is no simple solution. For example, many procrastination problems are not deep-seated; hence, simple solutions work. However, if the procrastination is due to a fear of failure, then the situation may be more deep-seated and other plans have to be put into place.

Other problems that students bring to a campus counseling center include test anxiety, over-scheduling, and personal problems. More students need to take advantage of counseling services. Don't let the name of the center stop you. That is, you may think that if you need counseling, something is wrong with you. Actually, something is right with you—seeing a counselor is a sign of strength, not weakness. College counselors are trained and experienced to work with college students, and have helped many of them. How lucky you are to have their services available to you, and at no cost.

Disability Resource Services. Some campuses offer aid to differently-abled individuals. Some of the services include early registration, course and classroom accommodations, tutorial referral and mentoring services, assistance at the computer center, note taking, and referral for auxiliary services such as interpreters for the deaf and taping texts for the blind. Instructors may be asked to make test-taking accommodations.

Student Ombudsperson. Your campus may have a student ombudsperson. This individual mediates any problems you might have with a student, faculty member, or administrator. You can make an appointment and express your concern. The ombudsperson may invite the two of you to meet with him or her to discuss the matter.

Student Grievance Policy. Your campus most likely has a grievance policy for students. This policy outlines the steps you can take to file a formal grievance against a student, faculty member, or administrator. A committee will meet to discuss the merits of your grievance. (There is also a grievance policy for faculty members. They can also file a grievance against any body at the college or university, including students.)

Admissions and Registration Offices. Information on programs and courses is available from the Offices of Admissions and Registration. Once again, the mathematics department is the best source for specific advice on mathematics programs and courses offered at your college.

You are ultimately responsible for obtaining accurate information

When getting advice on programs and courses, you have to make sure it is accurate. If you make bad decisions based on receiving incorrect information, your chances of having it excused will not be good. Consequently, get information and advice from multiple sources, and then put it all together to make your decisions. Advice is available concerning mathematics courses and programs from many sources including students, administrators, support staff, and written materials. Depending on the sources, you will get information and advice ranging from terrible to excellent. I advised students for many years and was privy to many horror stories resulting from bad information and advice students received. Using multiple sources to make an academic decision is no different from what you would do if you were buying a house or car, deciding on a doctor, choosing a spouse, or selecting a job. The more you know, the more you will be able to separate the wheat from the chaff, the truth from fiction, and the accurate from the inaccurate.

You have to make the decision to obtain help

Reading this section should convince you that plenty of help is available from a variety of places on your campus. You are the only person who can make the decision to get this help. Don't hesitate to do so.

4. Hiring a Personal Mathematics Tutor

A personal mathematics tutor, as defined here, is an individual who has contracted with you, formally or informally, to meet with you on a regular basis to help you learn mathematics. Typically this person has a very good to excellent mathematics background and appropriate teaching or tutorial experience. However, a student in the same course as you could qualify as your personal mathematics tutor, but only if it is understood by both of you that you are the tutee and he or she is the tutor. Having a classmate as a tutor can be effective, but it is not discussed here, per se. However, some of the comments in the ensuing sections do apply to these student tutors.

Should You Have a Personal Tutor?

It is not possible to lay out all the circumstances under which it is wise to have a personal mathematics tutor. Reasons range from high achieving students who want to do better in a course, to low achieving students who just want to pass the course. Addressed here are the circumstances for which a tutor seems necessary for you if you want a reasonable chance to succeed in your mathematics course.

Suppose you have, to an appropriate degree, the prerequisites for your mathematics course, and have been following most of the advice in this book on working in a mathematics course, including getting the help that your campus offers. Yet you still struggle in the course, receiving grades of less than C on your examinations. Then it is probably wise for you to obtain a personal tutor. A tutor's help will probably not be of immediate benefit to you, but could prove to be beneficial over a period of time. Thus, it is important not to wait too long to get a personal tutor. Having a personal tutor does not necessarily mean that you will succeed in the course; it just improves your chances of success. A personal tutor cannot compensate for weak knowledge of content that is prerequisite for the course you are taking. If you significantly lack prerequisite knowledge, then you need to drop the course and enroll in the proper prerequisite course. (Chapter 4 addresses mathematics prerequisites.)

The Cost of a Personal Tutor

If you are not provided a personal tutor through some program on your campus, then you most likely will have to pay for your tutor. If your tutor is an undergraduate college student, then the rate per hour can be significantly lower than the rate for a non-student such as a high school mathematics teacher. It is not unreasonable for a non-student tutor to charge in the neighborhood of $25–$40 per hour. Determine the going rate for private music lessons, and you have a good idea of the rate for a mathematics tutor. The money is worth it if the tutor is very knowledgeable about the mathematics you are studying, is reliable, is very concerned about your progress, and has excellent tutoring skills. More will be said on this shortly.

What a Personal Tutor Can Do for You

A personal tutor can only *support* your efforts to succeed in mathematics. A personal mathematics tutor, among other things, can help you—

- Help yourself
- Know and use good study techniques
- Be motivated and encouraged

- Learn prerequisite concepts that you did not learn as well as you should have
- Discern if your prerequisite knowledge is too weak to stay in the course
- Overcome difficulties you are having with specific content of the course, including your assignments

You and your tutor have to be in this venture together. Both of you have responsibilities, but ultimately your success in mathematics depends on you, not your tutor.

Selecting a Personal Tutor

The types of personal tutors include these:

Undergraduate or graduate students designated by the mathematics department. The mathematics department may have a list of students offering their tutoring services for specific courses. You may want to select a student as your personal tutor because they are typically closer in age to you, have the content of courses for which they are tutoring fresher in their minds, and charge less for their services. It is also easier to agree on a meeting time and place. One disadvantage is that they do not have the experience in tutoring of a more professional tutor. This lack of tutorial experience is not much of a problem for some student tutors, but it is for others. You should not expect student tutors to volunteer their time. High achieving students are generally very busy and may be willing to tutor if they are adequately compensated. Some will do it just for the money whereas others will do it for the money and the experience. There are very good students who do a creditable job tutoring in mathematics. I recruited students over the years to serve as personal tutors for my students and they performed well.

Undergraduate students recommended by your instructor. Ask your instructor if he or she can recommend students as tutors. You can also go to other instructors, friends, classmates, or acquaintances for their tutor recommendations. If a student is recommended who is currently in your course, you have the advantage of being tutored by someone working on the same content as you. A disadvantage is that this student tutor has not yet completed the course and therefore cannot tutor from that perspective.

Classmates you approach on your own. You can identify students in your mathematics class who you think you can relate to and who might be good tutors. Ask them if they would be willing to tutor for pay.

Current or former secondary or community college mathematics teachers. To find these professional tutors, (1) call or visit high schools, community colleges, or four-year colleges, and talk to professors or staff in the mathematics department; (2) ask your classmates, friends and acquaintances; (3) ask college mathematics instructors, especially those involved with the mathematics or mathematics education preparation of pre-service secondary mathematics teachers; or (4) ask the director of your campus support services center.

Professionals employed by a private tutoring service. Consult the advertising pages of a telephone directory or search online.

Qualities to look for in a personal tutor

A mathematics tutor has to realize that his or her major role is to *help you help yourself*. Tutoring is not just a matter of showing you how to do a problem or how a certain step in a proof follows from the previous steps. At times that has to be done. In the main, a tutor has to be able to diagnose why you are having difficulty with specific content, and then be skilled enough to help you help yourself progress from that point on. The tutor has to ask you a series of questions to determine where you are in your understanding of prerequisite knowledge. This is the starting point of your work together. If there is another way, I don't know of it.

To help you help yourself, your tutor has to let you know what is impeding your success. The tutor has to discern how you go about learning or studying mathematics, and then make suggestions for improvement. You would expect your tutor to address much of the advice in this book on how to learn mathematics successfully. Inexperienced tutors will address far less of this than experienced tutors will. You have to evaluate every now and then if you are benefiting from your tutor. If not, seek another.

How Often to Meet with Your Personal Tutor

Typically, a tutor cannot be of much benefit to you if you do not meet with him or her for at least one to two hours per week, over many weeks. The primary exception to this would be for something that can be handled quickly, such as help understanding a specific topic.

Preparing for Your Personal Tutor

Whether your tutoring session is of benefit to you also depends on how well you prepare for it, and how you use the session when you are there. You need to come to your tutor with a host of specific questions, and continue to ask questions throughout the tutoring session until you understand. These should be questions on your reading, on your class notes, on your attempts at the problem exercises, and on how you can work better in the course. Do many of the activities in preparation for your tutor session that you are advised to do when preparing for a visit to your instructor during his or her office hours. (See section 1 on obtaining help from your instructor.)

Remembering, Reviewing, and Summarizing Mathematics

A key benefit of reviewing mathematics is to remember mathematics; hence, it makes sense to discuss these two topics in the same chapter. This chapter is also a good place to discuss summarizing mathematics since it is a valuable technique for learning, reviewing, and recalling mathematics.

There are many facets of memory, which means that there are many facets of remembering mathematics. The first section is a discussion of these facets, including the major issues of the importance of remembering mathematics, and the means to do so.

1. Remembering Mathematics

Do you have difficulty remembering mathematics? Do you believe that you have a bad mathematics memory and that there is little or nothing you can do about it? If your answer to either of these questions is "Yes," then you can benefit from reading this section and implementing the suggestions given. It is unlikely that you have a faulty memory, present at your birth, which you can do nothing to improve. It is more likely that you need practice using it wisely and efficiently to remember knowledge, including mathematical knowledge. We can all improve our memory; hence, the information given here on memory is beneficial for all students, not just those who believe they have a poor memory.

In addition to the content of this chapter, considerable advice appears *throughou*t this book that will help you remember mathematics.

You often hear of the importance of memorizing mathematics with understanding. Remembering mathematics has a lot to do with understanding it, which mainly consists of knowing how mathematics ideas are related. (Considerable discussion on this appears in chapter 9.) Many of these ideas, and others related to memory that are scattered throughout previous chapters, are brought together in this section so you can read on the topic of remembering mathematics as a coherent whole. Many mathematics educators view rote memorization of mathematics negatively (i.e., learning it by mechanical repetition without much understanding of its meaning), and I am one of them. Unfortunately some instructors and students, *incorrectly* interpret this to mean that no plans need to be made for memorizing mathematics content. If this is your belief, I suspect you will change your mind after reading this chapter.

Short-Term versus Long-Term Memory

Our main interest here is with long-term memory. Short-term memory is what we momentarily remember. We all remember situations when we were introduced to someone and seconds later could not remember the person's name. Or, we are trying to locate someone's house, look at the address that we have on a piece of paper, and then after driving a half block or so, cannot remember it. Most of us, with a few suggestions and an increased focus, can improve our short-term memory. What we have more difficulty with is our long-term memory, which impinges greatly on our learning mathematics.

Remembering Details versus Generalizations

I would be hard pressed to say that it is more important to remember generalizations than it is to remember details. They are both important to remember, but particular circumstances may make one more important to recall than the other. One huge advantage of remembering generalizations, as opposed to details, is that many details can be quickly derived from generalizations; hence, there is less of a need to remember specific details. Details are more difficult to remember than generalizations because details are not as interesting as ideas conveyed by generalizations. In addition to this, details are primarily learned through repetition whereas generalizations involve understanding, making them easier to recall. For example, recalling the detail, $\csc 45^0 = \sqrt{2}$, is typically more difficult than recalling the Pythagorean

Theorem: The square of the length of the hypotenuse of a right triangle equals the sum of the squares of the lengths of the two legs of the triangle.

Importance of Memory to Learning

You may not like having to remember content that you think is detailed or trivial, but not doing so can have a negative impact on learning and solving problems that use this content. Paul Halmos reported a humorous incident involving the mathematician Tamarkin:

> On a PhD oral he asked the candidate about the convergence properties of a certain hypergeometric series. "I don't remember," said the student, "but I can always look it up if I need it." Tamarkin was not pleased. "That doesn't seem to be true," he said, because you sure need it now."[1]

When solving problems, you need to have a lot of knowledge at your fingertips, including knowledge that comes automatically to you. Flipping through books to recall what you might need to solve a problem interferes with the flow of information that you need to have within your grasp, and relate, as you reflect on the problem.

You Cannot Understand without the Use of Memory

Imagine trying to understand something without having short- or long-term memory. Suppose you are reading the statement of a problem. In your efforts to understand the key words of the problem and how they relate, you may have to integrate what is new in the problem with what you recall from your past experiences, including content from your class notes, what you read in the textbook, small group discussion, the last unit, your previous mathematics course, or an elementary or secondary mathematics course. Wouldn't you say it is difficult to understand without having adequate memory? To understand you have to carry over knowledge. Understanding is built on recalling a chain of related events and that requires recalling them when the demand is there to do so. This is not to say that you won't forget some things, but it does say that you will struggle understanding new things if your recall ability is weak. But don't despair because your recall ability can be improved. In summary, Erickson says that, "Without memory as an active agent there can be no accumulation of meaning persisting over time." This is an incredibly powerful statement, and one that should motivate you to employ the many means at your disposal to remember what you have learned, with as much understanding as you can acquire.

When You Begin to Forget

Unless you use special means to remember content, most of what you learn is forgotten soon after it is learned. For some students, forgetting may happen in a matter of minutes or a few hours; for others it may be a half-day or day. This is not very consoling, but you need to know about this so you can do something about it. Much of the mathematics you learn you want to retain, not just until the next mathematics examination is over, but perhaps for weeks, months, semesters, years, or a lifetime.

Why You Forget

There are a host of reasons why you forget mathematics, including many that you most likely don't know about. Fortunately, we know many primary reasons for not remembering and something can be done about most of them. The reasons we forget include—

- An inadequate knowledge base of mathematics
- Lack of initial learning
- No relearning
- Inappropriate review
- Not organizing for recall
- Interference from learning other things
- Not valuing what is learned
- No emotional investment in what is learned
- No intent to remember
- No plan for remembering
- Not restricting what needs to be remembered
- Lack of aspirations or ambition
- Undue anxiety

It is not difficult to surmise that these reasons will cause you to forget, and you need to have a plan to counteract them. We now turn to some counteractive measures.

Ways to Improve Remembering Mathematics

Measures to take to improve your ability to remember or recall mathematics are discussed here. Realize that they will work if you take them to heart and pursue them diligently.

Learn initially. Failure to *initially* learn mathematics content is most likely the greatest cause of not being able to retain it. You cannot retain something that you haven't adequately learned. And if you have adequately learned it, you are well on your way to retaining it. The key is to know what it means to learn something. You learn something when you understand it—that is, when it is meaningful to you—and it is meaningful to you if you can *relate it to other things you know*. (See section 2 of chapter 9 on understanding mathematics.) If you complain about having a poor memory in mathematics, you are mostly likely saying that you don't have a good understanding of mathematics. To improve your memory in mathematics, focus on understanding mathematics. *This is by far your best bet to retain it.*

Association. Memory experts use the term association when referring to connecting or relating details and generalizations to something else. Think about how you remember this information:

- Where you parked your car in a large parking lot. ["It's two aisles down from where I entered the lot."]
- The first name of the person you just met. ["Her first name is the same as that of my cousin Mary."]
- $\csc\dfrac{\pi}{6}$. ["I know that $\csc\dfrac{\pi}{6} = \dfrac{1}{\sin\dfrac{\pi}{6}} = \dfrac{1}{\dfrac{1}{2}} = 2$."]
- The graph of $\cos x$. ["It is the graph of $\sin x$, translated $\dfrac{\pi}{2}$ units to the left."]
- The graph of $\ln x$. ["It is the graph of e^x, reflected in the line $y = x$."]
- The period of $\cot x$. ["It is the same as the period of $\tan x$."]
- Whether $f(x) = |x|$ has a derivative at 0. ["Its graph has a "sharp corner" at 0; therefore it cannot have a derivative at 0."]
- Radian measure of $135°$. ["Since π radians $= 180^0, 1^0 = \dfrac{\pi}{180}$ radians $\Rightarrow 135^0 = 135 \times \dfrac{\pi}{180}$ radians $= .75$ radians."]

Memory is used to associate the knowledge in the brackets with what you are asked to find. It is a matter of *deriving* the results knowing this associated knowledge. Deriving the results is often done so quickly that it appears that you are recalling them. Some people say that *forgetting something is nothing more than not having the related knowledge to derive it.* The greater the amount of knowledge you possess, the greater the likelihood that you will be able to recall, or quickly reconstruct, a specific piece of knowledge. Your chances of relating one or more pieces of knowledge to what you need to recall is increased when you have a vast repertoire of knowledge to draw from. This related knowledge is then used to recall or reconstruct what you need.

Increase your knowledge base of mathematics. It is encouraging to know that the more mathematics you learn, the easier it is to recall what you have learned. The reason for this is that the field of mathematics can be viewed as one big story, or a series of smaller stories. A story is comprised of relationships. The more of the story you know (i.e., the more relationships in the story you know), the greater will be your understanding of any aspect of the story, and the greater will be your ability to recall it. Mathematically speaking, the more mathematics you know, the easier it is to learn other mathematics, and the easier it is to recall what you have learned. This is a win-win situation, one that should motivate you to learn as much as you can. If this book is about anything, it is about supporting you in building your knowledge base of mathematics. Suggestions are given throughout this book for building a knowledge base that will equip you to accomplish short-term goals in your course (e.g., completing an assignment). Fortunately, in the process of accomplishing short-term goals, you are building a knowledge base in mathematics that will stand you in good stead in the long term.

Relearn. Many things that you learn can be learned better. Hence, restudying what you have learned results in increased learning. Reviewing something you have learned helps you learn it better, and learning it better improves your chances of retrieving it when called upon to do so.

Appropriate review. There are so many facets to reviewing mathematics content that its treatment deserves a special section, and that is the content of section 2. Suffice it to say now that having an appropriate review strategy can make the difference between success and failure in your mathematics course. Meaningful review reinforces and enhances initial learning, hence increasing your ability to recall what you have learned.

Carefully select what you want to remember. You cannot remember everything, so don't attempt to do so. What you select to remember you want to learn well, and focusing on less content affords you this opportunity. But what should you select to remember? Lots of thought needs to go into deciding this, and included in making these decisions are your instructor's and textbook authors' comments on what they value. However, this advice is wise to heed: Remember information that can be used to easily derive other information. The more information that can be derived from a piece of knowledge, the more valuable it is to remember that knowledge. In general, it makes little sense to work on remembering information that you can easily derive from other information. Any attempt to do so involves remembering lots of details, which interferes with remembering more important information. It is important to remember *graphs of basic functions*, including the labeling of key values on those graphs. This knowledge allows you to retrieve considerable information. Suppose you were asked these questions:

- What are the domain and range of csc x?
- What is the period of csc x?
- What is the minimum value of csc x? Maximum value?
- Does csc x have vertical asymptotes and horizontal asymptotes? Where?
- On what intervals is csc x concave up? Concave down?
- On what intervals is csc x increasing? Decreasing?

Should you work diligently to memorize the answers to each of the above questions? Absolutely not. All of the above questions can be answered by knowing the graph of csc x. *Know well the graphs of these basic functions, including the labels of key values on the axes: linear, quadratic, trigonometric, inverse trigonometric, logarithmic, and exponential.* You have heard that a picture is worth a thousand words. In the same vein it can be said that a graph, which is a picture, is worth a thousand facts. Remember the graph, not the thousand facts.

Other important information to learn well, and hence have a greater chance to recall, are *definitions and general principles*. They are useful in answering or deriving a host of specific questions. For example, suppose you were asked to determine the intervals where the function $h(x) = \dfrac{x^5 - 7x - 8}{3x^2 - 2x - 8}$ is continuous. To answer this question for this and similar functions, you need to recall three things: (1) the definition of a rational function (note that $h(x)$ is a rational function), (2) the principle that a rational function is continuous everywhere it is defined, and (3) a rational function is not defined for values that yield a zero for its denominator. This is useful information to remember for it is used to answer many questions.

Knowing how *the graphs of specific functions can be derived from the graphs of related functions* is also worth remembering. This knowledge minimizes your work in sketching the graphs of functions. For example, suppose you were asked to sketch the graph of $g(x) = \ln(x - 3) + 7$. Recalling the effects that 3 and 7 have on $g(x)$ allows you to say that the graph of $g(x)$ is the graph of $f(x) = \ln x$, translated 3 units to the right and 7 units upward. As another example, the graph of $h(x) = -2(x + 5)^3$ is the graph of $g(x) = x^3$, reflected in the x-axis, translated 5 units to the left, and stretched vertically by a factor of 2 units.

You also need to be selective choosing the computational algorithms that you want to remember. For example, you would not want to remember an obscure computational algorithm that is seldom used. Once again, you cannot remember everything; therefore, be wise selecting what you want to remember, and then learn it well.

Have the intent to remember, and the attitude that you can remember. You are better served by having the intent to remember something than by expecting it to happen by chance. Tell yourself at the outset of a specific study that you want to remember what you are studying, and then take the necessary steps to see that it happens. This increases your chances of remembering it. When in the process of learning specific content, intent to remember is compromised by succumbing to distractions, or only wanting to learn it to complete an assignment or pass an examination. Having the intent to remember focuses your attention on the task at hand. You tell yourself that you want to learn it now, not later, and you use various techniques early on to quiz yourself on what you are learning. Have the intent to remember when you approach your reading or problem assignments, attend class, and discuss mathematics with others. You are working on the intent to remember when you orally recite to yourself what you are reading, ask questions in class, take good notes, etc. Having the intent to remember something only for the short term is not as beneficial to memory recall as having the intent to remember it permanently. As with most things, attitude also plays an important role in recalling content. If you believe that you can't remember no matter what you do, then that becomes a self-fulfilling prophecy.

Meaningful organization of content. Your ability to recall is helped by organizing content in a meaningful way and then studying the content using this organization. This can be done a variety of ways, depending on the content and what you find meaningful. Writing a summary statement of the development of a particular topic is one way of organizing content for recall. For example, if you have worked on a unit involving the inverse of a function, summarize in story form the main aspects of what you studied and the order in which it was studied. In doing this you are giving the story of inverse functions. In your story you define the inverse of a function, talk about how you can find the inverse of a function, and mention the theorems that involve inverse functions (for example, the graph of a function and its inverse are symmetric to the line $y = x$). If you can do this then you should have less difficulty recalling any part of it. Why? Because you have written a story and parts of stories are easy to recall since they are related to other parts of the story—your story is a continuum. A good summary statement of a topic, chapter, or unit reveals how well you understand the key aspects of what you summarized. Write your summary so that it is clear and concise. More is said on writing summary statements in section 2.

One way you can organize content is to group similar content. For example, you can group—

- Forms of the equation of a line, including a description of why one form is used over another
- Solutions to the various types of algebraic inequalities
- Basic derivative theorems
- Formulas for the area, circumference, and volume of basic geometric figures
- Fundamental trigonometric identities
- Graphs of basic trigonometric functions
- Basic limit theorems
- A variety of problems that use the same technique to solve them
- All the definitions in a unit of study

Have an investment in what you are to learn. What are your reasons for learning the content of your mathematics course? What is the payoff for your investment in the course? Do you value the content? Is it interesting to you? Will it help in your career? How well you remember is related to how well you learn it, which depends on the strength of your motivation to learn. Why are you in your mathematics course? Increasing motivation to succeed in mathematics is the topic of section 2 of chapter 11.

Use visualization. Your memory is assisted when you visualize what needs to be remembered since you are using a part of your brain that was mainly not used when you were reading or listening. Thus, whenever possible, visualize what you are learning. Ask yourself if it is possible to draw a picture or diagram, make a chart, or construct a graph. This can also be done with very abstract content, but you may have to be somewhat creative. To help remember the key steps in the proof of a theorem, make a flow chart of them. To solve a problem involving various relationships among elements of the problem, represent each element by a point and join two points with a line segment if the elements they represent are related. Visualization is a powerful learning and memory tool. When you need a piece of information, you can use the "mind's eye" to visualize it. If you have a visual preference for learning you will benefit more than those who don't if you use this study technique. The characteristics of those with a visual preference for learning are given in section 1 of chapter 13.

Remember exact statements. When it comes to the importance of remembering exact statements, like so many other things in the not-too-distant past, the baby was thrown out with the bath water. Allow me to explain. Awhile back many mathematics educators complained that there was too much memorization of mathematics at the expense of understanding mathematics. How true that was. Unfortunately, too many mathematics educators overreacted to this statement by downplaying memorization so much that any memorization work was viewed negatively. There is a need to memorize exact statements of many definitions and many theorems, as given by textbook authors and instructors. This does require some rote memorization or mechanical repetition, in addition to the understanding that should accompany it. Instructors who do not encourage their students to memorize exact statements of definitions and theorems are doing them a disservice because—

- Definitions and theorems have been carefully worded to say exactly what needs to be said. Many students have difficulty supplying their own words to come up with equivalent statements.

- Memorizing these statements compels you to read them carefully and hopefully, reflect on what is said. When you need to apply the information in a statement, and the statement is not in front of you, you will have difficulty doing this if you have not memorized the statement, and understood it.

- Since statements of definitions and theorems are carefully crafted, very precise language is used, containing key words such as "if," "then," "there exists," "for every," "or," "and," "at least," and "at most." Memorizing the exact wording of a statement forces you to pay attention to these words. It also helps you remember to use these words when you have to craft a precise statement. However, I never discouraged students who gave a definition using a formulation that was a little different from that given by me or textbook authors. However, I cannot tell you the number of times I saw students incorrectly apply a definition or theorem when working exercises. When I asked them, in class or in my office, to give me the statement of a specific definition or theorem, their response bordered on total nonsense. How can they properly apply a definition or theorem if they do not know what it precisely says? Memorizing definitions is discussed in section 1 of chapter 16.

Recitation. Recitation, as used here, is the act of saying aloud content that you are learning or have learned, using or not using your own words. This is a strong means of learning and promoting recall. That it is a viable means is supported by the following rationale:

- If you know before reading specific content that you will be reciting it after you read it, you most likely will pay more attention to it as you read. The *intent to remember technique* that was discussed earlier in this section is operational here. The role of recitation in answering questions that you construct based on reading your textbook is discussed in section 2 of chapter 15.

- Reading content and then reciting it requires the use of different parts of your brain.

- We often say silently the statements we say aloud; hence, this additional repetition helps you remember them. Some of our oral statements are chants or mantras. Recall how valuable knowing the Counting Chant ("one," "two," "three,") has been in your life. For those of you who have had an introductory calculus course, the chant for finding the derivative of the quotient of two functions (i.e., "It is the denominator times the derivative of the numerator, minus the numerator times the derivative of the denominator, all over the denominator square"), has stood you in good stead when applying it to find the derivative of a specific function. You most likely chant it silently. Do you often recite mathematics content to yourself? I suspect that if you do, you do it infrequently. Do you recite mathematics content with others? Find a partner and take turns doing so.

Consolidation. The term consolidation is one that may not be familiar to you. It is a term that appears often in memory studies. As used here, consolidation is the process in the brain that enables you to recall content. Perhaps the best promoter of consolidation is *time*. It takes time for content to assimilate in your brain. You have to plan for consolidation to take place. Much of what appears in this section, and in other parts of this book, promotes consolidation, which allows time for information to be cemented in your brain. One very effective technique to promote consolidation has not been discussed, per se in this book. This is the technique of distributive practice, which is now discussed.

Distribute your study over more study periods. It may seem to you by now that the techniques to increase your recall of mathematics content are endless (they probably are). Another key technique for learning and recalling is to distribute your study over more study periods, which means that each study period will be shorter in length. This technique is the technique of *distributive practice*. Reasons for using it may go counter to what you believe, but you may change your mind after reading this logic: (1) Spacing study periods allows more time for consolidation to take hold. (2) You remember more at the outset and at the end of a study session; hence, if more is studied between the outset and end of the session, more will be forgotten.

Let me give an example of the second reason for using distributive practice. It is not clear how much time has to pass for this time to constitute the outset of a study session, and the same can be said about the end of the session. For the sake of argument, let's suppose they are 20-

minute intervals. Now assume that you study for three consecutive hours. Most of what you remember occurs during 40 minutes of this three-hour session (20 minutes at the beginning and 20 minutes at the end). Consider a different situation in which you study for one-hour time blocks at three different times, each spaced by about a half or full day. Most of what you remember occurs during two hours of these three hours $[(20 + 20) + (20 + 20) + (20 + 20)] = 120$ minutes = two hours]. Hence, you have studied the same amount of time in each situation, but with the second scenario you could have retained approximately three times as much content. You can rightly question the preciseness of these figures, but they do indicate the *importance of distributed practice*.

Not only do shorter durations of study favor distributive practice and consolidation, but also it is easier to fit shorter durations of study time into your schedule. For example, you can study immediately after you rise in the morning and before you head off to school, between two classes spaced somewhat close to each other, or during your lunch hour. *A word of caution*: You can err thinking that all study sessions should be short. At times, you absolutely need longer study sessions to grapple with more complex content.

Using only one or two of the memory aids presented in this section will help you recall content or retard forgetting; however, imagine how adept you will be recalling content if you use most or all of them. This does not suggest that you will not forget content, for these memory aids are not cure-alls. If I had to choose only one of them to employ it would be to learn the content better.

2. Reviewing Mathematics

There is no doubt that the mathematics content you learn will fade from your memory if it is not used frequently. This, coupled with the knowledge that it is not possible to use this content very often in your courses, dictates that you need to have a *systematic review plan* for keeping it present. Bertrand Russell, the famous British philosopher who is well known for his work in mathematical logic captured this well: "What is best in mathematics deserves not merely to be learned as a task but to be assimilated as a part of daily thought, and brought again and again before the mind with ever-renewed encouragement."

It has come to my attention over the years that many students are learning specific content for the first time when they are reviewing for an examination. Enhancing one's learning during a review will almost always take place, but when you are reviewing content during your final preparation for an examination, your initial learning phase should be over with. Avoid mixing the time to initially learn a topic with the time to review the topic. There are distinctions among learning, reviewing, relearning, and overlearning.

Distinctions Among Learning, Relearning, Reviewing, and Overlearning Mathematics

When reviewing mathematics you are relooking at mathematics *that you have learned*. It is not to be confused with relooking at mathematics that you have partially learned. If you have not forgotten what you learned, then your review can be viewed as *overlearning*. Overlearning helps ensure that you will recall what you learned, upon demand. If in your review, you cannot recall all or part of what you learned, then you go to your sources to take another look at it (for example, your notes or textbook). This is called *relearning*, which puts you at a much better place than you were when you initially approached the topic. To ensure retention of content, you use overlearning and relearning.

Are you learning content when you conduct your review of content? You have to say, "Yes," to this question since you can always learn something better. However, if this learning is more than fixing up a few loose ends now and then, you are unwisely mixing the processes of learning and reviewing. The problem with this shows up in a major way when you are reviewing within a week or less before an examination. It is then that you may find out that there is content you don't understand, and there isn't adequate time to learn it. There is not enough time for consolidation to take place, and the time spent learning what should have been learned earlier is taken from the time that needs to be spent reviewing what was learned. When this takes place, and we have all experienced it, panic and stress set in, which raises havoc with the further preparation that needs to be done, and with performance on examinations.

How often have you seen an instructor lecturing to a group of students in his or her office, within a few minutes or hours of an examination? Perhaps you were in one of these groups. These students were trying to learn specific material within a few minutes or hours of an examination. The best-prepared students can be confused on a minor point or two, which may first come to light during a review. But if the problem is more severe, it is far too late to be in such a position shortly before an examination. There needs to be ample time to review close to an examination, and that will not happen if you are primarily using this time to learn new ideas from throughout the unit. You do not want to be in the position of being unable to apply a specific computational algorithm without making errors, not understanding key steps in the proof of a theorem, not being able to apply the results of a specific theorem to solve a variety of problems, not understanding a specific definition, and so on. You might say, "Better late than never," and that is true. But I say, "Better early than late." Having a systematic plan for reviewing throughout a unit includes finding out early in the unit (i.e., prior to a final review), the topics that you need to learn better.

Gaining Insight Through Review

One advantage of reviewing a lot of content within a short period of time is that, with this much knowledge before you, you are well equipped to discover new insights. Perhaps an unusually alert student arrived naturally at these insights while progressing through a unit of study, but if this is not you, you can also arrive there if you know how and when to review. Do you have a good systematic review plan? Perhaps you will be better equipped to answer this question when you read the next two subsections, which address key times to review and methods of review.

Key Times to Review

Any systematic plan that you construct for review must address key times to review. For example, you might know what to review and use appropriate methods of review, but the fruit of your review will be significantly reduced if not done at key times. The mindset of most students is that they think reviewing strictly applies to reviewing for an examination near the time of the examination. The most important thing for you to learn from this section is that reviewing needs to take place throughout a unit, *beginning the first day of the unit.*

Some of the key recommended times to review may surprise you. They include these:

As soon as you can after a class session ends. This is a most important time since forgetting takes place quickly. You will also discover that you are confused on some topics presented in class because your notes are incomplete or consolidation has not had time to take place.

Shortly before the next class session begins. Review what you did in the last class in preparation for the class session about to begin. This review will enable you to better understand the class session and to ask questions in class. It is often the difference maker between a class session being incomprehensible or very meaningful. It won't take much time, perhaps 10 to 15 minutes.

Just before beginning an assignment. Not only is this good to do to reinforce what you learned, but the content you are reviewing is most likely needed to complete the assignment.

Almost immediately after completion of a reading or problem assignment. I found that very few students looked over a reading or problem assignment they just completed. They might say to themselves, "Why should I do this, I just completed it?" You will be surprised at how well a quick review can reinforce what you worked on over one, two, or more hours. (The role review plays in reading a mathematics textbook is presented in section 2 of chapter 15.) Once again, realize that frequent short reviews pay more dividends than infrequent longer reviews.

Once a week, review all that you learned up through that week. This may take from a half hour to an hour, but it is time well spent. Record a time for doing this weekly review into your schedule. If possible, schedule this weekly review as the last thing you do in an evening for that is the best time to ensure a more lasting memory of what is reviewed. Do you conduct a weekly review?

Every week or so review content from past units. Include in this review a rereading of your notes and a reworking of a variety of problems. Students normally don't review from past units when studying a new unit; yet knowing some of this content is necessary for success in the current unit and in future units. I know instructors, including myself, who deliberately place a few items on a unit test that come from the previous unit.

Start your final review for a unit examination at least one week before the examination, and start earlier than that to review for the final examination. Realize that far less time needs to be spent reviewing for an examination if you have been reviewing throughout a unit or the semester. Do you think that your mathematics instructor has an obligation to conduct in-class reviews prior to an examination? If you said, "Yes," think again. Reviewing is your responsibility, not your instructor's. Your review for a unit examination starts the day you begin the unit, and your review for a final examination starts the first day of the semester.

You may miss some key review times due to planned or unplanned interruptions. This is not a problem unless it happens frequently. Make it a goal to review during the key times that have been mentioned here, and then strive to meet that goal.

Methods of Reviewing

In constructing a systematic plan for reviewing mathematics content, besides selecting key times to review, you have to select the content to review and methods for reviewing it. (Guidelines for choosing the types of content to review, or to recall, appear in section 1.)

Methods to use in your review include these:

Review actively, not passively. If I had to make a bet, it would be that the majority of students are passive in their review. That is a big mistake. Do you make this mistake? Does your review basically consist of reading over your notes and solutions to problems? This is not a bad way to review for examinations if your examinations consist of reading and answering questions on whether you agree with what you read. Examinations are very different from this. You will be asked to recall important content that can be used to answer questions that appear on your examinations. To answer some of these questions, you have to go through your repertoire of content, putting it together at times in creative ways. *This demands active review*, which mainly consists of reciting answers to questions that you have posed during your study of the unit, which relate to the content.

Write summaries of content, including that of a topic, section of the textbook, or unit of study. This technique is so important that it is the focus of section 3, which follows this section.

Work problems randomly selected from various sections in the textbook. One way to accomplish this is to write a variety of problems on 3×5 cards, shuffle them, and work them in the order they appear. Place your solutions on the backside of the cards, and then use these cards again in later reviews. When you go through the cards again, only use the solution on the card to *check* the solution that you just derived. Remember, you want to be an active reviewer and that is not accomplished by just reading the problem and its solution.

Write definitions and statements of theorems. Place the name of a concept or theorem on one side of a 3×5 card and the definition or statement on the other side. In your review, write the definition or theorem statement and then check the card to see if you wrote it correctly.

Work chapter review exercises. These exercises appear at the end of a textbook section, and are labeled as such. These are often good to work since they cover a variety of ideas and problems from the textbook chapters comprising the unit of study. Many instructors do not assign them, but that should not keep you from working them.

Work a study guide for a unit. You, a classmate, or your instructor can construct a study guide for a unit of study.

You might have other methods of reviewing mathematics content. Determine whether they are passive or active techniques. Then make a decision to keep them or discontinue using them.

To conclude this section, this perspective is placed on the importance of having a viable systematic review plan: Imagine studying two hours one day on a particular mathematics topic. In three weeks you would most likely forget much of it. Suppose you retained 20% of it. Contrast that with studying for only one hour on the given day, and then spending 30 minutes reviewing it later in the week, 20 minutes reviewing it near the end of the second week, and 10 minutes at some point during the third week. At the end of three weeks, with the same investment of time, you would most likely retain almost all of what you studied with the latter plan. I would be remiss if I did not mention this: For all intents and purposes, there is a form of built-in review that takes place for many mathematics topics because these topics are used to learn new content. Call this defacto review, if you will; however, it cannot take the place of planned systematic review. Never underestimate the power of frequent review. An analogy to the positive effects of frequent review is the effect that compounding interest has on your bank balance at the end of a compounding period.

3. Summarizing Mathematics

A mathematics study technique that is slighted by most students is *summarizing content* from their mathematics course at various times within a unit of study. Summarization is creating a shorthand version, a condensed form, of the main points of a development over a specific period of time. That is, it is creating an outline of the highlights of a body of mathematics content over that period of time. For example, a summary of a section of the textbook or a unit of study might consist of a listing of the statements of the definitions and major theorems that were introduced, the various techniques associated with them, and the uses for all of this. More generally speaking, a summary is a succinct informal accounting of a body of mathematics content, often including reasons for its existence. At various times in a unit of study, you (as well as your instructor) need to summarize content. Summarizing content is one of the more critical responsibilities of an instructor; however, it is also important for you to summarize.

Writing a good summary requires a good understanding of what you are writing. In the process of writing a summary you may find that your understanding is deficient and that you need to improve it before continuing your summary. Armstrong said, "A well-written summary is a test of how completely you have understood a paragraph, a chapter, an assignment, or a book—and how concisely and clearly you have been able to condense it without loss of meaning."[2] Writing summaries is germane to your development as a mathematics student. I cannot overemphasize the value of practicing this technique. If you have not been writing summaries, you will struggle somewhat in your initial efforts to do so.

Reasons for Writing a Summary

Once again it is my belief that you are more apt to subscribe to a specific learning technique if you know what it can accomplish for you. Objectives attained by writing summary *statements*, as opposed to other forms of summaries, include the following:

To attain a good overview of the key aspects of a body of knowledge. Writing summaries helps you see the big picture, or the important ingredients you have been exposed to in a day's, week's, or unit's work.

To see how content is related or tied together. If mathematics is about anything it is about relationships. Writing summaries highlights these relationships; which will help you learn new content and identify the content that can be used to work specific exercises.

To develop your ability to separate key content from supporting content.

To improve your ability to write mathematics stories, and to organize content.

To provide another review technique that helps in recalling the key features of content.

The importance of these objectives to your mathematics learning should provide ample motivation for you to make summarizing a key component of your study techniques.

Key Times to Summarize

The important times to summarize in your mathematics course are the same as the key times to review that were listed in section 2. You need to write summaries throughout a unit of study. There are very few things in your studying that will serve you as well as summarizing.

Types of Summaries

The type of summary we have been discussing is referred to as a *summary statement*, often consisting of a series of sub-statements. There are other ways to summarize content besides writing a summary statement, and these have their advantages and disadvantages over writing a summary statement. Other methods of summarizing include constructing a classification hierarchy, chronological outline, or flow chart. These summaries use far fewer words, are not stories to the extent that summary statements are, and may be less problematic for you. They are not discussed in this book. The focus here is on writing summary statements because of the many benefits you can accrue from writing them.

Examples of Summary Statements

A summary statement is written in paragraph form using proper grammar. In such statements, you use your own words and your sentences should flow nicely, avoiding any evaluative remarks. You are giving, in story form, the highlights of a body of content, including an indication of how its development progressed. An analogy would be giving an abbreviated account to a friend of how you came to enroll in the college you are attending. You would probably mention how you first became interested in the college, what you did to find out more about the college, how you eliminated other colleges, and who or what was instrumental in making your final decision. You would dispense with less

important details of your journey. Hopefully, your story would be coherent and unified. Your story would be different from another student's story. Similarly, summaries of the same body of mathematics
content written by each of two students would be different.

Three examples of summary statements follow:

Example 1. (This is a summary of a section of a precalculus textbook, which discusses methods of combining functions.)

> Five ways to combine functions are presented in this section.
> These combinations are the result of defining four arithmeti-
> cal operations on functions, called addition, subtraction, mul-
> tiplication, and division. A fifth operation is called
> composition. For each operation, two functions are chosen,
> the operation is applied, and the result is a third function,
> called the sum, difference, product, quotient, or composition
> of the two functions. For functions f and g, their sum, differ-
> ence, product, and quotient are denoted as $f + g, f - g, fg$, and
> f/g, respectively. The definitions of these arithmetical opera-
> tions follow: For x in the domains of functions f and g,

$$(f + g)(x) = f(x) + g(x).$$
$$(f - g)(x) = f(x) - g(x).$$
$$(fg)(x) = f(x)g(x).$$
$$(f/g)(x) = f(x)/g(x).$$

> For the quotient of two functions, x cannot be chosen if it
> makes $g(x) = 0$, for then $\dfrac{f(x)}{g(x)}$ is undefined.

> The composition of two functions f and g is denoted $f \circ g$. It is
> defined as follows: For x in the domain of g, such that $g(x)$ is in the
> domain of f,

$$(f \circ g)(x) = f(g(x)).$$

> One reason for defining the composition function is to express a
> more complex function as the composition of simpler functions.
> For example, the function $y = (2x + 3)^3$ can be viewed as the com-
> position of $f(x) = x^3$ and $g(x) = 2x + 3$.

Note. The development of this content was presented over four pages
in a textbook. Its essential features were captured in this (rather detailed) summary comprised of two paragraphs.

Example 2. (This is a summary of a class period in a calculus course on an introduction to the derivative).

> Today, the concept of the derivative of a function was introduced. A function $y = f(x) = 3x$ was graphed and two interpretations were given of the derivative of f at the real number a. The first interpretation, demonstrated graphically, is that it is the limit of the slopes of the secant lines through the points $(a, f(a))$, and $(a, f(a + h))$, as h approached 0. This limit was called the slope of the tangent line to the graph of f at $(a, f(a))$. The second interpretation is that it is the instantaneous rate of change of f at a. The slope of the line through the points $(a, f(a))$, and $(a + h, f(a + h))$ is the average rate of change of f over the interval $[a, a + h]$. As h approaches 0, the average rate approaches a value called the instantaneous rate, or rate of change of f at a. Several other graphical interpretations of examples were given to make these ideas more meaningful.

Example 3. (This is a summary of a section of a pre-calculus textbook on inverse functions.)

> The inverse of a function f, is a function and is denoted f^{-1}. The function f^{-1} takes a range value of f, call it y, and assigns it to the domain value from which it came, call it x. That is, if $y = f(x)$, then $f^{-1}(y) = x$. Stating this another way, f takes x to y and f^{-1} takes y back to x. For specific values, if $f(\frac{1}{2}) = 4$, then $f^{-1}(4) = \frac{1}{2}$. Notice that
>
> $$(f^{-1}\circ f)(\frac{1}{2}) = f^{-1}(f(\frac{1}{2})) = f^{-1}(4) = \frac{1}{2}.$$
>
> In general, when you compose a function and its inverse and apply it to x, you get x back again, regardless of the order of the composition. This is shown for $f\circ f^{-1}(4)$: $f\circ f^{-1}(4)) = f(f^{-1}(4)) = f(\frac{1}{2}) = 4$. Since this relationship characterizes a function and its inverse, the definition of the inverse of a function f is a function g, referred to as f^{-1}, such that
>
> $$g(f(x)) = x, \text{ for each } x \text{ in the domain of } g.$$

and

$$f(g(x)) = x, \text{ for each } x \text{ in the domain of } g.$$

The topic then turned to finding the inverse of a given function, say $y = f(x)$. The process for doing this is to solve for x in terms of y, replace y by x and x by y, and then solve for y in terms of x, yielding $y = f^{-1}(x)$. For example, given $y = f(x) = 1 - 2x$, we get, after making replacements, $x = 1 - 2y$, and in solving for y, we get $y = \dfrac{1-x}{2} = f^{-1}(x)$. Recalling previous knowledge that points (x,y) and (y,x) are symmetric to the line $y = x$, it turns out that the graphs of f and f^{-1} are symmetric to the line $y = x$. *Note.* If all points (a,b) satisfy $y = f(x)$, then all points (b,a) satisfy $y = f^{-1}(x)$.

Not every function, such as $y = x^2$ has an inverse function. The ordered pairs $(2,4)$ and $(-2,4)$ satisfy this function, but the ordered pairs $(4,2)$ and $(4,-2)$, obtained by interchanging the first and second numbers of each pair, cannot satisfy a function since 4 would be assigned to two values, 2 and -2, violating the definition of a function. Thus, for a function to have an inverse, it must be the case that each different domain values are assigned to different range values. Stating it another way, each range value of the function cannot be associated with two different domain values. Hence, a function has an inverse function if distinct domain values give distinct range values. Such a function is called a one-to-one function. Geometrically speaking, a function f is one-to-one (and hence has an inverse) if and only if each horizontal line intersects the graph of f in at most one point. This is called the Horizontal Line Test.

Notes. (1) This topic was developed over eight pages of the textbook from which it was taken. It was summarized here in a little more than one page, and *could have been shortened* if some of the remarks that contributed to understanding were eliminated. When writing summaries, you have to decide how many contributory remarks are too many or too few. In the main, you are writing summaries for yourself. You have to decide how much you need. (2) Imagine how much better prepared you are for an examination if you can write summary statements, as examination time nears, of all the sections of your textbook that are included in a specific unit of study, without looking at your notes, textbook, or problem assignments. This is not all you need do to review for an examination, but it is key to do. The summary statement in example 3 is very thorough and takes some skill to write. Don't be intimidated by it; yours will most likely be far less thorough. You are not expected to write summaries that are publishable. One way to shorten summaries is to leave out examples. A long summary can be viewed as a collection of smaller summaries. This view makes the work of writing a summary statement less daunting.

Self-Evaluation of Your Progress in the Course

You will be formally evaluated throughout your mathematics course. Included in these instructor evaluations of your performance will be your graded assignments and examinations. These evaluations provide important information on your progress in the course. However, you need to couple them with *your own evaluation*, called *self-evaluation*, which needs to begin the first day of class. You need to reflect frequently on how well you are doing in your mathematics course and what changes you may need to make. In this light, the following sections are addressed in this chapter:

1. Know When You Know and When You Don't Know
2. Determining Whether to Drop a Mathematics Course
3. Continuing to Work in a Course You Drop
4. Retaking a Mathematics Course
5. Learning the Hard Way

1. Know When You Know and When You Don't Know

You will receive a good grade in a mathematics course if you know the content and can show that you know it. This section helps you decide whether you know the content before you have to show your instructor that you do. You want to avoid initial awareness that you don't understand a unit's content to be when you receive your first examination back displaying a low grade. That could be as late as three to five weeks into the term, which might result in having to drop the course.

A statement expressed by many students who do poorly on an examination goes something like this: "I don't understand why I didn't do well on the examination. I was able to work all of the problems in the assignments." Or, a student would come up to me after an examination and say, "I liked the examination and think I did well on it," or "That was a good examination," obviously implying that he or she did well on it. After grading the examination, I found out that the student did poorly. These students' perceptions were out of touch with reality. They thought they understood the mathematics being tested when they did not. This was a serious problem for them.

You need to know when you know and when you don't know, at least to a degree that allows for success. There are times when the best mathematics students think they understand something and find out on an examination or graded problem assignment that they were mistaken. This happens to everyone. My concern is not with that situation, but with students who think they understand most of what they study, and in reality understand little of it. Perhaps they are experiencing a form of denial. The major problem with not knowing that you don't know is that you are unaware that you need to make changes.

Signs That Indicate You Don't Understand or Are Being Irresponsible Toward Your Course

There are many clues or signs that indicate you are confused or are being irresponsible toward your course, which will inhibit your success. In some instances there is enough time to take corrective action so you can succeed in the course. In other instances, due to an inadequate amount of time remaining in the course, you will have to drop the course or receive an unacceptable grade. Elaborations on the changes you may need to make to rectify the situation are addressed in related sections or chapters of this book. Some of the signs that indicate you don't understand or are being irresponsible toward your course are more problematic than others. How problematic they are is unique for each student. For example, missing class may cause you more problems than someone else. But don't get the wrong idea here—missing class causes almost everyone serious problems. If you are aware of troublesome signs and are honest in admitting that you display some of them, you have no excuse for saying, upon receiving a poor grade on an examination, "I didn't know that I was in trouble." I saw troublesome signs in students who performed poorly in my mathematics courses. Many of them were in denial about these signs.

Signs that indicate you don't understand or are being irresponsible toward your mathematics course, include the following:

Continual need to have your work verified by other sources, right up to when you take an examination. You are continually consulting verification sources such as classmates, instructors, answers in the back of your textbook, or a solutions manual. That is, you do not use internal checks to determine if you correctly performed a computation, understood a concept, applied a theorem, or constructed a proof.

Not understanding what your instructor is discussing with you during his or her office hours. That is, you leave your instructor's office hours almost as confused as when you arrived. This confusion may apply to questions on your reading, on your class notes, an exercise, or how your examination was graded. You may walk in with a classmate and wonder why you got fewer points on an examination question than your classmate did, and not understand your instructor's explanation. You nod your head in agreement even though you don't understand.

Having little or no idea what will be on your examinations. You are always surprised when you see the exercises that appear on an examination—they are not what you expected. This happens most frequently on a first examination in a course, even though your instructor may have handed out objectives or a study guide for the test, assigned specific problems and reading assignments throughout the unit, and given clues in class as to the content he or she views as most important. That is a lot of examination clues, and you have to recognize them. Unless your instructor's examinations are hardly representative of what is being revealed to you by his or her actions, you should have a good sense of the types of content that will be tested.

Not asking enough questions of yourself or others to achieve understanding. If you don't ask yourself and others appropriate questions, how are you going to know if you know? You need to test your understanding by reflecting on important questions such as these: Why can I make this step? How does this relate to what I just did? What uses are there for this concept? Is this a theorem or a definition? If you don't do this, how do you know that you know?

Thinking that working on the assigned exercises is basically all that you need to do. This may have worked for you in the past, or at least you thought it did; hence, you believe that it is all you need to do now. Beyond completing assigned exercises, you need to read your textbook, attend class, and work non-assigned exercises. These activities will help you know that you know.

Believing that understanding your instructor's development in class is the bulk of what you need to do with that content. Understanding your instructor's development of a specific topic does not mean that you have come close to mastering it. The thinking that has gone into it by him or her is at a higher level than what you currently possess. It is easy for you to agree with each statement made by your instructor who has explained it so well. For sure, you will gain some understanding of the problem, but it is often superficial and fleeting. You have to work with it before you can master it or before you know that you know.

Believing your assignments are thoroughly done when they are not. Many students who did badly on an examination told me that they worked all the exercises. Upon checking what they had done, I discovered that their work on the assigned problems was a hodgepodge. Some exercises were not tried, whereas others were in various stages of incompleteness. For whatever reasons, these students had a warped view of what it means to work all the exercises.

Thinking that obtaining the correct answer to a problem means you have a correct solution to the problem. You would be surprised how often you can have the correct answer but your work leading to it is defective. Once again, have internal checks that help you discern whether your process, at each step, is faulty. Otherwise, you won't know that you know.

Believing that reviewing is not that necessary to do. Reviewing throughout a unit helps you know that you know or don't know.

Not handing in some assignments that are to be graded, consistently handing then in late, or handing them in uncompleted. Is there any doubt that this tells you that you don't know?

Missing class at least once or twice every two weeks or so. This ranks at the top or near the top of a list of bad behaviors that indicate you are in serious trouble in the course.

You are consistently ill prepared for the class session because you delayed your work on the previous class period's reading and problem assignments.

You keep trying to understand many aspects of the course to

no avail, yet you believe that you should be able to understand it by yourself, which keeps you from seeking help. You view getting help as a sign of weakness, not strength, and do not consistently seek help from others.

Having had poor performances in mathematics courses before, yet have not substantially changed your study methods or attitudes. This quote says it all: "If you always do what you've always done, you'll always get what you've always got." Inappropriate study habits yield inappropriate results.

Considerable prerequisite content that surfaces in the course is confusing to you, even after it has been briefly reviewed by you, or in class by your instructor. Most likely you do not have the prerequisites for the course.

Having the attitude that the study techniques you employed in past mathematics courses, where you received good grades, will work for your current instructor and course. Receiving a good grade in a course does not necessarily mean that you were successful learning what needed to be learned. It depends on the appropriateness of the instructor's content objectives and grading standards. Receiving a good grade for your last course is not proof that you have an adequate background for your current course. You may have to increase the number of hours you study or change your method of study, which includes what you study.

Doing badly on the first examination. This is a very serious warning sign. You need to determine why it happened, and whether you can make the necessary changes as quickly as possible. Don't hesitate to seek the help of your instructor to determine this.

You continually blame your instructor for your shortcomings. If you believe that your instructor causes your problems in the course, then you will not focus on changing your behavior, which is most likely the problem.

How many of these signs, indicating that you may be in trouble in your course, apply to you? What do you to intend to do about them? In summary, all students, at times, know that they know or don't know something. A problem exists for those students who frequently don't know when they know or don't know something. Unfortunately, the behaviors that cause them to have difficulty understanding are the very behaviors that cause them to be unaware of what they know or don't know. If you are one of these students, the situation is not bleak if you reflect on the warning signs that were given here, and make the necessary changes. If it is too late for you to make the necessary changes to be successful in your current course, then your next step is to drop the course.

2. Determining Whether to Drop a Mathematics Course

A time may come in a mathematics course when it is wise for you to drop it before the official drop date set by your college. The number of weeks into the term when you can still drop a course without receiving a grade in the course varies from college to college. At my college it is about nine weeks into the 14-week term. At some community or junior colleges you can drop a course right up until the final grade is assigned. It is your responsibility to find out the official drop date at your college, as well as the amount of tuition you can receive back (if any), which is dependent on how early in the term you drop the course. This information can be found in your schedule of courses or college bulletin, and you should write it on your calendar.

I knew many students who had decided to drop a course a day or two after the official drop date, which was too late. In my college, if you applied to drop a course after the drop date, you could file a petition with the Academic Standards Committee explaining why you should be allowed to drop the course without penalty. There are a few legitimate reasons to drop a course after the drop date. Ignorance of the drop date or doing badly in the course are not legitimate reasons.

It is terribly unwise to receive a grade of D or E in your mathematics course because you (1) didn't know the drop date, (2) were careless about meeting this drop date, or (3) were unrealistic just prior to the drop date about the grade you expected to receive in the course. I saw many students receive unacceptable grades due to these failings. It never did and never will make any sense to me as to why they were so careless. Perhaps the various behaviors that led to their bad grades are the ones that caused them to be inattentive to the three reasons just listed for not dropping the course on time. It does not have to be that way for you. The solution is to take control of your life.

What Needs to Be Considered Before Deciding to Drop a Mathematics Course

You are the one who has to make the decision to drop or not to drop your mathematics course. Your decision becomes more difficult to make if you are on the borderline of receiving a C or D for the course. Your instructor, classmates, friends, or parents may give you advice, but don't ask them to decide for you. I had many students who wanted me to make the decision for them. I refused to do this, but did help them understand the pros and cons of making the decision. There are a host of things you need to consider before making the decision to stay in a course or drop it, and many of them are presented here, along with some unrealistic thoughts you may have that could cause you to make the wrong decision.

Having the prerequisites

If you discover that you don't have the prerequisites for your course, then you will not succeed in it and should officially drop it. Hopefully, you will discern this early in the course, within the first week or two, so you can still enroll in the prerequisite course for that term. (See chapter 4 on mathematics prerequisites.) The time period that you can drop a course and not lose any tuition is early in the term. It might also be the time period when, if you drop, no grade of W (indicating withdrawal) appears on your transcript. It may also be the time period when you can add a course. The longer you wait to drop within this add period, the less likely it is that you can get into the prerequisite course. Also, missing most of the first two weeks of a prerequisite course can be problematic.

Course conflicts

There are reasons, other than not having the prerequisites, for dropping a course early in the term. These include time conflicts with a job or another course, dissatisfaction with your instructor, transportation problems, or insufficient time to work adequately in the course. You have to be realistic deciding what needs to be changed; perhaps you need to make other changes in your life so you don't have to drop your mathematics course. Making a sound decision requires knowing what your priorities are or what they should be. Obtaining a college education most likely is at or near the top of your priorities. But just saying that it is a high priority doesn't necessarily mean that you make it a high priority. What actions are you taking that indicate it is a high priority for you?

Weighing the consequences of dropping or not dropping your course

Suppose you determine that you have the prerequisites for a course in which you are doing poorly. Your question becomes: "Can I significantly improve my performance before the course ends so as to receive an acceptable grade?" Many students agonize over this decision, and I suspect that you may also. Perhaps it is the first time you have had to drop a course; hence, you may view this as a failure. It is a setback, but one that may allow you to have a brighter tomorrow. Contrast these two possible actions: (a) Dropping the course and receiving a W on your transcript (which indicates a withdrawal from the course), which has no effect on your grade point average. (b) Staying in the course and receiving a grade of D or E on your transcript, adversely affecting your grade point average, and having the stigma of a bad grade displayed on your transcript. To place the second action in tangible terms, consider the following scenarios:

1. Suppose you are carrying four 4-credit courses, decide not to drop your mathematics course, and receive a course grade of E. Also suppose that your final grades in your other three courses were two C's and a B. With your grade of E, and a four-point grading scale, your grade point average for the term is 1.75. This puts you on probation and an E would appear on your transcript.

2. Suppose you receive a D in your mathematics course, with your other grades as stated above. Then your grade point average would be 2.0, just shy of what is needed to be placed on probation, and a D would appear on your transcript.

Which is the greater misfortune, receiving an E or D, or a W? If you had dropped the course no later than the drop date, your grade point average would be 2.3, you would not be on probation, and a W (indicating withdrawal) would have appeared on your transcript. There are times when one of the smartest things you can do is drop a course.

Making a decision based on faulty reasoning

Students who decide to stay in a course, or to drop it, often make their decisions based on faulty thinking. Here are examples of faulty thinking:

"It has been a waste of my time if I don't complete the course and get a grade, even if it is a D or E." If having been in a course is viewed as a waste of time if it is dropped, why will staying in the course make it not a waste of time? More will be said later on why the time spent in a course you drop should not be viewed as a waste of time.

"My instructor will not fail me or give me a D or E as a final grade even if that is what I averaged in the course." These students may be thinking back to their high school days when some of their mathematics teachers came up with the miraculous B or C for a final grade. By all means don't rely on this grading protocol in college.

"I will be able to do much better in the remaining four or five weeks of the course." This is unrealistic for many students who have done poorly up to that point, considering that there will be a comprehensive final examination, and about seventy percent of the content of this examination will come from the earlier part of the course (which they didn't understand). Furthermore, much of the earlier content has to be understood to make sense out of the content yet to come. If they don't make considerable changes in how they approach the course, with time running out, it is unrealistic to think that more of the same behavior will produce different results. Their learning of past and up-coming content in the course has to come during a time when the pace of this mathematics course, and other courses they are taking, has significantly quickened, papers and projects are coming due, and preparation needs to be done for hourly and final examinations. Many of these students think that doing well is a matter of luck and maybe they will get lucky on the few remaining examinations. "Maybe I will be lucky and be examined on easier content," "Maybe I will be lucky and have studied the right material for the test," or "Maybe by luck I will be able to work more of the problems on the examinations." Perhaps the final lucky statement that is in their minds is related to my earlier comment: "Maybe I will be lucky and my instructor will give me a higher course grade than the average of my grades up to this point." Don't bet on it, for that is a bet you most likely will lose. If you want to base important decisions on luck, stay away from casinos and mathematics courses.

"I can see that I won't get an A in the course and that is unacceptable to me." I had a few very good students drop my classes near the drop date, yet their grades in the course up to that time were B's or B+'s. I recall asking a freshman why she needed an A in the course. She said, "I want to get into law school upon graduation and a high grade point average is necessary for that to happen." How sad to go through college thinking that you need an A for every course you take. Another student, a woman pursuing a teaching degree, said that she was used to getting very high grades and would accept nothing less. Once again, how sad to see such an undue focus on grades.

I had a mathematics student who was a college junior and had not received a grade less than an A in college. It was clear to both of us that her grade would not be an A for the course. She struggled in making the decision to stay in the course, but did stay based upon some of my remarks. She discussed with me her trauma over the thought of receiving a grade of B because she had never received less than an A. She stated that her obsession with obtaining all A's was fostered by her mother many years ago. She received a B from me and mentioned to me several times in subsequent semesters how wonderful it was to receive this grade, although it took her awhile to accept it. She said, "My life now is so much more enjoyable."

Obtain assistance in making your decision

If you have doubts as to whether you should drop your course as the drop deadline approaches, make an appointment to discuss them with your instructor. This discussion can include—

- An overview of how you have done so far in the course, including the grade he or she would give you at the current time
- The work remaining in the course and how you will be evaluated on it, including the importance of the final examination to your course grade
- The study habits you may need to change in the remaining weeks and the likelihood of being able to do so to the extent that it will make a difference in your performance
- What you think your chances are to receive an acceptable grade in the course, and evidence that supports this
- Remarks made by your instructor about what he or she would do if in your position

You may also want to consult other people such as an advisor, classmate, or friend. Make your decision in a careful and thoughtful manner.

Live with your decision to drop a course

After dropping a course you may feel badly about this action, and believe that you wasted your time and money. Unless you totally ignored your course when in it, it is highly unlikely that it was a wasted experience. It is true that you don't have a grade for the course; however, you (1) learned whether you have the mathematics prerequisites for the course, (2) attained a roadmap of the knowledge that comprises the course that will help you the next time you take the course, (3) acquired an understanding of some of the content of the course, and (4) gained information about the changes in study habits you need to make that will help you be more successful in the course, if you retake it. Dropping a course may be the very thing that some students need to do to awaken them from their slumber. Isn't this important lesson, by itself, worth the cost of tuition for the course?

With more careful planning before you enroll in a course, you can minimize your chances of having to drop the course. Some of the circumstances that can cause you to drop a course can be dealt with prior to enrolling in the course (e.g., ascertaining whether you have the prerequisites for the course).

3. Continuing to Work in a Course You Drop

If you drop a course and your reasons for doing so do not rule against continuing to attend class and work in the course, then you need to give this serious consideration. It makes little or no sense to stay in the course if you don't have time for the course, or if you are so far behind it would be almost impossible to attain a creditable understanding of the remaining content. However, if this is not the case, then it might be in your best interest to unofficially continue in the course. But to do this you need to obtain the permission of your instructor. Some instructors will not only allow you to continue in the course, but they will also allow you to take examinations and receive feedback on them. Gaining this additional knowledge will improve your chances of succeeding the next time you enroll in the course.

I would be remiss if I left out this final caution: The vast majority of students in my courses who decided to stay in my courses after officially dropping them, either eventually stopped coming to class, or continued to come to class most of the time, but did no or very little work in the courses. They had the best intentions at the outset, but due to demands from their other courses and obligations, they were not able to fulfill their intentions. The cliché, the squeaky wheel gets the oil, applied. Their other obligations took precedence, as they should have. A disciplined and motivated student can keep this from happening, but it is not easy. It is unwise to think that showing up for class without carefully attending to the assignments will be of much benefit to you. It is wiser to stop attending class and use the time in a more constructive way.

4. Retaking a Mathematics Course

You may find yourself in the position of having to retake a mathematics course because of poor performance (i.e., you received an unacceptable grade or because you dropped it). It has been my experience that far too many students, who repeat a course due to poor performance the first time they enrolled in the course, don't improve their performance or perform as well the second time through. To have the best chance to succeed the second time around there are certain issues that need to be considered.

Issues to Address If You Are Retaking a Course

You may or may not know the reasons why you performed poorly in your mathematics course. If you know what they are, then you know what needs to be changed or corrected the next time you take the course. Experience tells me that you may not know what caused your poor performance or that you are in denial about one or more of the causes. The questions you need to address before reenrolling in the course include these:

Do you have the prerequisite knowledge for the course? This is the first thing you need to ask yourself. Don't assume that you do just because you were enrolled in the course before. Suppose, due to a bad performance, the course you are retaking is Calculus II. Also suppose you got a B or C in Calculus I from an instructor whose standards were low. This might have been another instructor's grade of D or E. For the sake of argument, let's assume it should have been a D or E. This poor performance in Calculus I may be due to an insufficient precalculus background. Let us assume this was the case. You then enroll in Calculus II with your grade of B or C in Calculus 1, and receive a Calculus II grade of D. Your first inclination likely will be to retake Calculus II. This is the not the correct decision to make. Most likely it is your inadequate precalculus background that led to an inadequate Calculus I performance, which led to an inadequate Calculus II performance. Is it wise for you to retake Calculus II without rectifying your inadequate precalculus and Calculus I backgrounds? It is frightening to consider retaking two courses to get the background you need, but do you have any other realistic choice? Hence, it is important to determine whether you have an adequate background for a course before you enroll in it, and if you fail to do this, to determine this as soon as you can when in the course. (See chapter 4 on mathematics prerequisites.)

If you find yourself in the position of having to retake a mathematics course due to a poor performance, you need to ask yourself these questions and get answers to them: (1) If I am retaking this course did I do badly in it because of a lack of prerequisite knowledge? (2) If so, do I have to retake the prerequisite course or courses, or can I compensate in other ways for what I am lacking? I had students who retook an introductory calculus course twice, and still didn't succeed because they did not remedy a poor precalculus background.

Are your study skills appropriate? If one of your problems leading to a poor performance in a course was how you worked in the course, then this can be easily determined by you. If you didn't work diligently in the course, then you know this has to change. For example, did you *minimally* spend from two to three hours outside of class for every hour in class? If you did work diligently in the course, did you work appropriately? That is, did your method of study include many of the suggestions given in this book? If your answer is "No," then you know what you have to change.

I would be remiss if I didn't mention that many students who retake a course mirror the behavior that they exhibited the first time through the course. For some reason they believe the results will be different the second time through, even though they changed nothing. Some of them are hoping that they will have an instructor who grades higher than their previous instructor. Even if they got such an instructor, this most likely would not be beneficial to them. These students need to focus on the changes they need to make to gain a better understanding of the course content. To do this they need to know the importance of changing, and be motivated and disciplined to make changes.

Do you confuse understanding with familiarity? The adage that a little knowledge can be a dangerous thing definitely applies when retaking a course. In retaking a course some students believe they should have an easier time because they have already been exposed to much of the content and understand a good portion of it. Hence, they spend more time studying other courses that they are taking for the first time. They are bored in class, or are prone to miss class because they were exposed to most of the content before. They are confusing familiarity with the content with understanding the content. Their poor performance the last time they took the course should tell them that they don't understand the content. This may be hard to believe, but many students who retake a course study no more or even less than before. The end result is that they either drop the course or obtain a grade as low or lower than before. I saw this happen time and time again. What these students need to understand is that their knowledge of the course is very meager, that they should view it as almost nonexistent, and that they need to work more diligently and smarter than they did before. Don't confuse familiarity of the content with understanding the content. The former can give you the false feeling that you know it. Recognizing content and being able to apply it or reproduce it when called upon to do so are two different things.

A major problem exists with students who took calculus in high school and didn't advance place out of it because it was not Advanced Placement Calculus, or because they performed poorly on the Advanced Placement Examination. Hence, they retake this calculus content in college. Many of them mistakenly believe that they know the calculus and consequently don't study, as they should, at the outset of their college calculus course. When they finally realize they don't understand, it is often too late to turn things around. They drop the course or receive a poor grade. Many of them resent this outcome, become discouraged, fault the college course, and don't choose, or drop out of a concentration that requires a few years of college mathematics.

Should you choose the same instructor? It is often wise to change instructors when retaking a course (if this option is available). However, changing instructors should not be done without considerable reflection. Do you want to change instructors just because your first instructor was too difficult, or because he or she was lacking in many ways? If you thought your instructor taught a good course, and you can come up with other reasons for your poor performance, then discuss these reasons with your instructor and see if he or she agrees with you. Ask your instructor whether he or she would recommend changing instructors. Unfortunately, a student's thinking often does not go beyond this: "I had the prerequisite course from Instructor X and received a good grade. I am receiving a bad grade from Instructor Y, so Instructor Y is the problem. When I retake the course I will choose Instructor X, and if not available, I will choose someone else, but definitely not Instructor Y." You need to consider much more than the grades you received from Instructors X and Y. It might be far better to retake the course with Instructor Y. However, a change of scenery can be a good thing, but not necessarily so.

There are many issues to consider when choosing an instructor, whether it is the first or second time through a course, including the teaching style of the instructor and how it relates to your learning style. (See chapter 13 on choosing a mathematics instructor based on his or her teaching style and your learning style.)

A Success Story on Retaking a Course

I have had success and failure stories with students retaking a mathematics course. An example of a success story is a student who took Calculus I at a major university near mine. She received a D in the course. Since she lived at home in the summer, near my university, she enrolled in my Calculus I course in a seven-week Spring Term. She was not sure why she struggled the first time around, but did indicate it was partly due to her instructor's struggles with English. Also, it was her first semester in college and she said she was somewhat naive as to how to function in a college mathematics course. She was very disappointed in her performance at the other university and questioned her ability to achieve in calculus. I quickly ascertained that she had the prerequisites for the course. She said she wanted to become a medical doctor. She was extremely motivated to succeed in my course and frequently came to my office to discuss the course content, as well as techniques of study. She ended up being my best student, receiving an A for the course. She was an impressive person, with great ability, integrity, and work ethic. Needless to say, she and I were thrilled with her performance and the confidence she regained in her ability to do mathematics. She wrote me several letters over the next few semesters thanking me over and over again, and telling me what was happening in her life since she left my course, especially in subsequent mathematics courses. I might add that her success was far more her doing than mine. Yes, teaching can be rewarding, and retaking a course after a poor performance can result in superior achievement if appropriate changes are made, not the least of which is choosing a supportive instructor.

5. Learning the Hard Way

It seems that some students have to fail before they take seriously the things that need to be done to be successful. That is, their poor performance in a course, to the extent that it needs to be retaken, is enough for them to make changes that allow for success. They learn not to miss class, to do their homework in a timely and smart manner, to not wait until just before an examination to study for it, etc. For others, doing badly in a course is not enough to motivate them to change. Some need to retake a course two times before they get serious about doing what needs to be done to be successful. Hopefully, they have not failed out of college by that time. This is a tough way to go, but seems to be the only way for some students. Is that you? Life can be a tough teacher! Learning the hard way can be financially costly, demoralizing, delay your progress, and raise havoc with your life in other ways. There is wisdom in this remark by the well-known mathematician Paul Halmos:

> I took a course in German the first semester, but I didn't like it. It was too easy, therefore I didn't study, and therefore it was too hard. In retrospect I can see the same pattern as I encountered in linear algebra: I didn't understand it, and therefore (a) I didn't like it, and (b) I had to work hard to learn it; having worked hard I came to understand it better than the subjects that seemed easy, and therefore ultimately I came to like it more.[1]

26

Preparing for Unit and Final Examinations

Your instructor evaluates your performance in your mathematics course, culminating in assigning you a course grade. This evaluation is partly on unit examinations, and most likely a comprehensive final examination. If you want your course grade to reflect accurately what you know about the content of your course you have to prepare diligently for these examinations. You may claim that you know more than your examinations show, but this subjective opinion will not be part of your instructor's determination of your knowledge of the course.

Preparation for an examination, as discussed in this chapter, is separated into two facets, *continuous preparation* (this is the ongoing preparation that you do as you are learning the material), and *final stage preparation*. The main focus of this chapter is on final stage preparation since aspects of continuous preparation are discussed throughout the book.

The sections of this chapter are as follows:

1. Continuous Preparation for a Unit Examination
2. Final Stage of Preparation for a Unit Examination
3. Preparing for the Final Examination

1. Continuous Preparation for a Unit Examination

Studying for a unit examination starts at the time you begin studying the unit. You cannot separate what you do while learning a unit of content from your preparation for an examination on that content. You should be continually preparing for a unit examination as you progress through the unit. This preparation for a unit examination will be called *continuous preparation*. The items you need to address *throughout* your study of a unit, which will influence your performance on the unit examination, include:

participating in class	writing mathematics
taking notes	writing about mathematics
reading the textbook	reviewing mathematics
working assignments	summarizing mathematics
seeking assistance	motivation to succeed
distributing your study	commitment to study
speaking mathematics	minimizing anxiety

All the items in this list have been thoroughly discussed in prior chapters of this book. Your performance on a unit examination, as well as on a final examination, is a function of how well you address these, and other items. This is the most important thought you can take from this section. In summary, effective study for a unit examination means meeting all of your responsibilities as a student *from day to day*, beginning with the first day of class. Learning takes time and you won't have enough time near the time of the examination to learn what you should have learned earlier. What you do in the final stages of preparation, say a week or more before the examination, plays an important role, although a significantly smaller one, than what you do throughout the unit, which we have labeled here as continuous preparation.

Continuous Preparation Versus Cramming for the Examination

What you need to do throughout a unit of study, as just outlined, requires a lot of work and sacrifice. But the payoff is great. Not only are you studying to learn content that is at hand and will be tested on your unit examinations, but you are also studying for longer-term goals such as doing well on a final examination, or having the background knowledge to succeed in subsequent courses. Continuous study is far less pressure-packed than responding to immediate learning emergencies. You can be so well prepared that a surprise examination would not be a difficulty, although some special study would have helped a bit.

Just think how your joy will build throughout your mathematics course and in subsequent mathematics courses if you sustain daily study in your classes. You will have lessened anxiety, and will feel good about meeting your responsibilities. Knowledge in mathematics begets more knowledge in mathematics. How comforting to know that your excellent knowledge of mathematics will snowball, culminating in superior mathematical performance, than it is to know that your lack of mathematics knowledge and skill will overtake you, resulting in poorer and poorer mathematical performance.

Let us now look at what is done to prepare for a unit examination when sustained daily study has not taken place. This will be characterized as *cramming* for the examination. You have to cram for an examination when you have not adequately planned your study throughout the unit, and consequently have to resort to unplanned study over a very short span of time. This time span varies from student to student, but for many it is only a few days, one day, the previous evening, or maybe only a few hours before the examination. A week can even be an insufficient amount of time to prepare for an examination if your preparation was faulty throughout your study of the unit. There are many downsides to cramming for an examination and they include:

High anxiety (due to insufficient study time). You find out that you need to spend more time than you have to learn content you didn't learn earlier, as opposed to reviewing this content.

Confusion or interference of ideas (due to too many ideas coming too fast). Signs of this are using the wrong content to answer specific questions, many silly mistakes, blanking out, and recalling how you should have responded to specific examination questions shortly after leaving the examination.

Setting aside almost everything else (which means missing classes, turning assignments in late, and not being prepared for other classes).

Focusing more on recalling rote or low-level knowledge (at the expense of not being prepared to respond to questions requiring a greater depth of understanding).

Restricting your study to what you think will help you pass the examination (thus missing out on tangential learning that comes from broader study).

Loss of sleep (causing weariness and a lack of alertness during the examination).

Little retention of knowledge for the final examination or later

courses (not enough time for consolidation of knowledge to take place; that is, for the process in the brain to take place that enables you to have a more lasting learning).

You do not have to be a genius to understand the harmful effects of late preparation for impending examinations. There are times when everyone has to do some cramming, but unless it is held to a minimum, it is a "high tension wire that burns." You should not treat the knowledge that an examination is on the horizon as an emergency. Avoidance of cramming for an examination demands that you organize for continuous study. Do you organize for continuous study? At the very top of the list of attributes you need to succeed in college is wise management of time. Without it, sufficient continuous preparation cannot happen.

In asking students for advice on what I should include in this book, a Calculus II student said:

> ❖ *"As for study techniques for calculus, the best advice to give anyone is not to cram for the exams. From personal experience, I've found this to be the most useless method of study. You get too much in your mind all at once and then on the exam day, the person either forgets it all or gets different concepts mixed up."*

If you cram for an examination in lieu of properly preparing for it throughout the unit of study, isn't it time to befriend yourself?

Remembering, Reviewing and Summarizing Mathematics

Since remembering, reviewing, and summarizing mathematics throughout a unit of study play a critical role in learning the content of a unit and in performing well on a unit examination, it is recommended that you take the time to read and reflect on chapter 24, where these items are discussed. The next subsection pertains to the final stages of preparation for a unit examination.

2. Final Stage of Preparation for a Unit Examination

There is a need for *final preparation* for an examination, even if you have been continuously preparing throughout the unit. It should take minimal time if your preparation throughout the unit was carefully attended to. You may find, as you complete final preparation for an examination, an idea or two that you do not understand as well as you thought. This has to be taken care of, but learning content for the first time during this stage should be a rare occurrence. We now look at other items that are part of the final stage of preparation for a unit examination.

Plan Your Final Preparation Activities and Then Schedule Them

You have to decide what you intend to do by way of final preparation and the priority you will give to each aspect of it. Then reserve time in your schedule to accomplish this. Avoid using large blocks of time; they are not needed at this stage of your preparation. It is far better to distribute your study over a week or more using small intervals of time. Four or five study sessions of one hour or less are superior to two study sessions of three or more hours each.

In planning these preparation sessions, you need to take into consideration other responsibilities you will need to meet during your final preparation time. The world does not stop for you because you have an examination pending. One huge advantage of continuous preparation is that you will have fewer things to study and review as part of your final preparation for the examination.

You cannot adequately schedule final preparation study time if you don't know when your examinations will be given. Mathematics instructors do not always know at the outset of a term when they intend to give a specific examination. If you miss class when an examination date is announced, and do not talk to someone who attended class, you may lose valuable final preparation time. Are there students who come to class and find out for the first time that an examination is being given that day or the next class session? I suspect you know the answer to this question.

Spend Your Time Doing What Needs to Be Done to Prepare Rather than Thinking about Preparing

Many students live with an incredible amount of self-inflicted worry because they spend too much time thinking about what they need to do, rather than doing it. It is so easy to be overwhelmed by the thought of having to work your final preparation time into your already busy study schedule, that you resist doing the work. Wouldn't you say that your time is better spent doing rather than worrying? Here is a solution to stop debilitating worry: Do nothing but worry for ten minutes. Then stop all your worrying (actually, it will naturally happen since you will tire of it—the uselessness of it becomes obvious), and get down to work. *Start with something easy,* just so something is being done. Once you get working, it is highly probable your worry won't come back and you will find out that you do have the time to do this final preparation, as well as most of your other work.

General Thoughts on Your Final Preparation

These ideas may help you with final preparation:

Final preparation for an examination has a somewhat different focus than your continuous preparation. An examination can only sample the many objectives that you are expected to learn in the unit. However, your instructor does not randomly choose this sampling. To the contrary, your instructor will test you on content that he or she deems most important for you to learn. To do the best job you can on the examination you have to determine what to expect on the examination. This will help you spend more of your preparation time in some areas rather than in others. You cannot, in a shortened preparation time, prepare yourself equally well in all areas of the unit. You have to be selective and determine the content—based on its importance and your understanding of it—that should receive the most attention in your final preparation. Here is a comment made by a Calculus I student upon being asked what she didn't like about her course.

> ❖ *"Not enough tests. Because there weren't enough tests, each test couldn't cover everything so some students studied the wrong stuff."*

What advice would you give students who study the wrong content?

Discerning what to expect on the examination. There are a variety of sources to consult to get a good sense of what will be on an examination, including the following:

1. Your instructor
2. Your instructor's study guide or review sheet for the examination

3. Your class notes
4. Previous examinations on the same unit

Each of these ways is now discussed.

Your instructor. Most instructors want you to know what they believe is important content, so focus your study on this content. Most of them are willing to respond to your request to know the *types* of content that will be included on the examination. They might be less specific than you want, but some information is better than none. Ask questions such as these:

- Which sections of the textbook are covered by the examination?
- What do I have to memorize?
- Will I need to give definitions or proofs of theorems?
- Will I have to write short essays demonstrating my understanding of concepts and principles, and their usefulness?
- Will there be objective portions of the examination (e.g., true-false, multiple choice)?
- What types of applied problems will I have to solve?

Also ask what you need to bring, or are allowed to bring, to the examination (e.g., bluebook, calculator, scrap paper).

It is not unusual to be surprised by questions or problems that appear on the examination and how long or short the examination is. These remarks by three Calculus I students on what they didn't like about their instructor support this:

- ❖ *"The instructor did not cover material in class that was on the test."*
- ❖ *"The instructor went through the material very quickly and a lot of material was involved in each exam."*
- ❖ *"The instructor's tests were unfair (sometimes) because he put problems on them that he had not explained in class. However, it did make me think and study harder."*

Don't assume that because a student doesn't like something an instructor did, that the instructor did something wrong.

Your instructor's study guide or review sheet for the examination. Some instructors distribute a study guide or review sheet for the examination. These guides are especially beneficial for learning if they include a host of questions or exercises that require you to write *about* mathematics, some of which will be on the examination. Be serious in your response to these study guides. There is one drawback to an instructor-constructed study guide: You are denied or at least hindered in having the valuable experience of choosing and organizing the content you will study. Upon asking students for advice to include in this book, a Calculus III student said:

❖ *"I prepare a review packet for an exam that includes all the main formulas and ideas. Example problems follow each main idea to make the topic clearer. Actually, this method uses a lot of my time."*

Your class notes. Instructors choose the content that comprises their lectures and it is this that they deem to be most important. (They can't lecture on everything.) It then stands to reason that they would construct examinations that mainly reflect their lectures. Good class notes provide excellent clues as to what you can expect on an examination, especially if you include in them your instructor's remarks on the importance of specific knowledge and what you are expected to do with it. For this reason, and others, reading through your class notes is normally part of final preparation for an examination.

Previous examinations on the same unit. Your instructor may let you see previous examinations, and if not, a student who has had the course may let you see his or her examinations. This does not mean that your examination will closely reflect a previous examination, but it most likely will. *A word of caution*: Deciding what your examination will be like based only on your instructor's previous examinations is an unwise thing to do. You may be unpleasantly surprised at examination time. You will also be compromising your learning because you were too restrictive in your preparation.

What you learn in your mathematics course is highly related to what your instructor expects you to learn, and you need to have a good sense as to what that is. However, regardless of what your instructor discussed in class, and what is presented in your textbook, students typically learn *what they expect to be evaluated on*—this primarily dictates the focus of their study. They know it is their graded problem assignments and examinations that comprise their course grade. Consequently, your instructor has a responsibility to construct important test items so you are motivated to study important content. If these items are poorly constructed, then you have to go beyond your instructor's expectations in your study, which requires discipline on your part.

Perhaps study with others. In my mind, the verdict is out on the benefits of studying with others during final preparation for an examination. As a student, I seldom did it, and when I did, it was only to clarify any confusion I had. One thing is certain; taking an examination is a solitary act. You don't want to let group work get in the way of preparing adequately. This can happen if you are in a group with students whose preparation has been inadequate, or if the group does not attend to the task when it meets. Your time is valuable and it is unwise to waste any of it by working with students who have not been responsible. Group work at this time should not be spent doing work that should have been done earlier. That being said, a minimal amount of group study for an examination can be beneficial. But unless you have time to spare, work only with students who are well prepared and stay on task, and do not expect the bulk of your final preparation for the examination to be done with the group.

Construct your own examination. Once you have a good sense of what will be on the examination, make up your own examination, place yourself in a testing situation (e.g., restrict your time, use no notes, have a variety of well-chosen problems), and then work the examination. Doing well on your examination should build your confidence. Just the act of constructing questions is beneficial. Perhaps you can have a classmate grade the examination. You can also ask a classmate to construct an examination for you to take (and have the classmate grade it). Return the favor. If you have access to one or more previous tests of your instructor's on the unit in question, work it under testing conditions. Also, work the review sections and chapter tests in the textbook. How beneficial this is depends on whether these tests match the expectations of you and your instructor.

Perhaps attend instructor-led extra review sessions. My cautions here are similar to some of those I mentioned when studying with others. Reviewing for an examination is something you have to do; classmates or an instructor cannot do it for you. You have to decide if the time spent attending an instructor-led review session is as beneficial as spending more time on your own review. One benefit of attending an instructor-led review session is that you have a greater sense of what will be on the examination.

Know when it is time to stop preparing. I cannot tell you when to stop preparing for an examination. However, assuming that you have positioned yourself so that you do not have to cram, this advice is worth considering: The interval of time to stop preparing ranges from a half day to a few days prior to the examination. The reason for this is that your mind needs time to work. You want to wake up your mind when taking an examination and working almost until the examination interferes with this. Without this interference your mind will work without continual prompting. Let go of your mind and it will stand on its own—allow it to work. Be engaged in activities the evening before an examination that are relaxing (e.g., reading a novel, going to a movie).

An excellent time to prepare for your examination on any given day is near the time you retire for the evening, but not right up to the time you retire. You will not want to study other subjects near this time since your mind continues working on the areas you just studied as you are sleeping, and you don't want interference from other subjects. This is not to say that you shouldn't also prepare earlier in the day, if need be.

Don't change daily living habits. You are likely to change your daily living habits if you cram for the examination. Cramming causes you to relax less, drink more stimulants, eat more hurriedly, skip meals, eat less nutritious food, sleep at unusual times, and sleep fewer hours. You also want to avoid oversleeping. Changing living habits can have a negative effect on your examination performance.

Relax as the examination nears. With proper preparation you will be confident that you know the content—you will know that you are prepared for the examination. Take solace in the knowledge that you have prepared as well as you can—nothing helps you relax as much as this. Knowing that you cannot ask anything more of yourself is incredibly freeing. Be your best friend and enjoy this freedom! Look forward to the examination as an opportunity to show what you know. See yourself as being successful, which is a credible vision if you have prepared well. The time just before the examination will be best used putting yourself in this state of mind, rather than doing further study. I always went for a walk outside an hour or so before an examination, weather and time permitting, and dismissed any thoughts that came to mind about the examination. This moderate exercise helped me relax and cleared my mind of detrimental distractions.

Specific Activities That Can Be Included in Final Preparation

The activities to include in your final preparation for a unit examination are decisions you have to make. What you do should be *based on your continuous preparation and what you believe is needed beyond that*. You need to focus your study where you are least prepared and on what will pay the greatest dividends. The components to consider include the following:

Seeing the big picture. Your ability to see the big picture will help you immensely on the examination. Seeing the big picture allows you to see how things are related, and this helps you fit an examination problem into that structure. This in turn helps you understand and work the problem. One of the more important things you can learn from reading this book is the value of seeing the big picture. Here are some items that relate to reviewing by focusing on bigger pictures.

1. Read through your class notes, focusing on the progression of ideas.
2. Read the highlights of the textbook sections that relate to the unit.
3. From memory:
   1) Outline the major content of each section of the textbook that comprises the unit, or review the important questions you constructed as you progressed through these sections.
   2) State the definitions of concepts that were introduced in the unit.
   3) Write summaries of the development of key concepts (e.g., inverse function, transformations of functions, derivative, definite integral).
   4) State the theorems that were introduced in the unit, and give their uses.
   5) Outline the proofs of specific theorems.
   6) List problem-solving skills.
   7) Write the rule(s) for performing operations on functions comprising specific categories (e.g., [a] find derivatives of (1) the sum, difference, product, and quotient of two functions, (2) the six basic trigonometry functions, and (3) composite functions; [b] find indefinite integrals of specific types of trigonometric and logarithmic functions).

8) Think about the types of application problems you have solved, what procedures you used to solve them, and the signs that triggered recognition of the procedures to solve them.

The operational phrase you are being asked to use with each of these eight items is *from memory*. You are being asked to recall this content, not to recognize it by reading it. On an examination you will have to recall it, not read it.

Working exercises. It is a given that you will have to work a variety of problems on your unit examination. Some of them will be routine, others not. Upon reading a problem you have to understand it, apply an initial strategy to work it, and then decide what to try next if that doesn't work. Hence, you have to prepare well to maximize your chances of doing well on the problems on the examination. These activities will help your preparation:

- Work a variety of exercises from the review exercises in the sections of the textbook or from assignments, that relate to the unit you are studying.
- Work problems that require specific use of a graphics calculator.
- Work problems that use tabular or graphical information.
- Work exercises out of sequence. That is, choose a variety of exercises from throughout the unit that come from section and end-of-chapter exercises, place them on cards, mix up the cards, and then work the problems as they come up.
- Work exercises out of the textbooks that have not been assigned. Work some that have answers in the back of the textbook or solutions in the solutions manual, and some that don't have answers or solutions. You have to decide which exercises you do not need to work out in detail, and those you do.
- Cover the solutions of problems in your class notes and examples in the textbook, work them, and then uncover the solutions and read them if you need verification of your work.

Knowing mathematics terminology. You have to understand the mathematics terminology that is used in your unit of study. For ideas on how to make this part of your final preparation, read section 2 of chapter 16 on reading mathematics terminology.

Constructing proofs. You may be asked on your examination to replicate part or all of the proofs of specific statements. You may be told which ones to study. Rather than read over the proofs of theorems, it is far better to see if you can construct their proofs. You may also be asked to justify statements on the examination that you have not seen before. Writing original justifications prior to an examination is good preparation for this. (Understanding logical arguments or proofs is discussed in section 4 of chapter 16.) Good problem-solving skills will also help you here. It is wise in your final preparation to review and apply problem-solving skills. See section 2 of chapter 22 on devising a problem-solving plan.

Writing mathematics. You should expect to write mathematics on an examination in an appropriate way. Perhaps, due to time limits, you can be somewhat less formal about it than if you were writing as part of an assignment. Technical aspects of writing mathematics is the subject of chapter 20.

Use 3×5 Cards to Accomplish Many of Your Review Objectives

I have mentioned in various places throughout this book the use of 3×5 cards to record specific information. A card should be made as soon as you gain important mathematics information as you progress through a unit. You will use your 3×5 cards in your continuous and final preparation. These cards are also useful in preparing for the final examination. It is not critical that you use 3×5 cards for these purposes—there are other means to accomplish similar objectives. However, their use lends itself nicely to accomplish much of what is good to do in studying mathematics. The following are examples of what can be written on these cards and how to use them:

Example 1. Write the verbal and symbolic name of a concept on one side of the card and its definition on the other side. Make a card for each new concept. In your study, read what is on one side of the card and see if you can state the information on the other side. At times, reverse the side of the card you begin with. Do this periodically as you progress through a unit, adding a new card for each new concept presented.

Example 2. Write each significant symbolic expression introduced in the unit on one side of a card, and on the other side write only one of these: its literal translations, interpretations (geometric and otherwise), uses, name or title, equivalent symbolic forms, and uses or applications. (Eleven fluency objectives on the symbolic language of mathematics are presented in section 2 of chapter 19.)

Example 3. Write the name or brief description of a theorem on one side of a card and its statement on the other side. When reviewing, look at one side of the card and see if you can write what is on the other side.

Example 4. In learning the proof of a theorem, a good technique is to place the statement of the theorem on one side of a card and an outline of the proof on the other side. The outline consists of the major steps of the proof. If you cannot write the outline of the proof when reading the statement of the theorem, read it on the backside of the card—study to remember it (with understanding)—and then write it down again from memory. Once you know the outline of the proof, use it to come up with the complete proof, filling in the necessary details. Knowing the outline of a proof should be enough information for you to fill in the missing details; hence, deriving the proof.

Example 5. Record the more important graphs or diagrams on cards, placing the graph or diagram on one side of the card with the question, "What do I learn from this graph?" Place your answer on the other side of the card by completing this statement: The purpose of this graph is to . . . When reviewing, look at either side of the card and see if you can write what is on the other side.

Example 6. Write a theorem or symbolic expression on a card, and a mantra or chant for it on the other side of the card. Look at one side and see if you can give the other side. (Using mantras or chants is the subject of section 4 in chapter 19.)

Example 7. Write the name of a theorem on one side of a card, and its uses on the other side. If the theorem gives a technique for working certain types of problems, you can also include some of these problems on the backside of the card. Look at one side of a card and see if you can give the information on the other side.

Example 8. Select problems from the various exercise sections of your textbook, your class notes, and graded problem assignments. Write a problem on one side of a card and a solution to the problem on the other side. Do this almost daily for each type of problem. Work these problems a few times weekly by going through your set of cards. If you struggle working a problem, look at the solution on the other side, but before doing that give yourself an adequate amount of time to figure it out. *A word of caution*: Don't look at the problem and then immediately read the solution on the other side of the card. You need to work on the problem and then read the backside, if need be. Only use the solution on the card to check your solution. You want to be an active reviewer, not a passive one. If you continue to struggle with a problem, go to your sources to find other problems of the same type and see if you can work those. When constructing these cards for problems out of the textbook, it is a good idea to note the page numbers of the text where information is found that will help you work problems of this type.

When using the cards, shuffle them and then work the problems in the order they appear. This is important since on an examination you will not know the section of the textbook from which similar problems come. Knowing the section a problem comes from is a crutch that can be used to solve the problem, and that crutch will not be available to you when taking the examination. Your deck of cards will expand as you progress through the course. When your final examination is near you can spend the bulk of your preparation time using these cards. They display problems that represent content from throughout the entire course. Great preparation for a final examination is to shuffle the cards from all the units of the textbook and then work the problems as they arise. When going through a set of cards, remove those that contain problems you have difficulty working. Do further study in these problem areas, using your textbook and class notes, and then work through these cards again. You may want to add more cards to your deck containing problems in these areas.

Your final study for an examination should consist of reviewing your problem notebook and class notes, rereading sections in the textbook, going through your 3×5 cards, working a good selection of problems from the chapter review exercises, and working the chapter tests. It should not be spent trying to work problems from the problem assignments that you could not get as you were working through the unit. It is most likely too late to do that; these problems should have been taken care of earlier.

3. Preparing for the Final Examination

It is almost a certainty that you will have an in-class final examination in a lower-division mathematics course. The length and comprehensiveness of this examination varies from instructor to instructor, although you could have a departmental examination. Most instructors give an examination covering the whole course, but the content on the examinations can vary. For example, one instructor will require you to justify statements whereas another will not. This section addresses, in the main, the value of preparing for a final examination, the role it can play in determining your final course grade, and how to prepare for it, including pitfalls to avoid. The section ends with a discussion on seeing your graded final examination, and what you can do about a final course grade that is lower than you anticipated.

Value of Preparing for the Final Examination

Preparing for a *comprehensive* final examination can be the most valuable learning experience in a mathematics course. However, it might not be a valuable learning experience for you. It depends on how you prepare, the time you devote to preparing, and how responsible you have been during the term. More is said on this later. It can be a valuable learning experience since it affords a wonderful opportunity to view the structure of the entire course (i.e., to see how the whole course fits together). Near the end of a course you are not the same student you were when you entered the course. You look at content introduced earlier in the course with a different set of eyes— you look at earlier content *knowing later content*. As you look back, you may deepen your understanding of some concepts and principles, or understand them for the first time. You find yourself saying such things as, "Now I understand what is being said here," "I now see how useful this is to understand that definition or that proof," or, "This type of problem is not as difficult to work as I first found it to be."

There are other reasons why instructors give final examinations besides it being a good learning experience for students. Some do it because they believe their students will work more diligently throughout the course to learn and retain the content, knowing that they will be examined on all of it at the end of the course. Many, if not most instructors, give a final examination to determine if their students can put it all together at the end. They believe that the best indicator of a student's performance in the course is how well they do on this examination. Consequently, this examination is almost always weighted more heavily than the unit examinations to determine a course grade. More is said on this later. Finally, a final examination affords you a last-chance opportunity to show what you learned in the course. If you have not been doing that well in the course, especially earlier in the course, perhaps you will have put it all together at the end, and a final examination can show this. However, if your unit examination grades have been consistently C's or D's, the chances of receiving as high as a C on a comprehensive final examination is a formidable task, to say the least. This speaks volumes about the importance of doing well on unit examinations.

Students Tend to Do Less Well on Final Examinations

Doing well throughout the course does not necessarily translate into doing equally well on the final examination. No doubt there is a positive correlation between these performances with most students, but a comprehensive final examination is unlike unit tests; hence, grades on a final examination tend to be lower than those on unit examinations. I have had students who averaged an A or B on their unit examinations, receive grades of C, D, or E on their final examination. There is little doubt that this happened primarily because of their inadequate preparation for the final examination.

Two things primarily dictate performance on a final examination: (1) how well you understand as you progress through the course, and (2) how well you prepare for the examination. It is one thing to do well on a unit examination, which you have studied for extensively over three to four weeks, and another thing to have to respond to questions related to that unit on a final examination that may occur a few weeks to *four months* later. Fortunately, mathematics knowledge builds on itself (i.e., in learning new knowledge you are using past knowledge). But content appears on final examinations from earlier units that might not have been addressed in the course, after those units were studied. To do well on this knowledge on a final examination, you have to study it carefully in your preparation for this examination.

How Your Grade on the Final Examination Affects Your Course Grade

Many mathematics instructors give considerable weight to a comprehensive final examination to determine course grades. For example, it may count the equivalent of at least two unit examinations. Suppose you had three unit examinations, received B's on all of them, and received a D on the final examination. Giving the final examination a weight equal to two unit examinations, results in your examination grades being the average of B, B, B, D, D, which is a C. How would you feel if this happened to you? It can easily happen if your unit grades of B did not reflect the strength of your understanding, or your final preparation for the final examination was faulty. Good grades on a unit test do not always reflect the *strength* of your understanding. You might understand specific content well enough to do well on a unit examination (which requires you to know far less content than what you need to know for a final examination), but not well enough to retain it for a few weeks or months when it appears on a final examination. An examination only samples certain objectives and can only ascertain to a limited degree the amount of understanding. The appropriateness of your unit examinations and the quality of their grading have a bearing on how well your unit examination grade reflects your knowledge of the unit's content. Your goal has to be to understand the content, not just to get a good grade on your unit tests. If you focus on getting good grades, then the understanding you need to do well on a final examination, or in a subsequent course, may not be there. *Focus on understanding and your grades will take care of themselves.*

Suppose you averaged a C on your unit examinations. This suggests that your knowledge of the course content is not strong. Will it be considerably more difficult for you to get at least a grade of C on a comprehensive final exam? In general, the answer is "Yes," for reasons that have been presented. However, it also depends on the order in which your unit grades were obtained. Consider these two very different examples of how unit examination grades can average out to a C, yet the chances are greater in one situation to receive a better grade on the final examination:

Example 1. Suppose you received grades of D, C, B, and B, in that order, on unit examinations. You started off poorly in the course, but finished well. It is not that unreasonable to expect to receive a grade of B on a final examination with sufficient time to prepare for it. It appears that you got your act together, and you most likely were able to eventually learn much of the earlier content since a large amount of it was used to develop the later content.

Example 2. Suppose you received grades of A, B, D, and D, in that order, on unit examinations. You started off well in the course, but finished poorly. It will be difficult for you to receive a grade as high as a C on the final examination for two reasons: (1) the content focus on the final examination is often on the last half of the course, which you do not understand, and (2) there are reasons for your drop-off in performance—perhaps a change in study habits—and these same reasons will most likely exist when you prepare for the final examination. I could easily see you receiving a grade of E on the final examination. I have seen this happen time and time again. It is not my intention to be grim, but it is my intention to help you see reality. Burying your head in the sand is not a productive course of action. The way to avoid this happening is to do this: As soon as your grades begin to slip, get help on the changes you need to make to turn this around. Your instructor is a good place to begin to seek advice. It is not by accident that grades slip two or more letter grades. It is too late to do much about this much slippage in grades (and understanding) during the final two weeks of the course.

Departmental Final Examinations

You may have to take a *departmental* final examination if you are involved in a multi-section course such as calculus. This is an examination usually constructed by a committee of instructors who divide up portions of the examination to grade. There are pros and cons to departmental examinations, but their purpose is to help ensure more uniform standards across sections of the course. They are a means of exercising control on what is being examined and how the examinations are graded. One drawback to them is that they stop an excellent instructor from giving a better examination. You can only imagine the compromises that have to be made by a committee of instructors to reach agreement on the content of the examination. The number of calculus students across the country taking departmental final examinations could be near fifty percent. Some mathematics departments also have departmental mid-term examinations. If you have to take a departmental final examination, ask your instructor how it will differ from one that he or she would give.

Length of Final Examinations

The time allotted for taking a mathematics final examination is not uniform across mathematics departments or instructors. For example, some departments have two-hour time blocks and others have three-hour time blocks. My department had the latter. I would construct a comprehensive final examination that a good student should be able to complete in about two hours, but my students had three hours in which to complete it. Taking such a final examination is a tiring experience. Imagine having to take another final examination of this length within an hour or two of that one. That is not an uncommon experience for many students.

Moving Final Examinations to Improve Performance

Preparing for a mathematics final examination is not a valuable experience for many students due to the lack of time they have to prepare for it. Other than some basic tinkering, there is not much you can do near the end of the term to find the necessary time to adequately study for a final examination if you have not been responsible during the term. You might get lucky and have only two final examinations in the four or five courses you may be taking. This will not help you much if your mathematics final examination is scheduled soon after your last class day. A day or two to study for this examination is hardly sufficient. Suppose you are even unluckier and your other final examination is on the same day. Then you have a day or two to study for two final examinations. Now suppose you have final examinations in all four of your courses and all have to be taken within the first three days of the designated examination time. Then again you might be fortunate and have them spread over a week or two. Even so, if you are responsible for new content in all of your courses that was developed during the last week of classes, learning this content can use up much of your time, to say nothing about reviewing past content and finishing up a term paper or two.

Can anything be done about the lack of time near the end of the semester to prepare for your final examinations? The answer is a resounding, "Yes," but you must take certain initiatives. Foremost among them is having most of your preparation for the final examination completed two or three weeks before the final examination. I have called this continuous preparation and it is discussed in section 1. Continuous preparation will lessen the demands on your time near the end of the course. Also, knowing the rules of your college related to final examinations can help eliminate or lessen time constraints that you may have during this time. For example, there are rules that instructors may have to follow including these:

- Unit tests cannot be given the last week of classes.
- Final examinations cannot be given before classes end.
- Final examinations cannot be moved to another time period during final examination week unless permission is granted by the college academic dean and departmental chair, or by every student in the course.
- Instructors cannot schedule classes or have assignments due on officially designated study days for final examinations.

You may be able, with the permission of your instructor or academic dean, to take a scheduled examination at another time if you have too many finals scheduled on one day (e.g., three finals scheduled on one day is viewed as a hardship). Nothing prevents you from asking your instructor if you can take the examination at another time. You may be granted permission to take it during the time your instructor is giving another final examination, or during the time a colleague of your instructor is giving a final examination. Granting your request depends on the reason for the request. I would grant a request for compelling reasons, but not many reasons were compelling. You need to know your rights as final examination time nears.

Get Hints from Your Instructor on What Will Not Be Tested

Quite often instructors will not test specific types of content on final examinations that they tested on unit examinations, and they most likely will tell you what that is. For example, you may not be expected to give definitions or prove theorems. Your instructor may make other statements such as, "About three-fourths of the examination will be on content from the last half of the course." Due to the wealth of material on which you could be examined, it is reasonable to expect guidance from your instructor on what should be the focus of your study. If nothing is said, ask.

What Constitutes Your Final Preparation

Other than employing many of the preparation ideas discussed earlier in this chapter, you can also review your unit examinations. Cover your solutions to the questions on these examinations and then answer the questions. Check your work by looking at your instructor's solutions keys (if they exist), or by whatever means you have. Analyzing the types of questions asked on these examinations will help you focus your study. By all means focus on those things that you had difficulty with on the unit examinations. If what you know now about them is the same as what you knew before, your results on the final examination will be, at best, comparable to your results on the unit examinations. Avoid making the same or similar mistakes. In conclusion, the best advice I can give you is this: *The better your understanding of the content of the course as you progress through the term, the better your retention of the content, and the less time needed for final preparation.* Just as cramming for a unit examination is beset with problems, it is even more problematic to cram for a final examination.

Looking at Your Graded Final Examination

You may not be able to keep your graded final examination, but you have the right to look it over. For example, all instructors in the mathematics department at my college are required to retain all materials, not previously returned, for a period of at least one regular term following the course. A student has the right to either have the material returned, or made available for viewing. If final examinations are not returned to students as a matter of policy, then they most likely will be restricted to viewing it in their instructor's office.

Getting a Lower Course Grade Than You Anticipated

If you receive a lower course grade than you expected, you have the right to have your instructor explain how it was calculated. This calculation has to meet the criteria that were set up by your instructor, which you should have obtained at the outset of the course. The criteria are most likely stated in the course syllabus. At my campus all instructors had to have a course syllabus, and statements had to be in it identifying the items used to determine course grades and their relative weights. If you are not satisfied with your instructor's explanation as to how your grade was calculated, and he or she will not change it, learn about procedures at your college for settling a grade dispute. In your efforts to have the grade changed, be polite and respectful in all your conversations.

27

The Mathematics Examination

You may be well prepared, mathematically speaking, to take an examination, but that does not necessarily mean you will be successful on the examination. Most examinations are given in a limited time period, which dictates that there are things you need to know to move through the examination in a meaningful and expeditious way. These items are presented in this chapter along with ideas on how to handle anxious moments that may arise before and during examinations, and how to learn from your examinations. The sections of the chapter are titled:

1. Taking Examinations
2. Coping with Test Anxiety
3. Learning from Your Examinations

1. Taking Examinations

Would you celebrate if there were no examinations in your mathematics course? If learning is a priority for you, your answer should be "No." Having to take examinations compels most students to overcome the human frailties of spending too much time in the coffee shop, watching too much television, sleeping too long, or playing too many online games. Examinations will be around for a long time and your grades in mathematics courses are highly dependent on how well you do on them. Rather than fretting over having to take them, your time is better spent dealing with them. To be successful in your mathematics courses you have to know the content and *show* that you know it. The purpose of many of the past chapters has been to give advice on learning mathematics content. This section gives ideas on how to show, through your course examinations, that you have learned it. Knowing the content, knowing that you know it, and showing that you know it are three different things.

Since it is my belief that certain statements cannot be said enough, I repeat a seemingly obvious statement that is ignored by many students: *The best thing you can do to succeed on examinations is to know the content well.* (See section 1 of chapter 26 on continuous preparation for unit examinations.) There are many student complaints about examinations and how they felt while taking them. They include statements such as: "It was too long," "The problems were too difficult," "I couldn't understand some of the questions," "I was so nervous on the exam that I panicked," and "I could not recall much of what I knew." Knowing well the mathematics that is to be examined will eliminate many of these complaints, but not all of them. There are many examination-*taking* issues that you need to address, many of which are now presented.

Plan Before an Examination How You Will Tackle It

There are many ways to work through an examination and you need to decide, prior to the examination, how you will work through it. The decisions you make ahead of time can be revised or modified as you progress through the examination.

Read Through an Examination at the Outset

Spending time at the outset of an examination to read through it is time well spent. You may feel that you don't have this time, but this actually saves time. The maxim, "Haste makes waste," applies here. The time-saving techniques you can use as you spend several minutes reading through the examination, include the following:

Note the number of problems and how the various types of problems are distributed throughout the examination. To have a ballpark idea of the average amount of time you can spend on a problem, you need to know the number of problems. For example, if it is a 50-minute examination comprised of 10 problems, you have, *on average*, about 5 minutes per problem. You have to allow more time for the more difficult problems.

Note the difficulty of each problem and use a code to characterize its difficulty. It makes good sense to have an overview of what is required in the way of problem difficulty; otherwise, you cannot make good decisions on how to move through the examination. Code the difficulty of a problem and use this code to decide the order in which you will work the problem. For example, you can use "1" for the problems you think you can do quickly, "2" for those that may take a little longer, and "3" for the most difficult ones. You won't have much time to determine the code to assign to a problem—you assign it based on a quick impression of the problem's difficulty as you initially survey the examination.

Note the point value of each problem. Knowing the point value of each problem allows you to choose the problems to work on that help ensure you get the most number of points in the time you have available. It makes no sense to spend considerable time on a difficult problem, and not have the time to work several easier problems with comparable or higher point values.

Decide on the best problems to begin with. Some students like to progress through the examination by beginning with problems they are certain they can do, move to those that they are less certain about, and end up with those that are most problematic. It is wise to do this if you want to avoid panic and build your confidence. Furthermore, it allows your mind to work subconsciously on the more difficult problems before you begin working them (another reason for reading them at the outset of the examination). Some students would just as soon take the problems in order. Others will begin with the problems that are assigned the most number of points. You have to decide what is best for you.

Write comments on some problems as you survey. A thought or two on how you can work some problems may cross your mind as you read through the examination. You may want to jot down a word, short phrase, or formula by the item to remind you of these thoughts. You may also want to do this if a thought comes to you about working a problem as you are working another problem.

Give Yourself Time to Interpret a Problem and Plan a Method of Attack before Beginning to Write

You don't want to begin working on a problem until you understand the statement of the problem. You might say, "Of course"; however, vast numbers of students cannot solve a problem due to misreading the problem. You won't get any credit on a problem if you work the problem you think is given, instead of the one you are given. *Take time* to understand the statement of the problem before you begin writing a solution. You have more time than you think to understand it, but you don't have forever. A minute or so may seem like a very long time to interpret a problem when you may only have about three to four minutes to interpret and solve it. You have no other choice but to spend the necessary time. (See section 1 of chapter 22 on understanding the problem.) I now take great liberty in paraphrasing a remark by Einstein that maintains the spirit of his remark: "If I had an hour to solve a problem, and my life depended on reaching a solution, I would spend 55 minutes to understand it, and 5 minutes to solve it." Your life does not depend on your examination performance; nonetheless, resist the urge to begin writing almost immediately upon reading a problem.

In the long run, you save time by doing more thinking before you begin writing. I am convinced that the vast majority of students feel that, unless they are writing, they have not yet accomplished anything on the examination. Writing too soon causes false starts, which takes up time. It delays the necessary thinking that needs to take place on how to respond to an examination question. Note that I am not saying that you should always know how to completely solve a problem before you begin writing. The nature of the problem may be such that what you need to do to finish the solution can only be determined by first deriving partial results. What I am saying is that you need to give yourself enough thinking time at the outset *to understand the problem* and how to initially attack it. That being said, there are times that an understanding of a problem only comes about through your attempts to solve it.

When you are taking an examination with your classmates it can be anxiety-provoking to see others writing when you feel you don't have a clue on how to solve the problem. You hardly received the examination when you notice that most of your classmates are furiously writing. In a panic-stricken manner you say to yourself, "They know what to do, but I don't." If you are having difficulty understanding a problem and how to attack it, you can be sure you are not the only one. The paper that gets handed in first is often from someone who has given up, or has done poorly on the examination. Rarely in my classes was an examination handed in early by someone who received an "A" on it. Many of those writing almost immediately upon receiving the examination will give an incorrect response, or are working a problem that is different from the one you are working. Work within yourself and ignore others.

Underline or Circle Key Words in the Problem Statements

There are important words in the statements of problems that indicate what you are to do. Sometimes students overlook these words when working a problem; hence, underlining or circling them may further imprint them in your mind. Examples of these words follow: simplify, verify, argue, prove, show, denote, summarize, explain, develop, discuss, derive, formulate, and compare. You may also want to underline or circle connector words such as "and" and "or," and quantifying phrases such as, "at most," "at least", "if and only if," and "if…then."

Moving on to Another Examination Problem

You cannot spend forever trying to solve an examination problem. If you see that no solution is forthcoming then it behooves you to move on to another problem, unless you only have a problem or two remaining and time left to work them. However, you want to allow yourself a reasonable amount of time to solve a problem. Moving too quickly from incomplete problem to incomplete problem is an inefficient use of time. You may be on the verge of a solution so don't leave a problem too soon. For example, if you have about five minutes to work the problem, don't leave it after two or three minutes, unless you have a good sense that any more time spent on it will not bear fruit. How much time you spend on it depends on how much time you have already spent on other problems. For example, if you have just completed a problem in two minutes for which you had five minutes, you can spend more time on another problem. Obviously you cannot monitor this precisely, but you need to have a decent sense of whether you are gaining or losing time as you progress through the examination.

Use a Code to Characterize the Difficulty of Problems That You Leave Unfinished

You may want to revisit problems that you have left and not solved, time permitting. Be wise establishing the order you will follow to revisit these problems. Where you place a problem in the order depends on the likelihood that you are able to progress further on it, and how long it may take to do so. Code the problem with "a" or "b," indicating whether you are likely or not likely to solve it in this amount of time, respectively. Revisit "a" problems first.

Vary Your Work Pace as You Progress Through the Examination

It is nice to use the same working pace throughout an examination, but it will have to be a faster pace than your pace when working assignments. Working too hurriedly leads to mistakes, and working too slowly means you won't complete the examination; hence, it is a balancing act. There typically comes a time on an examination when your pace has to quicken. If you have one-third of the time left and you have more than one-third of the problems left, you have to work faster if you want to complete the examination. I had students who worked the first three pages of a five-page examination without error, and did not get to the last two pages. I also had students who spent 15 minutes on a 10-point problem and did not receive any points on it. Because of the time they spent on the problem, they were not able to get to less difficult problems; thus, forfeiting any chance to gain 20 or 30 points from working these easier problems.

Are you a student who knows how to solve a problem but finds it difficult to move quickly in writing a solution? With practice you can increase your speed. Prior to an examination, time how long it takes you to write solutions to specific problems and then work to improve these times by working comparable problems. Reflect on what seems to be getting in the way of moving faster and on ways to overcome these impediments.

What to Do If Time Is Running Out

When time remaining on an examination is getting scarce and you do not have time to complete a solution, *outline* the steps you would take to arrive at it. Your outline may contain a sketch, a formula or two, or the name of a theorem that you would use to solve the problem. Here is an example of the form of an outline: "First, I would use . . . to find . . . Then I would see if. . . . Finally, I would find . . . to arrive at the answer." This shows your instructor that you have the knowledge to work the problem. The complete solution giving the details may take five minutes, but an outline could be given in less than 20 seconds. If your instructor assigns partial credit, you may obtain anywhere from 0 to 100 percent on an outline of the solution. I typically gave from 50 to

75 percent of the point value of the problem, depending on how detailed the outline was.

What to Do If a Problem is Poorly Worded, Is Missing Information, or Contains Mistakes

My experiences tell me that many students do nothing about a poorly worded problem. It was not unusual to have only a few students come up to me and ask me about one or more difficulties they perceived with the statement of a poorly worded problem. And if they came it would often be near the end of the examination period when it was too late for me to make the corrections for them or other students. The only conclusions I could reach for this inaction was that (1) they did not feel they had the time to do this, (2) they were not sure whether the problem was poorly stated, or (3) their former mathematics teachers had not allowed them to make comments or ask questions about a problem during examinations.

There are clarification questions you might ask your instructor concerning a problem that he or she will justifiably *not* answer. It may be something you are expected to know and that is part of what is being examined. However, you will not know what your instructor will say unless you ask, so take the time to ask. You most likely have the time to ask, even if you don't think so. It is difficult to construct a good examination; hence, it is not unusual for there to be one or more problems that need fixing.

Another reason you might want to approach your instructor during an examination is to obtain information that you need, but cannot recall in your efforts to solve the problem. For example, you might need the number of feet in a mile, a formula for the volume of a cone, or a trigonometric identity. Your instructor may believe that giving you this information is a reasonable thing to do so you can continue working the problem. The instructor's decision may depend on whether it is incidental information from a previous course, or germane information from the course you are in. If it is the latter, and your instructor is reluctant to give it to you, ask if you can have this information if he or she denotes on your examination that you received it. In grading this problem the instructor can decide how many points to deduct for giving you this information. It is reasonable for your instructor to deny any request that you make of this nature.

Writing Too Much or Too Little

If you look at your instructor's solutions key to a mathematics examination, you will see that the solutions are concisely written. What he or she needs to say is said, and nothing more. Similarly, it is unwise for you to pad your answer with extraneous information in order to have a long response. You will be graded on how well you solve the problem, not on all that you know that relates to the problem. You waste valuable time by writing what is not needed. Furthermore, the grader of your examination may deduct points for your verboseness, even if a complete solution appears somewhere in your writing.

Most mathematics instructors give partial credit; hence, it behooves you to show most of your work, displaying key statements. If you are to show computations, only show the more complex ones. If you make an error in solving the problem, your instructor can find it and give you credit for what you have done, up to (and after) that point. It is difficult, if not impossible, to give partial credit if little is written.

Organize Your Solution

It is important to organize your work on an examination. This most likely will not happen unless you are in the habit of organizing your written work on your assignments. On an examination there is less time to think about how you should write down your solution to a problem; by this time it has to be instinctive. I have graded many tests in my day and most students' solutions are reasonably well organized; but there are grim exceptions to this. I have seen solutions that are correct, but it took me a long time to place all the parts in an order that made sense. I have taken points off for lack of organization. In some instances, I gave no points when the organization bordered on being incomprehensible, even though a solution could be found somewhere in the work.

Suppose there are seven steps to solving a specific problem, and the solution progressed in the order of step 1 through step 7. Here is an acceptable way to display the solution vertically:

Step 1	Step 5
Step 2	Step 6
Step 3	Step 7
Step 4	

Here is another acceptable way to display the solution horizontally:

Step 1	Step 2	Step 3	Step 4
Step 5	Step 6	Step 7	

Here is an unacceptable way to display the solution (believe me, I have seen solutions similar to this):

Step 7	Step 3	Step 1	Step 6
Step 4	Step 2	Step 5	

The recording of this solution causes more problems than evident at first glance. The instructor does not know where the various steps are since the student has not identified them. No examination grader should have to take the time to figure out the order of these steps and is perfectly justified giving no credit for this solution, even if the thought process is correct. Now compound the grader's problem because the handwriting is poor or the steps are incomplete. Well-written communication of mathematics has to be a goal of any mathematics student. See chapter 20 on the technical aspects of writing mathematics.

Give Yourself More Space to Write Your Solution

Even though your solution should be concisely written, you still need enough space to display it. Fitting your solution into a small space, whose size is determined by your instructor, can be problematic and interfere with its quality and readability, and frustrate you as well. If you need more space and are not writing in a bluebook where space is not an issue, refer the grader to the backside of a page by writing, "Continued on the backside of this page," or "Continued on the backside of the previous page." Once again, don't write too much—write only what you need to solve the problem.

Avoid Erasing an Incorrect Solution

You have no doubt worked on an examination problem and realized that the work you have done thus far is not going to solve the problem. Hence, you decide to start over. It is a waste of time to erase your erroneous attempt. Instead, cross it out and place your second attempt on the backside of an examination page, if you need more room. There have been times that I gave students partial credit for some of the work they crossed out, because the work related in a meaningful way to a solution of the problem. I could not have done this if their initial work had been erased.

Some students work a problem and arrive at a solution, and then realize that they did the problem incorrectly. They then give a correct solution and leave both solutions, not crossing out the incorrect one. How does your instructor know which solution you want graded? It is not the instructors job to determine this. Most, if not all instructors, will deduct points if they must decide which solution you want graded.

Identify Your Answer

If a problem is computational, circle or underline your answer, or write the words "answer" next to your result. Again, you don't want your instructor to have to look for your answer, or make a guess as to what you view as the answer. This is especially important to do if your work is disorganized on the page.

If You Know You Made an Error Let the Grader Know

You may know that you made an error in your work and derived an incorrect answer, but do not have time to find your error. Let the grader know since that creates a favorable impression, and can lead to receiving more points on the problem. Your remark to the grader could take this form: "I know my answer is in error; however, I do not have time to locate the error." Also, if you need to use the incorrect answer you derived to work another part of the problem, indicate this to your grader so you can continue working the other parts. You don't want a mistake in one part of a problem to keep you from working other parts of the problem. Also, if you are not able to come up with an answer to a part of a problem, write a note on the examination similar to this: "I could not complete this part of the problem, so I am making up an answer in order to proceed with the other parts." Or, instead of using an answer you made up so you can work the other parts, *outline* how you would solve the remaining parts.

Look Back Over the Statement of a Problem to See If You Did All You Were Asked to Do

You would be surprised how often I saw students omit parts of a problem due to an oversight. They may have noticed the various parts of a problem when first reading it, but when they finished working the first part, they forgot there were other parts to it. They think their work on the problem is finished and move on to another problem. Here are three ways to help avoid this: (1) Upon first reading a problem, say to yourself, "There are parts to this problem, and I need to be aware that I have to work all parts." (2) Number the parts of the problem (if they are not already numbered). (3) Always reread a problem as soon as you have completed it.

Don't Work a Problem in Its Entirety If You are Only Asked to Set It Up

If you are asked on an examination to *set up* a problem, don't do any more than that. Some students will work it in its entirety, which is a colossal waste of time. You will get no more credit for working it than you would have gotten for setting it up. What may have taken you a minute or so to set up, may take five minutes to work. For example, you may be asked to set up a definite integral that will yield a certain area, given specific conditions. Suppose the integral you set up is

$\int_{\frac{\pi}{4}}^{\frac{\pi}{2}} (\sin x - \cos x)dx$, which may have taken you 30 seconds. It may take

you a few more minutes to evaluate this integral (which you were not asked to do). In the meantime, you may have been able to work a few other problems. Don't do any more than you have to on an examination.

Reread the Problems and Your Solutions If Time Permits

It isn't very often that you get a chance in college to carefully go over a mathematics examination after you have completed it. It is probably more likely that you won't have time to complete it. Nonetheless, if you have some extra time, use it. It makes no sense to leave an examination early, as some students do, without rereading the problems and your solutions to them. Any product can be improved with additional work and an examination is no exception. Problems can be misread, errors can be made in solutions, solutions can be written better, and you may be able to work a problem that you couldn't work earlier in the examination. Have confidence that you can work a problem that you struggled with earlier. Sometimes, with the pressure of the examination lessening as you near the end of the examination, ideas can suddenly surface. Continue working on your examination until you are told to leave the room.

I recall a mathematics examination that I had in graduate school. After reading through the examination, I didn't think I would be able to work any of the 10 problems. I kept my cool, started working the examination, and ideas began to surface. I ended up with a perfect score, and was the only one to do so, even though there were students in my class who had far more ability than me. This was an eye-opening moment in my life as a mathematics student. It taught me a valuable lesson in persevering on an examination and trusting in myself. Whatever you do, don't panic and don't give up. Trust that your preparation will come through for you; however, it is another matter if your preparation is poor.

Gain Speed on an Examination by Slowing Down

I mention once again, but more comprehensively, that what may seem like delay to you as you take your examination is not delay, but a means of moving faster and better through your examination. *Often the best way to gain speed on an examination is to slow down.* I do not mean to minimize the importance of working quickly, but you need to know when not to rush. It will not be easy for you to slow down if you are a person who likes to rush, perhaps due to anxiety. But you can train yourself to slow down and may be willing to do so when you realize its benefits. A key way to do better on examinations is to take more time doing things that seem to delay working the examination. Take more time to—

- Look over the whole examination at the outset.
- Read and think about a problem before attempting to write.
- Stay with a problem before leaving it uncompleted to begin working another one (but you need to learn when to leave a problem).
- Ask your instructor about the statement of a problem.
- Reread the statement of a problem you have just solved.
- Write a solution to a problem that is easy to follow.
- Allow an anxiety attack to subside (which is addressed in the next subsection).

It is just as important in your final preparation time for a unit examination, to think about how you will work through a unit examination, as it is to review the content of the unit. Many of the issues you need to think about have been presented in this section.

Examination Logistics

You need to be aware of specific examination logistics. They include the following:

Know when and where the examination will be held. You may have missed a class session when your instructor announced the time of an examination, postponed it, or changed the room in which it would be held. You need to use the means you have available to learn this information in a timely manner.

Bring the supplies you are allowed to bring (e.g., pencils, blue-book, eraser, calculator, watch, scratch paper). You may find this difficult to believe, but there are almost always more than a few students in an examination room who do not bring a watch (and the room does not contain a wall clock), or a calculator. Not being able to check the time results in improper pacing of your work.

Allow yourself plenty of time to arrive at the examination room. You do not want to be rushed making it to the examination room on time. Allowing just enough time to get there can be problematic if you have unexpected delays. Comments on how early to *enter* the examination room appear in the next section.

Select a seat in the examination room where you will be most comfortable, considering distractions, lighting, work space, proximity of other students, and open space.

2. Coping with Test Anxiety

Don't feel that anxiety is necessarily your enemy. It is well established in psychology that we all need some anxiety to perform well on an examination. Anxiety increases our adrenalin level, which motivates us to prepare better and to be more alert during an examination. What we don't need is a level of anxiety that is debilitating to the extent that our examination performance is severely affected.

Many different labels are used to describe specific anxieties. As examples, you hear of social anxiety, separation anxiety, natural disasters anxiety, test anxiety, and mathematics anxiety. Test anxiety is addressed in this section. Test anxiety and mathematics anxiety are not the same thing. If you suffer from test anxiety you get overly anxious when you take examinations, regardless of the subject matter. If you suffer from mathematics anxiety, you get overly anxious when you have to address matters that involve mathematics, and that includes mathematics examinations. However, these two maladies are both anxieties; hence, they share many characteristics. The characteristics they do or do not share will be left to psychologists and are not addressed here, per se. Addressed here are the symptoms and causes of test anxiety, and suggestions for alleviating it. (This may be a good time to read section 4 of chapter 11 on decreasing mathematics anxiety.)

Symptoms of Test Anxiety

Most likely you know if you have test anxiety. However, it is useful in combating test anxiety to know what the symptoms are. Identify the following symptoms you possess:

- Unpleasant physical sensations when taking, or thinking about taking, examinations (i.e., sweaty palms, increased heart rate, panic attack, stomach flutters, nausea, loss of appetite, headaches)
- Stress or anxiety when thinking about the important role examinations play in determining your course grade
- Incessant worry over how well you will perform on impending examinations
- Frequent thoughts on the consequences you *will* suffer if you do badly on examinations
- Frequent thoughts on the consequences you *have* suffered due to poor performances on examinations
- Stress over taking examinations has been with you a long time
- Worry that almost all of your classmates will do better than you on examinations
- Difficulty focusing during examinations
- Forgetting knowledge during examinations that you knew shortly before taking the examinations
- Content you couldn't recall during the examination comes to you shortly after leaving the examination room
- Thoughts all jumbled during an examination
- Freezing on examinations

- Pessimistic feelings about the fairness of examinations based on negative past experiences

It is natural, and not of much concern, to have some of these symptoms to the degree that they interfere minimally with your performance on an examination. If you have many of these symptoms to a greater degree, there is a good chance you have test anxiety, which will interfere with your performance. If so, you are not alone for it is a malady of many students. Perhaps knowing that there are others is not that consoling to you; but knowing that something can be done about it should be consoling. Before making suggestions for alleviating test anxiety, we look at some of its causes.

Causes of Test Anxiety

You will not find here a large list of items that contribute to test anxiety. This is not because a large list of these items doesn't exist, but because many of the items on such a list are not that relevant to alleviating test anxiety in college students. Here are three examples of items that contribute to test anxiety in college students—the *last one is most important to you since you have significant control over it*:

Teachers who are insensitive, lack appropriate subject matter or pedagogical knowledge, or overly rely on examinations to evaluate students.

Key courses in which receiving good grades are a must for entrance to, or dismissal from, certain programs. Examples include a statistics course that is required for a concentration program in psychology, or a calculus course that is required for entrance to a business management concentration.

Inadequate preparation in subject matter. There can be little doubt that the major cause of test anxiety for the majority of college students afflicted with this malady is inadequate preparation in the content being examined. Most likely their test anxiety began early in their pre-college years. Regardless of when it began, inadequate preparation for an examination positions them to experience test anxiety in college.

Alleviating Test Anxiety

It was implied in the last subsection that a major way to alleviate test anxiety is to be better prepared in the subject matter. Inadequate preparation is caused by inappropriate study techniques, which includes insufficient time spent studying. Hence, for most students, the chief way to combat test anxiety is to have more viable means of learning, which is a major objective of this book. *The chances of examination panic raising its ugly head in a significant way are slim if you know the content and know that you know it.*

The following are other suggestions for alleviating test anxiety:

Don't expect to cure your test anxiety

Controlling debilitating test anxiety is a realistic goal. Expecting to completely rid yourself of debilitating test anxiety is an unrealistic, perilous, and unnecessary goal. It is unrealistic, because as a human being you are not perfect, psychologically or otherwise, and never will be. It is perilous because anything short of perfection may be demoralizing to you and viewed as failure. You can function very well by *controlling* examination anxiety. Your goal should be to lessen anxiety so it no longer has devastating effects on your examination performance.

Know how to take an examination

Knowing and using good test-taking procedures usually helps reduce anxiety. For example, you need to know which problems to begin working, what to do if a problem is confusingly worded, how much to write on a problem, and to outline the solution to a problem if you are running out of time. There is no doubt that good test taking procedures can lessen anxiety. (See section 2 on taking examinations.)

Prepare yourself psychologically

Preparing yourself psychologically can minimize test anxiety. Here are some ways to do that:

Expect to see problems that you do not know how to work. It may happen that you know how to work every problem on the examination, but most likely not. Hence, if you see a problem on an examination that you can't work, you won't be surprised and won't panic if you expect to see such a problem. It is unrealistic to think that you should always be able to work every problem. Say to yourself as you are reading a problem on an examination, "This is one of the problems I can't work that I expected to be on the examination." Isn't that better to say than, "I can't understand this problem, what is wrong with me?" The former statement is telling you that sometimes it is okay not to know, which allows you to calmly spend more time on the problem, or move on to another. The latter statement is telling you that it is never okay not to know, and elicits panic.

Look forward to showing improvement on an examination and determined to apply and show what you have learned. Contrast this attitude with dreading an examination because of what might be on the examination. The former attitude feeds your strengths and starves your weaknesses. The latter attitude starves your strengths and feeds your weaknesses.

Do not expect wholesale improvement in test anxiety to occur quickly. Your debilitating test anxiety didn't surface overnight and it won't greatly improve overnight. Be patient and be consoled by any improvements, as small as they may be. (See sections 3 and 4 of chapter 10 on basics of change and the process of change, respectively.)

Maintain a journal of your thoughts and feelings. It helps to write about your concerns, frustrations, failures, successes, and plans. They become less vague and less anxiety-provoking when you can write about them. In the hidden recesses of our minds our fears are vague and exaggerated. In the brightness of day they are made more visible, realistic, and can be more adequately handled. The benefits of journaling on your thoughts and feelings can be compared to the benefits that come with the brightness of day.

Focus on your examination, not on yourself or your classmates. These four things are present at your examination: The examination, your classmates, your instructor, and you. You can only focus on one of these at a time. Occupying your thoughts with yourself or with your classmates robs you of the focus you need to place on the examination. There is no time for these thoughts: "Woe is me," "I knew I should have spent more time studying that topic," "She is already on the second page of the examination," or "He is already finished with the examination," or "The instructor is so unfair." Quickly dismiss these thoughts if they arise. We cannot control the thoughts that may arise but we sure can control whether we entertain them. Remind yourself that you are being tested on the examination, not on what you feel about yourself and others in the room. It is not the time or place to worry, doubt, blame, or wonder. It is the time and place to focus on the task before you.

Long examinations are not to be feared

Many students get very upset and anxious when an examination is too lengthy for them. Some of them will blurt this out for everyone to hear:

"This examination is too long." This might happen at anytime during the examination, but is most likely to happen at the end when students are told to hand in their examinations. These anxious and vocal students would like their instructor and classmates to think that they are speaking for almost everyone taking the examination; sometimes they are correct, but most likely not. Almost any test is too long for some students, especially when they spend 10 minutes or more on each of a few problems that, with proper preparation, should have taken a few minutes each. No matter how reasonable my examination was in length, some students would say it was far too long, even though most students finished it in the allotted time.

Instructors would love to muzzle test-anxious students who vocally condemn the length of the examination. The only purpose their comments serve is to trigger the feelings of other anxious students. Remember, "misery loves company," and soon after someone comments negatively on the length of the examination, several sounds of agreement can be heard throughout the room. If you hear these remarks, you need to stay calm and realize that, in the main, test-anxious individuals are vocalizing their anxieties. That being said, an examination may be too long for many students in the course. If you get overly anxious about this, then reflecting on these comments may help you to handle it better:

- Everyone in the room is taking the same examination.
- Your instructor most likely knows or will know if it is too long, either after constructing or grading it. Leave the grading to your instructor. Hopefully, the length of the examination will be reflected in the grading scale that is used. If you have a valid complaint about the way it was graded based on your perception of its length, discuss this with your instructor.
- It gives you more opportunity to find problems that you can work.
- Some instructors do not believe that every examination should necessarily be completed by most of their students. They just want to give their students the opportunity to show what they know.
- Being upset about an examination's length can cause you to lose focus and perform poorly. You have two alternatives—

choose the better one, which is staying calm. It is a matter of "mind over feelings."

Avoid being with classmates near examination time

It is generally unwise for a test-anxious person to be around test-anxious or hyper classmates within a few hours before an examination. As the examination approaches, students make anxiety-provoking comments about their lack of preparation or they will ask you questions on what you did or did not study. Their comments and questions can raise your level of anxiety at a time when you should be relaxing and clearing your mind. These comments can stimulate these thoughts: "I forgot to study this," "I thought I knew that," "I didn't think we would be tested on that," "She is really prepared for the test," or, "I wish I was as smart as he is." If you are around these students at this late date you have much to lose and little to gain. Being around them during this time can also interfere with the knowledge that you do possess. For example, if you are able to answer a question posed to you by a classmate, you may not be able to recall it when you are asked the same or analogous question on the examination, especially if your knowledge is not as strong as it should be. If a problem could talk as you work on it in an examination, it might say, "I cannot give you at this time what I gave you just before you entered the examination room. Call on me later for this information." The mind works in mysterious ways. Allow it to rest with what you know, rather than continually working it just before entering the examination room.

It is not wise to arrive at the examination room too early. You are apt to hear many anxious or panic-stricken comments as you sit near classmates. On the other hand, you don't want to arrive at the classroom as you are about to receive the examination. You have to strike an appropriate balance that works for you between arriving too early or too late.

Relax before the examination

How wonderful it is to look forward to relaxing before an examination. At some point in advance of the examination, perhaps a half day or more, you have to say to yourself, "It is time to end my final preparation for the test. I need to relax and rest my mind on this content so I can function well on the examination." Take a walk an hour or so before the examination if you can, preferably outside, enjoying nature, and being at peace with your preparation and the impending examination. You can have your notes with you, which should give you a sense of security in case you want to look at them, but have the confidence that you don't need to. Don't think about what will be on the examination, or how you may do on it. If these thoughts surface, quickly dismiss them. You may want to relax in a comfortable lounge chair, being at peace with yourself. In other words, just be. I frequently make it a point to do this, whether I am to give a special presentation, chair an important meeting, or spend time on an important conference call. I have no doubt that this improves my performances significantly. As you relax, feel the calm coming over you. It is a blessing indeed.

Plan for rewards

It is always encouraging to look forward to a reward after an examination. Perhaps you can plan to attend a movie or sporting event, visit a friend, or have a special meal out. Have the mindset that you will reward yourself for having been faithful to your preparation and because the examination is over, *regardless of how you do on it*. Looking forward to this reward will buffer any trials and tribulations that may come your way related to the examination.

Seek help from your instructor, academic counselor, or counseling center

If your stress level is more than you can handle, don't hesitate to speak with your instructor, academic counselor, or someone in the counseling office. Test anxiety is a common occurrence in college, and these individuals should be able to help you lessen it, or refer you to someone who can. If a person you see does not help you, go to someone else. You have to work at getting the assistance you need. Consider attending a workshop conducted by your college on reducing test anxiety.

Employ a technique for desensitizing stimuli that arouse anxiety

You can artificially create an examination environment, and by repeated reflection on key parts of it, *desensitize* your emotional reactions to them. Begin this desensitization process as far in advance of the examination as you desire, and repeat it as needed as the examination approaches. I am a strong advocate of this technique and continue to use it for desensitizing purposes.

Here is an *example of a technique for desensitizing test anxiety*, which employs good test-taking procedures:

Sit in a comfortable chair away from others. Close your eyes and contract and relax your muscles, feeling the tension leave your body. For each of the following scenes described below keep your eyes closed and visualize yourself in the scene. *Stay with a scene until you experience it with a minimum amount of anxiety.* Add intermediate steps to the scene if you are experiencing anxiety; then move on to the next scene:

1. You are walking into the examination room looking forward to the examination.

2. You are seated in a chair waiting for your instructor to enter the room. Your classmates are making various anxiety-provoking comments on the impending examination and their preparation for it. You see yourself remaining calm through all of this, not paying attention to what they are saying.

3. Your instructor begins passing out the examination. You look forward to receiving yours.

4. You notice that some students are already writing on their examination, yet you do not have yours. You continue to remain calm and focus on what you need to do when you receive it.

5. You read through the examination with deliberate speed, noticing how long it is, the nature of the content of each item (including whether you think you can work it), the point value of each item, and the degree of difficulty of each one, which you indicate by writing a number near it. You are calm and collected during this activity even though you have not begun writing. You understand the benefits of delaying your writing, including knowing where you will begin, how you will move through the examination, and the average amount of time you have for each item.

6. You look at your watch and notice that about three minutes have passed since you received your examination, and are cognizant that you have not yet begun to write. Also, you happen to notice that some students are already working on page 2. You continue to remain calm, focusing on what you need to do.

7. You carefully work on an item that you know you can work, and then progress to other less difficult items as noted by the code you placed by each one. You see yourself progressing carefully and confidently. You also see yourself writing no more than what needs to be written and in an organized way.

8. You are now working on an item that you have difficulty understanding. You have read it a few times and are not sure what to do. This does not alarm you since you realized this could happen. You also notice that you have about a quarter of the time remaining with a little more than a quarter of the problems remaining (and they are the more difficult ones for you).

9. You decide to spend a little more time trying to interpret the problem you are struggling with. You decide that the problem is badly worded and wonder if you should take the time to approach your instructor to get clarification. You calmly decide to do this, and to leave the problem for another if your instructor's remarks don't help.

10. You notice that some students are turning in their examinations, but you quickly turn your thoughts back to the problem you were working on, not allowing their departure to interfere with what you have to do.

11. You make some progress on the problem after talking to your instructor, but decide to leave it unfinished and work on the three remaining problems in the 10 minutes that are left.

12. You have a sense of how to work two of these three problems, but not enough time to write out the details. You know that you can save time by outlining solutions to them, so that is what you do.

13. You are almost done outlining these solutions when you are asked to hand in your examination. You would have liked more time, but you feel good since you did the best you could in the time allotted.

I encourage you to create your own test-taking scenarios that you can use to desensitize the stimuli that provoke anxiety in you. You know best what triggers your test anxiety. Create scenarios that involve good test-taking procedures, and go through the scenario as often as you need to until your anxiety subsides. (You might want to use all or part of the list just given. It is a long list, so cut out those items that do not arouse your anxiety.) Repeat your test-taking scenario several times in one sitting (being sure to stay with any part of it until you feel the anxiety subsiding), do it again that evening, or come back to it the next day. You will find out that it does desensitize test anxiety.

Handling a panic attack during the examination

If at any time during an examination you feel panicky to the point that it is difficult to think rationally or recall content, stop working, put your pencil down, and "step outside of your body." That is, be aware of both sides of yourself, namely, the rational and calm side, and the panicky and irrational side. Tell your panicked and irrational side to calm down and relax. Give it a minute or two by taking deep breaths, relaxing you muscles, and thinking pleasant thoughts. Trying to rid yourself of your anxiety while continuing to pursue the very thing that triggered it, such as a problem you cannot work or your fear of the consequences of a bad grade on the examination, is a no-win situation. Continue the conversation between your two sides until the side you don't want decides to flee the premises. When that happens get your calm and rational side back to work. Go through the same process if it happens later on.

You may think you don't have the time it takes to let your anxiety pass. Yes you do—*what you don't have is the time to continue working on the examination in a panic-stricken state*. You can practice this scenario prior to the examination by visualizing it happening and seeing your rational side getting back in control. This may seem like a bit of hocus-pocus to you, but it does work if you let it. In this situation it really is "mind over anxiety."

Be Your Best Friend

I would be remiss if I didn't leave you with this most important thought: You have to place your performance on examinations in its proper perspective. No doubt your performance on examinations is important, and you need to do what you can to properly prepare for them. But *you are more important to you than the grief you inflict on yourself* by accommodating your test anxiety. Your best friend has to be you! If you can honestly say to yourself that you have prepared diligently for an examination, including reflection on good test-taking procedures, then you have fulfilled your obligation to yourself and no one, including yourself, can ask any more of you. Take on the attitude that what happens on the examination will happen. You are being your worst enemy if you always give yourself permission to get extremely anxious before and during an examination. In other words, the anxiety you create makes your life miserable. You deserve and should want better. Resolve to treat yourself better; that is, do not allow any anxious feelings that may surface to linger. Life is too short to make it that unpleasant. Once you make this decision, you have created an environment for anxiety to lessen and success to flourish.

Finally, to conclude this section, avoid the trap of thinking that you are a failure if your efforts to alleviate test anxiety result in small positive changes, or seemingly no changes at all. Your test anxiety almost certainly developed over a long period of time, resulting in well-established patterns. Any positive change to long-standing anxiety, no matter how small it is, is a significant one indeed! Relish it, and look forward to making more positive changes.

3. Learning from Your Examinations

There are many aspects of students' behavior that get in the way of their learning. Right up there with the major ones is allowing their lack of understanding to snowball out of control. Mathematics is a subject that builds on itself as you progress through your mathematics courses. This dictates that future progress in a mathematics course depends heavily on past progress. *Your success in subsequent units is problematic, at best, if you continue to remain confused on key content of past units.* Most students have no trouble accepting this intellectually, but many of them do very little about past confusion. They do not take the necessary steps to overcome their confusion of past content, and they pay a heavy price for this dereliction. A big step in overcoming confusion of past content is to learn from your graded unit examinations. A graded examination is a wonderful diagnostic tool for determining ways you can improve as a mathematics student. It will help you identify the study habits you need to change to better understand past content, as well as improve your test-taking ability. However, its value as a diagnostic tool is severely limited if it is a poorly constructed examination; that is, if it tests inappropriate objectives, is very limited in scope, or is poorly graded.

This section mainly discusses how you can use your graded examination to:

- Diagnose your difficulties with the content
- Learn the content that you didn't understand
- Improve your study methods
- Improve your test-taking skills

Diagnosing What You Didn't Understand

A unit examination *samples* your understanding of a unit's objectives; hence, it has limitations as a diagnostic tool. However, if appropriately constructed, it reveals your knowledge of most key objectives of the unit. Therefore, it is crucial for you to use your examination results to identify and take steps to learn content that you did not learn prior to the examination.

What Your Instructor May Require You to Do with Your Examination

Some instructors have their students redo the examination, allowing them to use whatever means they have at their disposal (for example, textbook, notes, completed assignments), and requiring it to be turned in to be graded. The grade received may be viewed as an assignment grade, or there may be a scheme for using the points received on it to improve your original examination grade. Many of these instructors will require their students to rework only those problems for which they did not receive full credit, and they will expect a more complete response than what they expected during the examination. These instructors may or may not hand out a solutions key to the examination after students turn in their redo examinations. Other instructors will recommend that their students redo the examination, but they do not ask them to turn it in to be graded. After a fixed amount of time passes, they may hand out a solutions key. A third set of instructors may just hand out a solutions key to the examination, not saying anything about redoing it. I suspect there are instructors who do not say anything to their students about reworking the examination, and also do not hand out a solutions key. Which instructor policies are in your best interest?

Regardless of your instructor's examination policies, you are in the course to learn the mathematics of the course as best you can. The best way to do this, minimally, is to rework all those problems for which you did not receive full credit, using the resources you have available. Do not go to a classmate or your instructor for help until you have worked on it by yourself. Do not look at a solutions key prior to redoing the examination. Incidentally, some instructors include a few problems on the next examination that came from the previous examination. Take advantage of the opportunity to gain points on the next examination by correcting your mistakes on the previous examination.

Reflect on the Grader's Written Remarks

Hopefully, your examination grader will write short messages on your examinations expressing his or her concerns, suggestions, and praise. There are a variety of things these messages address, including how you organized your work, the incompleteness of your solution, errors in your thinking or computations, a request to show more of your work, or encouragement to see the grader. These remarks are typically very constructive. You may not fully understand the remarks since they are often very quickly and briefly expressed, and perhaps even unreadable. See your grader if you need clarification and advice on these comments. You may want to journal on the more important ones and reflect on them periodically so you rectify the situations that need changing. Ignore them and your examination corrections at your peril.

Go Beyond Redoing Examination Questions

You most likely need to do more to improve your understanding of the content of the unit covered by the examination, than just redoing specific examination questions. An examination question that you had more than passing difficulty with is probably symptomatic of other problems that you have with the ideas that relate to the question. It behooves you to spend more time rereading the sections of the textbook and your class notes that discuss these ideas, and working additional exercises that use them.

Do not dispense with remediation. You may say, "I don't have time to remedy my deficiencies." I say, "You have to find the time, as difficult as that may be." Granted, your workload in mathematics will increase substantially, since not only will you be spending time on remediation, but you also have to work on the new unit. This new unit cannot be viewed as just another unit to learn if your knowledge of the previous unit was significantly lacking. Its difficulty for you has significantly increased if you had considerable difficulty with the last unit. Perhaps you can now better understand why so many students dispense with remediation, which is an enormous mistake on their part. Remediation takes time and that is often in short supply. Students want to move on with their work, not spending time to rework past content; however, this behavior is terribly unwise. What will your decision be?

Consult Your Classmates on Your Graded Examination

After completing your examination corrections, you may want to meet with a classmate, who has also made corrections, to discuss what you both have done. This affords you another opportunity, and a valuable one at that, to increase your knowledge of the unit's content. You may also increase your knowledge of how to take an examination after seeing your classmate's graded examination. In doing this, you most likely will look at the partial credit each of you received on the same problems. It is not uncommon for you to think that you should have received at least as many points on a particular problem as your classmate, but did not. You may or may not be in error having this opinion, but there are two things you need to know before complaining to your instructor:

The chances of you not being able to discern whether your solution is equivalent to your classmate's solution are reasonably high. Your instructor is in a much better position to judge the merits of solutions. I cannot count the number of times I had a student come to me with a classmate's examination in hand, saying something like this: "Mary got five points on the problem and you gave me zero points, yet our solutions are the same." After looking at each solution again, I would not change the assignment of points, and was able to explain why. Most students who came in and asked me to relook at their solution to a problem, understood my explanations for the credit I gave them. Those students who didn't were either in denial or they lacked the knowledge to understand my explanation. This is not to say that I never made mistakes assigning partial credit to the degree that it needed to be changed, but it didn't happen very often.

Your instructor has the option to only give full credit or no credit. A remark that many instructors use when students unjustifiably complain about the partial credit they received is similar to this: "It is not easy to assign partial credit, but I do the best job I can. It is in your best interest that I do give partial credit; however, if you don't like how I do it, then I will give full credit or no credit." This seems to quiet students down.

Consulting Your Instructor on Your Graded Examination

Your instructor is in the best position to help you with any issues concerning your graded examination. He or she can respond to your concerns about how your examination was graded, and what you can do to improve your performance on subsequent examinations. If you believe that you deserve more partial credit on a problem, mention why you believe this as opposed to just expecting your instructor to look it over again. Avoid referring to how a classmate was graded on the problem. A word of caution here: The final arbiter is your instructor, not you. Be careful not to convey the attitude that your opinion carries at least as much weight as your instructor's. It is unwise to virtually demand that you receive more credit. Do some students do this? Does the sun set in the west? Be tactful in your conversation, but don't be shy asserting your case.

There are students who ask their instructor to change partial credit that was given only because they want to improve their examination grade, not because they feel strongly that a problem or two did not receive enough credit. They come in asking their instructor to look at a specific problem. Then they progress to asking about another problem, then another, and another again. In a real sense, they are asking the instructor to regrade the complete examination. You do not want to do this; at best it will be poorly received and you most likely will be told that.

Students often make an appointment to see their instructor if they did less than satisfactory on the examination. They inquire as to whether they should drop the course or ask what they can do to improve their performance on subsequent examinations. This book has considerable information on the latter so nothing more will be said here. For information on the former read section 2 of chapter 25 on dropping your mathematics course.

What You Can Learn from Analyzing Your Graded Examination

Your graded examination can provide you with a gold mine of information. The amount of gold you find there is a function of the quality of the examination and how it was graded. An analysis of your graded examination can reveal (1) the specific content that was addressed by the examination that you need to learn better (you can then ascertain whether you addressed it in your studies, and where it appeared in your studies), and (2) the test-taking procedures you need to improve.

Identifying Types of Test Objectives You Need to Learn Better

A key thing I did with my students who came in after receiving a poor grade on their examination, was to help them analyze their performance by categorizing the problems for which they received various partial credit. My examinations tested a variety of cognitive objectives; hence, this analysis identified the types of objectives for which the student's performance varied. For sake of illustration, assume that each problem was worth 5 points. On a student's examination we jointly looked at the problems that I assigned (a) 4 or 5 points, (b) 2 or 3 points, or (c) 0 or 1 point. This told me a lot about the student, including the type of objectives he or she handled easily, moderately well, or with considerable difficulty. It also conveyed information on how devoted the student was to studying various types of objectives. Through discussion with the student, specific deficiencies in the student's study methods were identified.

I often saw students' graded examinations that were characterized by problems receiving either partial credit of 0 or 1 point, or 4 or 5 points, with almost no problems assigned 2 or 3 points. The 4- or 5-point assignments told me that the student was capable of doing well in the course, and I let the student know this. My comment was especially creditable if I saw that the student did well on the more high-level cognitive objectives, even though he or she did poorly on many of the lower-level ones. I also knew that there had to be a problem with the student's study skills since the student either knew how to work a problem or could not work it at all; there was no middle ground. The student and I analyzed why he or she was doing badly on lower-level cognitive objectives. A variety of reasons were possible and we looked for them. For example, lower-level objectives are more computational in nature and to do well on them they have to be practiced. A poor performance on the higher-level cognitive objectives implied that the student might not be completing these more difficult objectives in his or her assignments. If that was the case, reasons for this were identified.

I often saw students' graded examinations that were characterized by the predominate assignment of partial credit of 2 or 3 points. These students had an incomplete, but moderate understanding of almost all the test items. These students were capable of a better understanding, but either chose not to expend the effort to gain it, or needed to be more cognizant that they only had a partial understanding. They could be helped by getting more feedback on their work by working with classmates or making more use of their instructor's office hours.

To conclude, you can learn a lot from looking at the number and types of objectives for which you received a specific number of partial credit points. You can ask yourself what this tells you about your attitude toward these types of objectives and how you prepared for them. You can ask a classmate to help you analyze your graded examination, but by all means don't hesitate to ask your instructor to help you since he or she is much better equipped to perform this analysis.

Discerning What Was Asked on the Examination and Where It Came From

There are three basic questions pertaining to examination content that you can ask yourself, and get answered by analyzing your examination: (a) What was asked? (b) Where did it come from? (c) Why wasn't I prepared for it?

To get answers to these three basic questions, ask yourself these, and other questions:

1. What content was on the examination that surprised me?
2. What content was left off the examination that surprised me?
3. Was I expected to do things with the content that I was not expected to do in my assignments?
4. What was the requisite knowledge that I did not have?
5. Was there content I deliberately decided not to learn but was needed for the test?
6. Did I forget specific formulas or use incorrect ones?
7. Was there content I thought I understood, but found out I didn't?
8. Do I need to work more problems from my assignments?
9. Were there things I did not commit to memory that I should have?
10. Did I make simple errors?

11. Did I confuse one definition, theorem, or application with another?

12. Should I have started my final preparation earlier?

13. Was most of the content based on the lectures? On the textbook? On the assignments?

14. Was there content that came from my instructor's handouts?

15. Were the problems worded differently from those in my notes, textbook, or assignments?

16. What do I need to change so that I would have been better prepared for what was asked?

It is your answers to these questions and others, which dictates what you need to change in your approach to the course, including how you study for examinations.

Identifying Test-Taking Procedures That You Need to Improve

You will want to know if there is anything you need to change as far as how you functioned during the examination. What test-taking techniques do you need to change? These questions can stimulate answers:

1. Did I read the entire test before writing?

2. Did I note the difficulty of each problem and the points assigned to it before I began writing?

3. Did I work problems in their entirety that I was only asked to set up?

4. Did I write a note on the test indicating that I knew I had an error in my solution to a problem?

5. Did I waste time erasing an incorrect solution?

6. Did I check my work if time was available?

7. Did I understand what I was asked to do before responding to a problem?

8. Did I give too few or too many details in my responses?

9. Did I answer all parts of a problem?

10. Were any of my responses disorganized?

11. Did I circle key words in the problem statement?

12. Did I ask my instructor to clarify the wording of any problems?

13. Did I outline solutions to problems if time was running out?
14. Do I know why I lost points on a problem?
15. Did I read over my solutions to see if I made an error?
16. Did I have enough time?
17. Did I allow enough room for my responses?
18. Did I leave the examination room too early?
19. Did I get overly anxious at any time?
20. What was my major test-taking problem?

Wouldn't you agree that it is quite amazing what you can learn from your graded examinations? Taking time to do this is time well spent. How much time you take to do this is a function of how valuable you think it will be for you. I end this section by repeating an adage, which gives a reason to learn from your graded examinations: "If you always do what you've always done, you'll always get what you've always got."

Main References Cited

Chapter 5

1. Small, Don. Report of the CUPM Panel on Calculus Articulation: Problems in Transition from High School Calculus to College Calculus. *American Mathematical Monthly* 94, No. 8 (1987): 776-785.
2. Halmos, Paul R. *I Want to be a Mathematician*. New York: Springer-Verlag, 1985.
3. Small, Don. Report of the CUPM Panel on Calculus Articulation: Problems in Transition from High School Calculus to College Calculus. *American Mathematical Monthly* 94, No. 8 (1987): 776-785.

Chapter 6

1. Light, Richard J. *Making the Most of College: Students Speak Their Minds*. Cambridge: Harvard University Press, 2001.
2. Ibid.

Chapter 7

1. Light, Richard J. *Making the Most of College: Students Speak Their Minds*. Cambridge: Harvard University Press, 2001.
2. Ibid.

Chapter 8

1. Lewis, Robert, H. Mathematics: The Most Misunderstood Subject. Retrieved August 15, 2007, from http://www.fordham.edu/mathematics/whatmath.html.
2. Halmos, Paul R. Mathematics as a Creative Art. *American Scientist* 56 (Winter 1968): 375-389.
3. Danzig,Tobias. *Number: The Language of Science*. Republication of 4th edition, Joseph Mazur editor. New York: Pearson Education, Inc., 2005.
4. Halmos, Paul R. Mathematics as a Creative Art. *American Scientist* 56 (Winter 1968): 375-389.

Chapter 9

1. Meiland, Jack W. *College Thinking: How to Get the Best Out of College Thinking*. New York: The New American Library, 1981.
2. Bruner, Jerome S. *The Process of Education*. 2d rev. ed. Cambridge: Harvard University Press, 2004.

3. Halmos, Paul R. *I Want to be a Mathematician*. New York: Springer-Verlag, 1985.
4. Ausubel, D.P. *Educational Psychology, A Cognitive View*. New York: Holt, Rinehart and Winston: 1968.
5. Meiland, Jack W. *College Thinking: How to Get the Best Out of College Thinking*. New York: The New American Library, 1981.
6. Whimbey, A. and Lochhead, J. *Problem Solving and Comprehension*. New Jersey: Lawrence Erlbaum Associates, 1999.

Chapter 10

1. Brown, George, and Atkins, Madeline. *Effective Teaching in Higher Education*. London: Routledge, 1999.
2. Meichenbaum, Donald. *Cognitive-Behavior Modification: An Integrative Approach*. New York: Plenum Press, 1977.

Chapter 11

1. Branden, Nathaniel. *The Power of Self-Esteem*. Florida: Health Communications, 1992.
2. Burka, Jane B, and Yuen, Lenora M. Mind Games Procrastinators Play. *Psychology Today* (January 1982): 32-44.
3. _____. Procrastination and Group Treatment in a College Population. Paper read at the American Psychological Association Convention, August 24, 1981, Los Angeles, California.
4. Ibid.
5. Ibid.

Chapter 12

1. Meiland, Jack W. *College Thinking: How to Get the Best Out of College Thinking*. New York: The New American Library, 1981.

Chapter 13

1. Gardner, Howard. *Frames of Mind: The Theory of Multiple Intelligences*. New York: Basic Books, 1983.
2. Greyer, Peter. The MBTI. Retrieved July 27, 2007, from *http://www.petergreyer.com.au/mbti.php*.

Chapter 14

1. Halmos, Paul R. *I Want to be a Mathematician*. New York: Springer-Verlag, 1985.

Chapter 15

1. Cohen, David. *Precalculus: A Problems-Oriented Approach.* 4th edition New York: West Publishing Company, 1984.
2. Ibid.

Chapter 18

1. Hurley, James; F., Kuehn, Uwe; and Ganter, Susan L. Effects of Calculus Reform: Local and National. *American Mathematical Monthly* 106, No. 9 (November 1999): 800-811.
2. Ibid.

Chapter 19

1. Steenrod, Normand E; Halmos, Paul R; Schiffer, Menahem; and Dieudonne, Jeanne A. *How to Write Mathematics.* American Mathematical Society, 1975.

Chapter 21

1. Schoenfeld, A. H. Some Thoughts on Problem-Solving Research and Mathematics Education. F. K. Lester and J. Carofalo (editors). In *Mathematical Problem Solving: Issues in Research.* Philadelphia: Franklin Institute Press, 1982.
2. *Principles and Standards for School Mathematics*, Reston: The National Council of Teachers of Mathematics, Inc., 2000.
3. Schoenfeld, A. H. Some Thoughts on Problem-Solving Research and Mathematics Education. F. K. Lester and J. Carofalo (editors). In *Mathematical Problem Solving: Issues in Research.* Philadelphia: Franklin Institute Press, 1982.
4. Halmos, Paul R. *I Want to be a Mathematician.* New York: Springer-Verlag, 1985.
5. Schoenfeld, A. H. Some Thoughts on Problem-Solving Research and Mathematics Education. F. K. Lester and J. Carofalo (editors). In *Mathematical Problem Solving: Issues in Research.* Philadelphia: Franklin Institute Press, 1982.
6. Ibid.

Chapter 22

1. Polya, G. *How to Solve It: A New Aspect of Mathematical Method.* 2nd. edition. Princeton: Princeton University Press, 1988.
2. Schoenfeld, A. H. Some Thoughts on Problem-Solving Research and Mathematics Education. F. K. Lester and J. Carofalo (editors). In *Mathematical Problem Solving: Issues in Research.* Philadelphia: Franklin Institute Press, 1982.

3. Polya, G. *How to Solve It: A New Aspect of Mathematical Method*. 2nd. edition. Princeton: Princeton University Press, 1988.

Chapter 23

1. Light, Richard J. *Making the Most of College: Students Speak Their Minds*. Cambridge: Harvard University Press, 2001.
2. Ibid.

Chapter 24

1. Halmos, Paul R. *I Want to be a Mathematician*. New York: Springer-Verlag, 1985.
2. Armstrong, William H. *Study is Hard Work*. 2nd edition. Boston: David R. Godine, Publisher 1995.

Chapter 25

1. Halmos, Paul R. *I Want to be a Mathematician*. New York: Springer-Verlag, 1985.

Index

About the Author

Professor Richard M. Dahlke has an extensive formal academic background in mathematics and mathematics education. He graduated from the University of Wisconsin, Stevens Point, with a Bachelor of Science Degree in mathematics, and a teaching certificate in secondary school mathematics. He holds a Master of Arts degree in teaching mathematics from California State University, San Diego, and a Master of Science degree in mathematics and a Doctor of Philosophy degree in mathematics education from the University of Michigan, Ann Arbor. He taught high school mathematics in Wisconsin for three years and college mathematics for the next thirty-six years. His first two years of college teaching were at the State University of New York, College at Buffalo.

As a college professor of mathematics for thirty-four years at the University of Michigan, Dearborn, he taught all the courses in the calculus track lower-division mathematics program to students pursuing degrees in mathematics, engineering, natural sciences and business. At the upper-division level he taught linear algebra, abstract algebra, geometry from an advanced viewpoint, and specialized courses for students pursuing a major or minor in elementary school mathematics. For many years he taught courses on problem solving in mathematics, and techniques of teaching secondary school mathematics.

His professional accomplishments are many. He has been an active conference presenter and workshop leader at local, state, and national conferences. He served two terms as chair of the Michigan Council of Teachers of Mathematics (MCTM) Teacher Preparation Committee, and served for many years as the National Council of Teachers of Mathematics representative to MCTM and the Detroit Area Council of Teachers of Mathematics. His writings and research have focused on issues associated with the learning and teaching of mathematics at all levels of education. He has published widely in academic journals on a variety of mathematics and mathematics education subjects.

Professor Dahlke is a strong advocate for college students, including their right to receive quality mathematics instruction. He believes that learning is a partnership among the student, the student's class-

621

mates, and the instructor, and that students' efforts to organize and manage their time is critical to success in college. He further believes that students should enjoy college. In order for that to happen students shouldn't get stressed out or too alarmed when things don't go well. They need to focus their energies on making necessary changes that will improve their college experience and lessen stress and anxiety. The end result will be successful college students with contented spirits.

He served on the Board of Directors of CRISPAZ (Christians for Peace in El Salvador), and is a certified spiritual director. His current volunteer activities include giving spiritual direction and coordinating annual educational pilgrimages to El Salvador. His favorite recreational activity is camping and hiking in state and national parks with his wife, Mary.

"The Scholar" pictured on the back cover is a photograph of an Amish boy from rural Pennsylvania.* This professional photograph hangs in the author's home and reminds him of the time he spent as a boy on the porch of his family's farmhouse in Halder, Wisconsin, working arithmetic problems for his homework. He periodically visits Amish communities to remind himself of the work ethic, sense of responsibility, perseverance, importance of family and community, pastoral life, and faith he experienced growing up on a farm. These influences have served him well as a student, professor, and person. Professor Dahlke epitomizes the oft-repeated comment that you can take the boy off the farm, but you cannot take the farmer out of the boy.

---

*Used with permission from Bill Coleman Photography